Soil Water Erosion

Soil Water Erosion

Editor

Csaba Centeri

MDPI • Basel • Beijing • Wuhan • Barcelona • Belgrade • Manchester • Tokyo • Cluj • Tianjin

Editor
Csaba Centeri
Department of Nature
Conservation and Landscape
Ecology, Szent István University
Hungary

Editorial Office
MDPI
St. Alban-Anlage 66
4052 Basel, Switzerland

This is a reprint of articles from the Special Issue published online in the open access journal *Water* (ISSN 2073-4441) (available at: https://www.mdpi.com/journal/water/special_issues/Water_Erosions).

For citation purposes, cite each article independently as indicated on the article page online and as indicated below:

LastName, A.A.; LastName, B.B.; LastName, C.C. Article Title. *Journal Name* **Year**, *Volume Number*, Page Range.

ISBN 978-3-0365-3240-0 (Hbk)
ISBN 978-3-0365-3241-7 (PDF)

Cover image courtesy of Csaba Centeri.

Contents

About the Editor

Csaba Centeri is an Associate Professor at the Dept. of Nature Conservation and Landscape Management at the Szent István Campus of the Hungarian University of Agriculture and Life Sciences in Gödöllő. He has published more than 18 papers in Q1 and Q2 journals and has served as a Chief Editor in the Hungarian Journal of Landscape Ecology, Guest Editor of Special Issues in the Journal of Water, Sustainability, and Remote Sensing. His main research interest is soil water erosion; soil erosion modelling; land-use change; ecosystem services (related to the other research interests); and soil–plant, soil–wildlife, and soil–animal interactions, with special emphasis on nature conservation related issues.

Editorial

Soil Water Erosion

Csaba Centeri

Department of Nature Conservation and Landscape Management, Institute for Wildlife Management and Nature Conservation, Szent István Campus, Hungarian University of Agriculture and Life Sciences, 2100 Gödöllő, Hungary; centeri.csaba@uni-mate.hu

Abstract: The purpose of this Special Issue is to provide an opportunity for researchers to publish novel results that could help landowners, land-users, farmers, politicians, and other representatives of our global society to protect, and, if possible, improve the quality and quantity of our precious soil resources. Authors were encouraged to submit papers related to new ways of mapping, showing more detailed input data, new modeling results, areas that had never been mapped before, etc. The Special Issue provides novel results on the state of soil water erosion mapping and offers insight into new or easier ways to mitigate and reverse soil degradation. Papers from this Special Issue cover a good range of the world, from India through to Pakistan, Russia, China, Syria, Iran, Ethiopia, Italy, the Czech Republic, Germany, and Hungary. Several models or parts of them were analyzed, including USLE, MUSLE, USLE-M, RUSLE, EPIC, WEPP, WaTEM/SEDEM, GULTEM, and GYNDUL. Besides soil erosion modelling, machine learning was also used in one of the articles for the evaluation of gully development. One of the main subjects of the published research was sediments, which are related to one of the most interesting questions in the topic of soil erosion: "Where is the huge amount of soil being lost going?"

Keywords: soil degradation; soil erosion modeling; gully erosion; sedimentation; erosion models; freeze–thaw effects; infiltration; preferential flow; badland; rainfall simulation

1. Introduction

Soil erosion by water is considered to be one of the major forms of soil degradation (other than soil erosion by wind, acidification, salinization, desertification, etc.) and causes the majority of problems related to the degradation of soil resources, leads to the largest amount of soil loss, and covers the greatest extent of areas affected worldwide [1]. This subject is very wide, and there are even many forms of soil water erosion, from splash erosion to gully erosion and sedimentation. It would be worth running a Special Issue just one the numerous soil erosion models now available. As the final product of soil water erosion, sedimentation has been investigated by many interested parties. Each one of these related issues would be worth a separate Special Issue, as all keywords listed have been investigated by many researchers and have been the subject of many papers published in various journals. There have been efforts to provide a wider view of problems related to erosion, such as non-agricultural uses—e.g., industrial pollution and contamination; the disposal of wastes; the restoration of polluted and degraded areas; and recreational uses [2]. The IPCC Report [3] also describes numerous issues relating to desertification, land degradation, sustainable land management, and their interlinkages in some of its chapters—e.g., land–climate interactions, land degradation, etc.

This is the reason why this Special Issue does not aim to serve as a collection on one very specific topic but rather to provide insight into a number of related topics, including soil erosion modelling using the classic USLE as well as more recent models; field research, including rainfall simulations; and mitigation technics and their related problems—e.g., the use of dams as sediment traps.

Citation: Centeri, C. Soil Water Erosion. *Water* **2022**, *14*, 447. https://doi.org/10.3390/w14030447

Received: 19 January 2022
Accepted: 20 January 2022
Published: 1 February 2022

Publisher's Note: MDPI stays neutral with regard to jurisdictional claims in published maps and institutional affiliations.

Water **2022**, *14*, 447. https://doi.org/10.3390/w14030447

This Special Issue was also a subject of finding interested researchers. Who is submitting manuscripts and what types of papers are submitted and from which countries? On the one hand, the time of researchers is limited and, on the other hand, most researchers wish to reach the widest possible audience, so it was rather difficult for someone to decide if they wanted to publish in this Special Issue or another. We feel that the purpose of this Special Issue was achieved, as various publications submitted showed a good variety of information on soil water erosion and novel analyses.

I hope that this Special Issue will provide new and useful information on the mitigation of soil water erosion-induced problems across the entire world. I also believe that, regardless of the localities used (which is an aspect that is often criticized when a paper is under review), many other countries under different climatic conditions as well as people in other geographical regions can learn from the presented results and methodologies and thus move closer to achieving degradation-free land use, as this it is the core interest of the population of Earth.

2. Summary of this Special Issue

Nowadays, as resources such as time and research personnel are limited, computer-based methodologies are receiving more attention and greater popularity. Perhaps this is also the reason why the first paper submitted to this Special Issue used machine learning techniques [4]. These techniques were used to investigate gullies—more precisely, gully head-cuts. A total of 119 gully head-cuts were identified and mapped in N.W. Iran. The authors found that, based on goodness-of-fit and AUROC (Area Under the Receiver Operating Characteristic is a performance metric that we can use to evaluate classification models) of the success rate curve (SRC) and prediction rate curve (PRC), the results indicate that the bagging-ADTree ensemble model had the best performance.

Following this approach, another gully erosion investigation was performed for land management purposes by another author from Russia [5]. Four models were used: two models calculated gully depth and width along with its longitudinal profile, while another two models were used for the novel modification of the area-slope approach, which gives the most probable position of possible gullies. The author suggested the use of the dynamic gully erosion model GULTEM (GULly erosion and Thermoerosion Model) was recommended for calculations of gully geometry transformation in time and space for the most detailed projects of land management.

The third article related to gully erosion identified and assessed the areal and temporal changes of badlands in Italy [6]. The case study area was in the Modena Province (Emilia Apennines), which was important because there was no previous detailed investigation. The authors revealed a general stabilization trend of the badlands in the study area due to an intensified revegetation process around the badlands. This trend is mainly the result of intensive land-use changes—namely the increase in forest cover and the reduction in agricultural land (occurring in the study area from the 1970s onwards). There was another article related to gullies, but its main purpose was to handle sediment-related issues [7] which is also an important part of the whole issue.

Soil erosion models were used by various authors and some of the model details were also analyzed. Kreklow et al. [8] compared rainfall erosivity (R-factor) estimation methods with the utilization of weather radar data in Germany. Their results showed that R-factors have increased significantly due to climate change and that the current R-factor maps need to be updated. One of the possible options to accomplish this is using more recent and spatially distributed rainfall data.

Almohamad [9] used the RUSLE model to show the impact of armed conflicts on soil erosion through the land cover change in Syria. Land cover was decreased due to the increase in forest fires as a result of armed conflict. Damage to coniferous forest and transitional woodland and scrub, especially on steep slopes, had the biggest impact on factor C after the fire. The soil loss was 200% to 800% higher than it had been in the pre-fire situation.

Keller et al. [10] performed a rainfall simulation for soil erodibility calculation using various models, such as USLE (Universal Soil Loss Equation), RUSLE (Revised USLE), USLE-M (USLE Modified), and EPIC (Erosion-Productivity Impact Calculator). Based on the soil loss measured during the rainfall simulation, the authors found that the RUSLE model resulted in the best performance in event soil loss estimation.

Another type of field experiment was conducted by Li et al. [11], who were interested in the effects of infiltration on preferential flow characteristics and solute transport in a Chinese area. The impact of three precipitation amounts of 20, 40, and 60 mm was investigated. Solute concentration was found to peak around the end of the preferential flow path and when preferential flow underwent lateral movement. The results indicated that the infiltration volume and transport capacity of preferential flow had important effects on the distribution of Br^- and NO_3^- concentrations. The results can help in the management of protected forests in China.

Su et al. [12] analyzed the effects of thawing on the slope erosion and hydraulics in the sand-covered Loess Plateau in China. The authors conducted laboratory experiments on meltwater flow to quantify the temporal and spatial distribution of the hydraulic parameters of sandy soils concerning runoff and sediment yield under a constant flow on unfrozen and frozen slopes of variable sand thickness. Their results showed that sand can prolong the initial runoff time, and that unfrozen and frozen slopes have significantly different initial runoff times. A significant linear relationship was found between the cumulative runoff and the cumulative sediment yield.

Lu et al. [13] went further than Su et al. [10] by investigating the effects of freeze–thaw cycling on the soil detachment capacities of three loamy soils on the Loess Plateau of China. After 20 freeze–thaw cycles, the degree of decline of silt loam was the greatest (77.72%), while sandy loam (63.18%) and clay loam (39.77%) showed smaller degrees of decline. The soil detachment capacity of silt loam and sandy loam was positively correlated with the freeze–thaw cycle and negatively with that of clay loam. The authors believe that their results can provide references for further studies on the mechanism of soil erosion in seasonal freeze–thaw regions.

The RUSLE model was used in numerous articles for estimating the sediment yield of certain areas. One of these estimations was conducted in the watershed of the Chenab River at the border of Pakistan and India [14]. The 30-year average annual sediment yield was estimated as 4.086, 6.163, and 7.502 million tons based on different approaches.

Gurmu et al. [15] simulated sediment influx using the RUSLE model and compared it to the amount of sediment removed during desilting campaigns. They found the sediment deposition rate to be 308 m^3/km and 1087 m^3/km, respectively, for the Arata-Chufa and Ketar schemes. The spatial soil losses amounted to up to 18 t/ha/yr for the Arata-Chufa scheme and 41 t/ha/yr for the Ketar scheme. The authors concluded that overland sediment inflow is not considered properly related to canal sedimentation, and this is a major cause of excessive sedimentation problems (among others) in Sub-Saharan Africa.

Jáchymová et al. [16] used the RUSLE-based WATEM/SEDEM model in their research. This modelling approach allowed for the modelling of sediment fluxes at 127,484 < risk points, meaning that a good coverage for the entire Czech Republic could be achieved. Risk points were defined for the study as follows: outlets of contributing areas 1 < ha, wherein the surface runoff enters residential areas or vulnerable bodies of water. The authors found that the most important factor for risk definition is a combination of morphometric characteristics (specific width and stream power index), followed by the watershed area, the proportion of grassland, the soil erodibility, and the rain erosivity (described by PC2).

Szabó et al. [17] analyzed the use of various rainfall simulators in the determination of the driving forces of changes in sediment concentration and clay enrichment. The authors compared a field and a laboratory rainfall simulator. They found that the slope gradient is an effective regulator only in the laboratory. Rainfall intensity was also more effective in the laboratory than in the field simulations. These findings suggest that soil-related properties play a prominent role in driving sediment concentration in the field, whereas, in the laboratory, slope and rainfall intensity were found to be driving factors independent of soil-regulated sediment concentrations.

As we can see, on the one hand, there is a great deal of information on soil water erosion-related issues; on the other hand, as we go into more detail, there are always new findings. After all, we can conclude that a lot more research is still needed [18] in order to find the relations between the factors leading to water erosion. These findings can help in mitigation. We highly encourage the creation of more Special Issues in this same field of research.

Funding: This research received no external funding.

Acknowledgments: I wish to thank all of the authors who contributed to this Special Issue. The other group of people who contributed a huge amount of time and effort were the anonymous reviewers and the editorial managers—many thanks for them. I was very satisfied with the review process and management of the Special Issue, and all those involved are very much appreciated.

Conflicts of Interest: The author declares no conflict of interest.

References

1. García-Ruiz, J.M.; Beguería, S.; Nadal-Romero, E.; González-Hidalgo, J.C.; Lana-Renault, N.; Sanjuán, Y. A meta-analysis of soil erosion rates across the world. *Geomorphology* **2015**, *239*, 160–173. [CrossRef]
2. Lal, R. (Ed.) *Soil Quality and Soil Erosion*; CRC Press: Boca Raton, FL, USA, 2019. [CrossRef]
3. IPCC, 2019: Climate Change and Land: An IPCC Special Report on Climate Change, Desertification, Land Degradation, Sustainable Land Management, Food Security, and Greenhouse Gas Fluxes in Terrestrial Ecosystems. Available online: https://www.ipcc.ch/srccl/cite-report/ (accessed on 19 January 2022).
4. Arabameri, A.; Chen, W.; Blaschke, T.; Tiefenbacher, J.P.; Pradhan, B.; Tien Bui, D. Gully head-cut distribution modeling using machine learning methods—A case study of N.W. Iran. *Water* **2020**, *12*, 16. [CrossRef]
5. Sidorchuk, A. Models of gully erosion by water. *Water* **2021**, *13*, 3293. [CrossRef]
6. Coratza, P.; Parenti, C. Controlling factors of badland morphological changes in the Emilia Apennines (Northern Italy). *Water* **2021**, *13*, 539. [CrossRef]
7. Bai, L.; Jiao, J.; Wang, N.; Chen, Y. Structural connectivity of sediment affected by check dams in loess hilly-gully region, China. *Water* **2021**, *13*, 2644. [CrossRef]
8. Kreklow, J.; Steinhoff-Knopp, B.; Friedrich, K.; Tetzlaff, B. Comparing rainfall erosivity estimation methods using weather radar data for the State of Hesse (Germany). *Water* **2020**, *12*, 1424. [CrossRef]
9. Almohamad, H. Impact of land cover change due to armed conflicts on soil erosion in the basin of the Northern Al-Kabeer River in Syria using the RUSLE model. *Water* **2020**, *12*, 3323. [CrossRef]
10. Keller, B.; Centeri, C.; Szabó, J.A.; Szalai, Z.; Jakab, G. Comparison of the applicability of different soil erosion models to predict soil erodibility factor and event soil losses on loess slopes in Hungary. *Water* **2021**, *13*, 3517. [CrossRef]
11. Li, M.; Yao, J.; Yan, R.; Cheng, J. Effects of infiltration amounts on preferential flow characteristics and solute transport in the protection forest soil of Southwestern China. *Water* **2021**, *13*, 1301. [CrossRef]
12. Su, Y.; Li, P.; Ren, Z.; Xiao, L.; Wang, T.; Zhang, Y. Slope erosion and hydraulics during thawing of the sand-covered loess plateau. *Water* **2020**, *12*, 2461. [CrossRef]
13. Lu, J.; Sun, B.; Ren, F.; Li, H.; Jiao, X. Effect of freeze-thaw cycles on soil detachment capacities of three loamy soils on the loess plateau of China. *Water* **2021**, *13*, 342. [CrossRef]
14. Ali, M.G.; Ali, S.; Arshad, R.H.; Nazeer, A.; Waqas, M.M.; Waseem, M.; Aslam, R.A.; Cheema, M.J.M.; Leta, M.K.; Shauket, I. Estimation of potential soil erosion and sediment yield: A case study of the transboundary Chenab River Catchment. *Water* **2021**, *13*, 3647. [CrossRef]
15. Gurmu, Z.A.; Ritzema, H.; Fraiture, C.d.; Riksen, M.; Ayana, M. Sediment influx and its drivers in farmers' managed irrigation schemes in Ethiopia. *Water* **2021**, *13*, 1747. [CrossRef]
16. Jáchymová, B.; Krása, J.; Dostál, T.; Bauer, M. Can lumped characteristics of a contributing area provide risk definition of sediment flux? *Water* **2020**, *12*, 1787. [CrossRef]
17. Szabó, J.A.; Centeri, C.; Keller, B.; Hatvani, I.G.; Szalai, Z.; Dobos, E.; Jakab, G. The use of various rainfall simulators in the determination of the driving forces of changes in sediment concentration and clay enrichment. *Water* **2020**, *12*, 2856. [CrossRef]
18. Poesen, J. State of science. Soil erosion in the Anthropocene: Research needs. *Earth Surf. Process. Landforms* **2018**, *43*, 64–84. [CrossRef]

Article

Gully Head-Cut Distribution Modeling Using Machine Learning Methods—A Case Study of N.W. Iran

Alireza Arabameri [1,*], **Wei Chen** [2,3,4], **Thomas Blaschke** [5], **John P. Tiefenbacher** [6], **Biswajeet Pradhan** [7,8] **and Dieu Tien Bui** [9,*]

[1] Department of Geomorphology, Tarbiat Modares University, Tehran 36581-17994, Iran
[2] College of Geology & Environment, Xi'an University of Science and Technology, Xi'an 710054, China; chenwei0930@xust.edu.cn
[3] Key Laboratory of Coal Resources Exploration and Comprehensive Utilization, Ministry of Land and Resources, Xi'an 710021, China
[4] Shaanxi Provincial Key Laboratory of Geological Support for Coal Green Exploitation, Xi'an 710054, China
[5] Department of Geoinformatics–Z_GIS, University of Salzburg, 5020 Salzburg, Austria; thomas.blaschke@sbg.ac.at
[6] Department of Geography, Texas State University, San Marcos, TX 78666, USA; tief@txstate.edu
[7] Centre for Advanced Modeling and Geospatial Information Systems (CAMGIS), Faculty of Engineering and Information Technology, University of Technology Sydney, Sydney 2007, New South Wales, Australia; biswajeet.pradhan@uts.edu.au
[8] Department of Energy and Mineral Resources Engineering, Choongmu-gwan, Sejong University, 209 Neungdong-ro Gwangjin-gu, Seoul 05006, Korea
[9] Institute of Research and Development, Duy Tan University, Da Nang 550000, Vietnam
[*] Correspondence: a.arabameri@modares.ac.ir (A.A.); Buitiendieu@duytan.edu.vn (D.T.B.)

Received: 27 October 2019; Accepted: 16 December 2019; Published: 19 December 2019

Abstract: To more effectively prevent and manage the scourge of gully erosion in arid and semi-arid regions, we present a novel-ensemble intelligence approach—bagging-based alternating decision-tree classifier (bagging-ADTree)—and use it to model a landscape's susceptibility to gully erosion based on 18 gully-erosion conditioning factors. The model's goodness-of-fit and prediction performance are compared to three other machine learning algorithms (single alternating decision tree, rotational-forest-based alternating decision tree (RF-ADTree), and benchmark logistic regression). To achieve this, a gully-erosion inventory was created for the study area, the Chah Mousi watershed, Iran by combining archival records containing reports of gully erosion, remotely sensed data from Google Earth, and geolocated sites of gully head-cuts gathered in a field survey. A total of 119 gully head-cuts were identified and mapped. To train the models' analysis and prediction capabilities, 83 head-cuts (70% of the total) and the corresponding measures of the conditioning factors were input into each model. The results from the models were validated using the data pertaining to the remaining 36 gully locations (30%). Next, the frequency ratio is used to identify which conditioning-factor classes have the strongest correlation with gully erosion. Using random-forest modeling, the relative importance of each of the conditioning factors was determined. Based on the random-forest results, the top eight factors in this study area are distance-to-road, drainage density, distance-to-stream, LU/LC, annual precipitation, topographic wetness index, NDVI, and elevation. Finally, based on goodness-of-fit and AUROC of the success rate curve (SRC) and prediction rate curve (PRC), the results indicate that the bagging-ADTree ensemble model had the best performance, with SRC (0.964) and PRC (0.978). RF-ADTree (SRC = 0.952 and PRC = 0.971), ADTree (SRC = 0.926 and PRC = 0.965), and LR (SRC = 0.867 and PRC = 0.870) were the subsequent best performers. The results also indicate that bagging and RF, as meta-classifiers, improved the performance of the ADTree model as a base classifier. The bagging-ADTree model's results indicate that 24.28% of the study area is classified as having high and very high susceptibility to gully erosion. The new ensemble model accurately identified the areas that are susceptible to gully erosion based on the past patterns of formation, but it also provides highly accurate predictions of future gully development. The novel ensemble method introduced in this research is recommended for use to evaluate the patterns of gullying in arid and semi-arid environments and can effectively identify the most salient conditioning factors that promote the development and expansion of gullies in erosion-susceptible environments.

Keywords: gully head-cuts; machine learning modeling; soil erosion; Iran

1. Introduction

Gullies are common features in arid and semi-arid regions, and they are major causes of sediment erosion; they supply from 10 to 94% of the total sediment yield in some watersheds [1]. High erosion rates undercut agricultural sustainability and necessitate the search for (usually expensive) solutions in the context of costly governmental policies. However, studying and predicting gully erosion is difficult [2–4]. In terms of the ecosystem effects and environmental damages from gully erosion, studies have focused on the influential factors and on identification of susceptible areas using geographic information systems (GIS) and remote sensing (RS) [5–8]. This study develops a new model to detect and predict gully locations with high spatial accuracy to reduce gully erosion damages.

One method that many have used is gully-erosion susceptibility mapping (GESM). This approach can provide useful and easy-to-understand information to planners and hazard managers [9], but there is no standard procedure for producing these maps. In recent decades, researchers have devised and experimented with many GESM techniques and various traditional data-driven approaches, including logistic regression (LR) [10,11], weights of evidence (WoE) [12,13], conditional analysis (CA) [14,15], certainty factor (CF) [16], index or entropy (IOE) [17], analytical hierarchy process (AHP) [18,19], and frequency ratio (FR) [12].

One of the difficulties in the regional GESM process is that the factors influencing gully erosion require data usually derived from various sources at different spatial scales, which may contain uncertainties and imprecisions. Traditional data-driven approaches cannot be used to determine the relationships between geo-environmental factors and gully erosion occurrence because of the limitations caused by imbedded statistical assumptions about variables' independence and data distributions in susceptibility analyses [20,21]. New modeling methods are needed that go beyond traditional data-driven approaches, and methods that can deal with the above issues and can enhance model performance.

Recently, machine-learning (ML) techniques have become popular for the spatial prediction of natural hazards like wildfires [22], sinkholes [23], groundwater depletion and flooding [24–38], droughts [39], earthquakes [40], land subsidence [41], and landslides [42–48]. ML is a type of artificial intelligence (AI) that uses computer algorithms to analyze and forecast information by learning from training data. ML algorithms that have been used for GESM include random forest (RF), boosted regression tree (BRT), support vector machine (SVM), classification and regression trees (CART), artificial neural networks (ANN), stochastic gradient tree-boost (SGT), maximum entropy (ME), and multivariate adaptive regression splines (MARS) [13,49–58].

Ensemble models have been used in GESM due to their novelty and their ability to comprehensively assess gully-erosion parameters for discrete classes of independent factors [51,52]. Although some studies have been conducted on the spatial prediction of gullies, a standard framework considering all influential factors for achieving a reasonable and reliable prediction has not been established. Some studies and techniques should be used in different hydro-geomorphological environments to devise a global framework for gully-erosion modeling. Additionally, some factors contribute to gullying that are either difficult to recognize (and measure), or they are difficult to convert to raster formats for modeling. Therefore, one of the future fields of gully modeling should focus on the detection and application of the unknown factors that influence gully formation. This may be achieved by combining gullying research with GIS and data-mining tools to create a tool or technique that can map future, unknown factors. This could help planners, decision makers, and environmental managers to prepare gully erosion maps of the highest quality with the best possible accuracy to better manage gullying in erosion-susceptible areas.

The main difference between this study and previously conducted studies is that this study explores a new ensemble-intelligence approach that employs bagging as a meta-classifier with an alternating decision tree (ADTree) as a base classifier to spatially predict gully erosion. The results produced by this new ensemble-intelligence approach are compared to the results generated with a single alternating decision tree, a rotational-forest-based alternating decision tree (RF-ADTree), and benchmark logistic regression (LR) to assess and improve the accuracy of GESM. These ML modes haven't been used for GESM, so we assess the performance of the new ensemble model using a variety of statistical metrics and the area under the curve (AUC).

The Chah Mousi watershed (northeastern Iran) is an arid region very prone to gully erosion. Gullies are widespread throughout the region and cause land degradation and economic damages every year. This study illustrates and compares individual and ensemble machine learning models to assess gullying susceptibility. We test the efficacy of these models and compare them to find the most suitable model for land use planning. The main objectives of this study are identifying and mapping the extant gullies in the Chah Mousi watershed by (a) creating an inventory; (b) mapping, modeling, and predicting the locations of gully head-cuts; (c) characterizing the roles of various geo-environmental features as factors that control the distribution of gullies; and (d) evaluating gully erosion susceptibility in the study area.

2. Materials and Methods

2.1. Study Area

The Chah Mousi watershed is in Semnan province, Iran, and is located between 35°15′05″ and 35°37′12″ N and 54°35′44″ and 55°23′05″ E (Figure 1). It is a relatively small area of approximately 2176.02 km². The greatest change in elevation is along a NE to SE axis. The average elevation in the northeastern quadrant is 2123 m.a.s.l. In the southeastern quadrant, it is 672 m.a.s.l. As the region is relatively small, the slope degree varies significantly from flat to 67.8°, although the average is about 3°. Due to the predominance of flat landscapes, standing and slow-moving water is more typical than runoff. The mean annual precipitation ranges from 48 to 206 mm, principally during the wet season from January to March [59]. Temperatures typically reach a peak of 41 °C during summer, especially in the south, and a low below 0 °C during winter in the northern parts of the region; though average temperatures during the rest of the year range from 13 to 23 °C [59]. Together, these numbers indicate the potential for meteorological stress on the land surface with high thermal and precipitation variations and local spikes that may cause freezing and thawing of soils and expansion and contraction within the regolith [13].

The land covers include agriculture, bare land, kavir (barren sandy and rocky desert), rangeland, rock outcrops, salt lakes, wetlands, and salt lands. The latter are particularly vulnerable to dissolution processes during the wet season as the salt crust is easily weathered, giving rise to pores that promote changing groundwater levels and erosion of soils [60]. The distribution of salt crusts is evident in the regional soil map (primarily in areas featuring aridisols and entisols and where the outcropping lithologies are also reported). The main lithological units in the study area are marl, gypsiferous marl and limestone, shale, sandstone, granite, conglomerate, and salt flat [60].

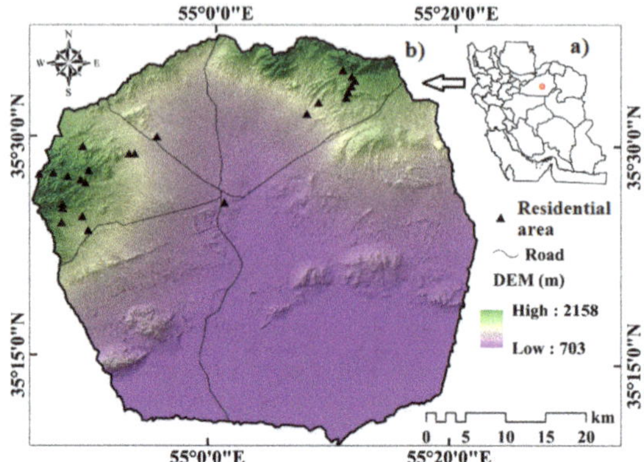

Figure 1. Study area. (**a**) Location of the study area in Iran and Semnan Province. (**b**) Elevation and Hillshade model of the study area.

2.2. Gully Mapping

Archival records containing reports of gully erosion that have been compiled by the Semnan Agricultural and Natural Resources Research and Education Center were used as the first source of locational data. Upon this historical foundation, gully locations and dimensions were identified and measured using remotely sensed data viewed through Google Earth. Finally, a field survey was conducted in the study region to update and refine the inventory (Figure 2). Sites of gully head-cuts were geolocated with a DGPS (Differential Global Positioning System) device. The survey yielded 119 gully head-cuts (Figure 2) to be used for modeling. Of the overall dataset, 75 gullies (63.02%) were identified from archives, 19 gullies (15.96%) collected using Google Earth, and 25 gullies (21.008%) were collected in a field survey. All gullies were checked and mapped using DGPS with millimeter accuracy. The universal transverse Mercator (UTM) coordinate system was used. The models described above were applied to the locations of 83 head-cuts (70% of the total). The models were tested (or validated) with the remaining 36 gully locations (30% of the total). As the models selected in this study correspond to a family that predicts the presence or absence of a phenomenon, an equal number of locations (36 no gully locations as validation data and 83 no gully locations as calibration data) were selected and tested as well [52]. In turn, this procedure creates a balanced dataset for the subsequent analyses, although it should be noted that the geomorphological features still debates whether balanced or unbalanced datasets should be created prior to a susceptibility analysis [19,58]. Some of mapped gullies are shown in Figure 3.

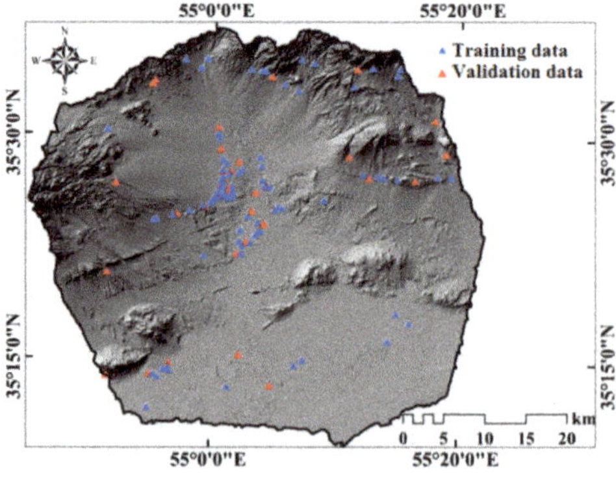

Figure 2. Location of training and validation gullies in the study area.

Figure 3. Gullies in the study area.

2.3. Gully Erosion Conditioning Factors

Several factors affect a location's susceptibility to gully erosion [17,19]. After completing a study of the gully-erosion literature, and considering local conditions and data availability, 18 variables were selected for inclusion in the modeling process. These include elements of topographical, geological, and hydrological conditions.

The following topographical factors were considered: elevation, slope gradient, aspect, plan curvature, convergence index (CI), slope length (LS), topographic wetness index (TWI), topographic position index (TPI), and terrain ruggedness index (TRI). Each was calculated using PALSAR DEM with 12.5 m spatial resolution applying the basic terrain analyses in SAGA GIS. A detailed explanation of the equations used to calculate LS, TWI, and SPI is available in Arabameri et al. [19].

The description of the lithology was acquired from a geological map at a scale of 1:100,000 (Geological Survey Department of Iran, [59]). The map was digitized and 6 geological classes were identified in the study area: A (including marl, gypsiferous marl, and limestone; dacitic to andesitic volcano sediment; well-bedded green tuff and tuffaceous shale; dacitic to andesitic volcanic; dacitic to andesitic volcano breccia; andesitic volcano breccia, sandstone, marl, and limestone; granite, pale-red polygenic conglomerate, and sandstone), B (including phyllite, slate, and meta-sandstone; Jurassic dacite to andesite lava flows), C (including Cretaceous rocks, in general), D (including light red to brown marl and gypsiferous marl with sandstone intercalations; red marl, gypsiferous marl, sandstone, and conglomerate), E (including fluvial conglomerate, piedmont conglomerate, and sandstone), and F (salt flat, high-level piedmont fan and valley terrace deposits, low-level piedmont fan and valley terrace deposits, and salt lake) (Figure 4p).

The hydrological gully erosion factors that were included in the modeling process are drainage density, distance-to-stream, mean annual rainfall, and stream-power index (SPI). Drainage density and distance-to-stream were calculated using the stream network information developed from the PALSAR DEM in ArcGIS 10.5. Raster maps of these factors were prepared using line-density and Euclidean-distance tools in ArcGIS 10.5. The SPI was calculated as follows:

$$\text{SPI} = \text{As} \times \tan \beta \tag{1}$$

where As is the specific catchment area, and β is slope (°).

Annual precipitation data were obtained for the period from 1984 to 2014 recorded at the Toroud, Razveh, Moalleman, and Hosseinan weather stations operated by the Iran Meteorological Organization (IRIMO, 2014). The rainfall data were interpolated using the kriging interpolation tool in ArcGIS 10.5. Gully erosion is also influenced by soils, land use, and vegetation [19]. Therefore, these factors are represented by soil types, land use/land cover (LU/LC), and normalized difference vegetation index (NDVI), and were used as conditioning factors. Soil type data were based on the information from the Soil Conservation Section of Agricultural and Natural Resources Research Centre of Semnan Province. LU/LC and NDVI data were obtained from Landsat 8 images (15 August 2017) with a 30 m resolution.

The LU/LC map containing eight classes (agriculture, bare land, kavir, poor range, rock, salt lake, salt land, and wetland) was prepared using the supervised classification method and maximum likelihood in ENVI4.8 software. The map was verified using the kappa coefficient with 459 ground control points (GCP). The kappa value of the resulting map was 0.976. The NDVI was calculated using Landsat 8 bands 4 (red) and 5 (infrared) data in ArcGIS 10.5.

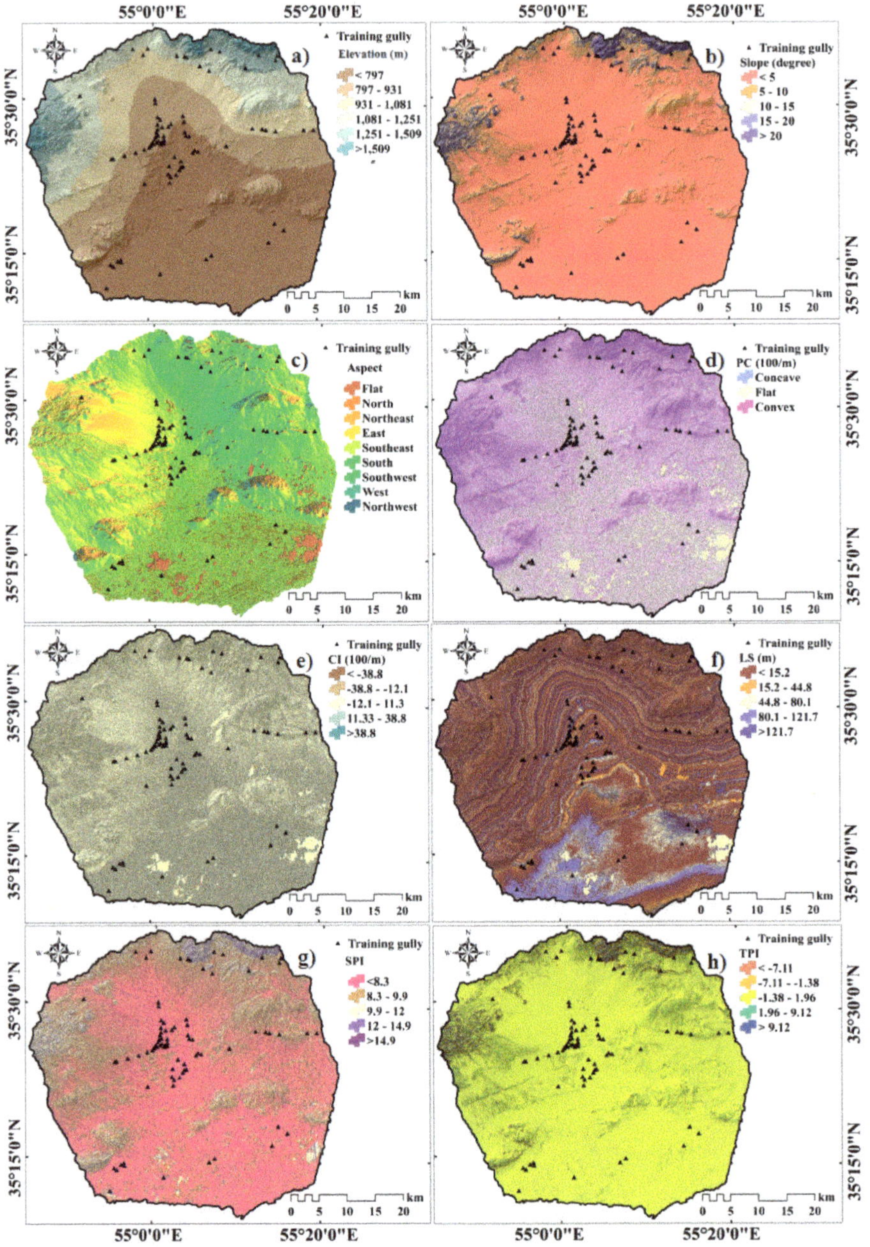

Figure 4. *Cont.*

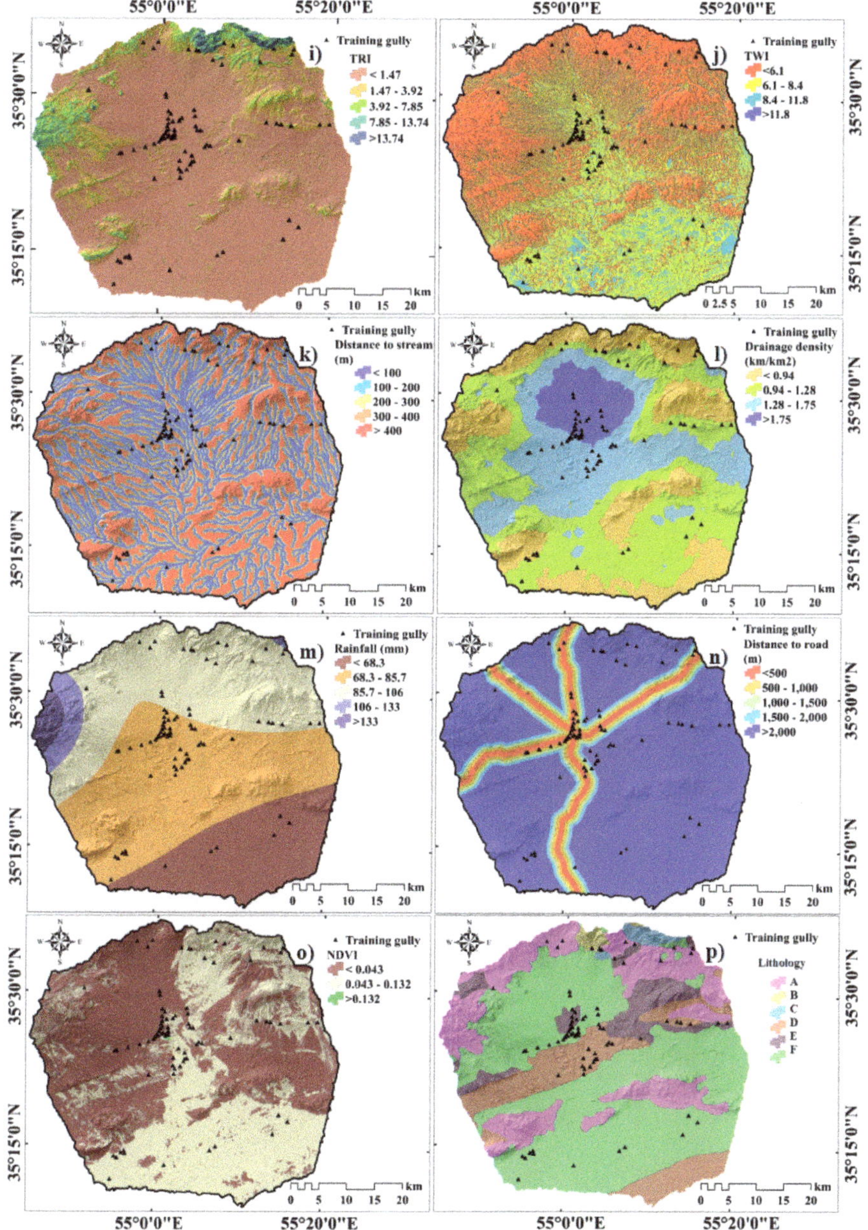

Figure 4. *Cont.*

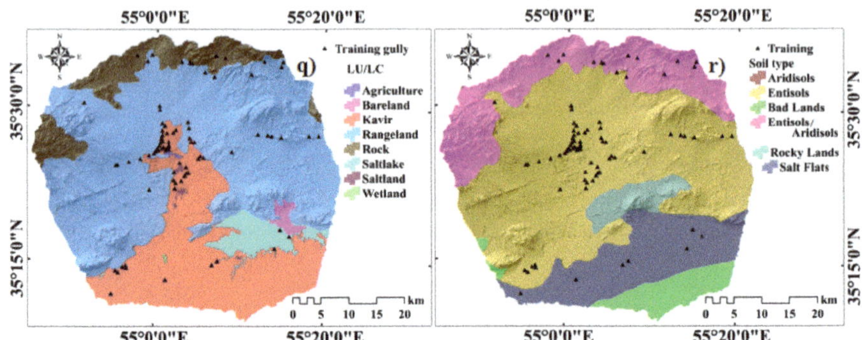

Figure 4. Gully erosion conditioning factors. (**a**) Elevation, (**b**) slope, (**c**) aspect, (**d**) plan curvature, (**e**) convergence index (CI), (**f**) slope length (LS), (**g**) stream power index (SPI), (**h**) topography position index (TPI), (**i**) terrain ruggedness index (TRI), (**j**) topography wetness index (TWI), (**k**) distance to stream (**l**) drainage density, (**m**) rainfall, (**n**) distance to road (**o**) NDVI, (**p**) lithology (**q**) land use/land cover (LU/LC), and (**r**) soil type.

Roads also affect gully erosion as they intercept and concentrate overland flow [17]. This factor is represented by the distance to road in gully and non-gully locations, which is determined by vectorizing topographic maps and Google Earth images, and then transforming the data to a raster map using line density tools in ArcGIS 10.5.

2.4. Models

2.4.1. Rotational Forest (RF)

RF modeling is a relatively new ensemble algorithm that increases the accuracy and diversity of base classifiers, and it was first proposed by Rodriguez et al. [39]. The success of RF modeling depends on the rotation matrix generated by transformations and base classifiers [61,62]. The basis of RF modeling is principal component analysis (PCA), which can extract features and create training datasets for learning base classifiers [63]. RF has been applied to classification problems, such as landslide-susceptibility research, land use mapping, and flash flood susceptibility research [64–66].

Suppose $x = (x_1, x_2, x_3, \ldots, x_n)$ is the vector of the landslide conditioning factor, $y = (y_1, y_2)$ is the vector of landslide or non-landslide class, X is the training dataset, $A_1, A_2, A_3, \ldots, A_L$ are the classifiers in the ensemble, and B is the landslide conditioning factor set. The steps of training classifier A_i are as follows. The rotation matrix R_i^a generated by the matrix of R_i is shown in Equation (2).

$$R_i = \begin{bmatrix} a_{i,1}^{(1)}, a_{i,1}^{(2)}, \ldots, a_{i,1}^{(Q1)} & 0 & \cdots & 0 \\ 0 & a_{i,1}^{(1)}, a_{i,1}^{(2)}, \ldots, a_{i,1}^{(Q2)} & \cdots & 0 \\ \vdots & \vdots & \ddots & \vdots \\ 0 & \cdots & \cdots & a_{i,1}^{(1)}, a_{i,1}^{(2)}, \ldots, a_{i,1}^{(Qk)} \end{bmatrix} \quad (2)$$

R_i is produced by the following four steps:

(i)   Divide B into K subsets, and the number of gully conditioning factors of each subset is $Q = n/K$.

(ii)   In case of the classifier A_i, let $B_{i,j}$ be the *j*th, where $j = 1, 2, 3, \ldots$ and K is the subset of gully conditioning factors. $X_{i,j}$ is the gully conditioning factor of $B_{i,j}$ from X. $B_{i,j}$ is randomly selected from the $X_{i,j}$ with the 75% size by bootstrap algorithm. Then, $X_{i,j}'$ would be transformed to achieve coefficient $a_{i,1}^{(1)}, a_{i,1}^{(2)}, \ldots, a_{i,1}^{(Qi)}$, the size of $a_{i,1'}$ is Q × 1.

(iii) Arrange a sparse rotation matrix R_i with the obtained coefficients.

(iv) The confidence of each class is calculated by the average combination method in the given test sample χ,

$$\mu_k(\eta) = \frac{1}{L}\sum_{i=1}^{L} \gamma_{i,k}(\eta R_i^a), \quad k = 1,2,3,\ldots,c \tag{3}$$

where $\gamma_{i,k}(\eta R_i^a)$ is the probability produced by the classifier A_i to the hypothesis that η belongs to the class k.

2.4.2. Alternating Decision Tree

The alternating decision tree (ADTree) model is an ensemble model that consists of a boosting algorithm and a decision tree [67]. It is a generalization of a decision tree in which each node is replaced by a splitter node and a prediction node [68,69]. The base rule mapping from an instance to real number involves a precondition c_1, a base condition c_2, and two real numbers a and b. If $c_1 \cap c_2$, the prediction is a, and the prediction is b when $c_1 \cap {}^-c_2$; $^-$ means negation. The values of a and b are determined by Equations (4) and (5), respectively.

$$a = \frac{1}{2}\ln\frac{W_+(c_1 \cap c_2)}{W_-(c_1 \cap c_2)} \tag{4}$$

$$b = \frac{1}{2}\ln\frac{W_+(c_1 \cap {}^-c_2)}{W_-(c_1 \cap {}^-c_2)} \tag{5}$$

where $W(p)$ is the total weight of training instance. The best c_1 and c_2 values are obtained by minimizing the $Z_t(c_1, c_2)$, which is defined as Equation (6).

$$Z_t(c_1, c_2) = 2\sqrt{W_+(c_1 \cap c_2)W_-(c_1 \cap c_2)} + \sqrt{W_+(c_1 \cap {}^-c_2)W_-(c_1 \cap {}^-c_2)} + W({}^-c_2) \tag{6}$$

Suppose that R is a set of base rules. Then, a new rule can be defined as $R_{t+1} = R_t + r_t$, $r_t(x)$, which shows two prediction values (a and b) at every layer of the tree. x is a set of instances. The classification of instances is the sign of the sum of all predicted values in R_{t+1}:

$$Class(x) = sign(\sum_{t=1}^{T} r_t(x)) \tag{7}$$

The algorithm first finds the best constant prediction for the whole data set [70]. Cross validation is often used for selection [71].

2.4.3. Bagging

Bootstrap aggregation or bagging (BAG) was introduced by Breiman in 1996 [72]. The bootstrap technique randomly selects and replaces samples to generate multiple samples to form a training dataset. Every subset generated is used to build a decision tree, and they are later aggregated in the final model. The accuracy of classification is improved by reducing the variance of classification error [73,74]. In recent years, BAG has been widely applied in landslide susceptibility research and has performed well [75–77].

2.4.4. Logistic Regression

Logistic regression (LR) is one of the most popular multivariate statistical analysis methods [78–80]. It can make a multivariate regression correlation between a dependent variable and several independent variables [81,82]. The advantage of LR is that the variables can be continuous, discontinuous, or a combination of the two [83,84]. In this study, the main purpose of using an LR model is to determine the relationships between landslide occurrence and gully conditioning factors, calculated using Equation (8).

$$P = \frac{1}{1+e^{-Z}} \tag{8}$$

where P is the probability of gully occurrence and ranges from 0 to 1. Z is a linear sum of constants, and its range is $(-\infty, +\infty)$. The calculation equation of Z can be defined as Equation (9).

$$Z = \alpha + \beta_1 x_1 + \beta_2 x_2 + \beta_3 x_3 + \ldots + \beta_n x_n \tag{9}$$

where α is a constant, β_i ($i = 1, 2, 3, \ldots n$) is the coefficient of the model, and x_i ($i = 1, 2, 3, \ldots n$) is the independent variable.

2.4.5. Frequency Ratio

The ratio between the frequency of occurrences and non-occurrences at a location within a given causative factor class is called the FR [19]. Larger ratios suggest that those factor classes are more important determinants of the occurrence (in this case, gully-erosion proneness or susceptibility. As there are numerous pertinent factors at play in each location (or area defined by a pixel in our digital map), the potential for gully erosion can be computed as the sum of all ratios for the predisposing factor classes [19]. FR is empirical. It is, in fact, not a statistical method; it is not based on statistical distributions.

2.4.6. Random Forest (RAF)

RAF uses multiple trees to classify locations based on a single conditioning factor [85]. The RAF algorithm continuously replaces the factors affecting each pixel space, thereby creating numerous decision trees. A combination of all decision trees in a study area provide the information to support decision making [85]. An RAF contains 3 user-defined parameters: (1) the number of variables used to construct each decision tree, which indicates the power of each independent tree; (2) the number of trees included in the RF; and (3) the minimum number of nodes within the trees. The prediction power of RAFs increase as the strength of independent trees increases and as the correlation between them decreases. Sixty-six percent of the data (the testing data) are used to grow a tree, and the result is called a bootstrap. A randomly introduced predictor variable splits a node in the tree's construction during the growing process. The remaining third of the data is used to evaluate (or validate) the fitted tree. The average of all predicted values produced during several iterations of the algorithm creates the final modeled prediction. In this model, two factors—the mean decrease accuracy and the mean decrease Gini index—are used to prioritize the effective factors. Comparing the mean decrease accuracy to the mean decrease Gini index determines the relative importance of the effective factors, especially the relationships between environmental factors. RAF analyses were carried out in R 3.3.1 using the "Randomforest" package [85].

2.5. Multicollinearity Assessment

In GESM, testing for collinearity among the effective factors in gullying is very important, because the collinearity reduces the accuracy of the GESM [86–89]. The variance inflation factor (VIF) and Tolerance (TOL) are very commonly used indicators for checking multicollinearity among parameters [90,91]. TOL values less than 0.1 or 0.2 and VIF values greater than 5 or 10 indicate collinearity between the parameters [17,19,86,89,92]. In the present study, the multicollinearity test of gully erosion conditioning factors (GECFs) was done using Equations (10) and (11) in SPSS software:

$$Tolerance = 1 - R^2 J \tag{10}$$

$$VIF = [\frac{1}{Tolerence}] \tag{11}$$

where $R^2 J$ is the regression coefficient for determining independent variable j.

2.6. Methodology

As described above, an inventory of gullies was created, and the gully-erosion conditioning data were compiled in a GIS to provide input for the modeling process (Figure 5). The gully sites were divided into two datasets: 70% were used for training, and 30% were used for validation of the models. An assessment of multicollinearity among the conditioning factors was performed. The relative weights of the GECFs were determined using an RAF model, and an analysis of the spatial relationships between GECFs and gullies was conducted with FR. GESMs were created using each of the four models: ADTree, RF-ADTree, Bagging-ADTree, and LR. Finally, the models were evaluated and validated using the receiver operating characteristic (ROC) curves and by calculating the area under the ROC curve (AUC) for each model [93–95]. The AUC values are between 0 and 1, which can be interpreted following these categories: 0.6–0.7 have poor, 0.6–0.7 medium, 0.7–0.8 good, 0.8–0.9 very good, and 0.9–1 excellent accuracy [9,17,19]. The four models used were objectively compared to determine the most effective approach.

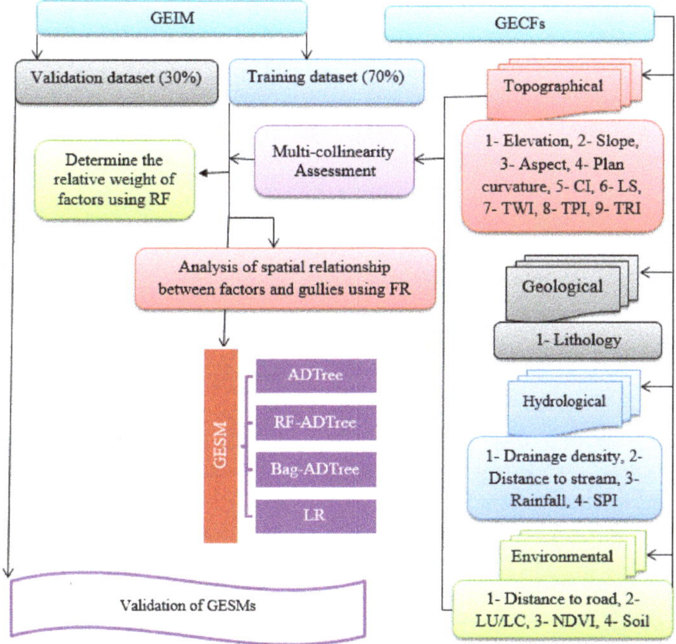

Figure 5. Flowchart of modeling procedure, where GIM is gully inventory map, GECFs is gully erosion conditioning factors, GESM is gully erosion susceptibility map.

3. Results

3.1. Multicollinearity Assessment

A multicollinearity analysis of the GECFs was performed (Table 1). The analysis revealed that TOL and VIF values for all factors are >0.1 and <5, respectively, indicating that the variables are not significantly correlated and that they can be used in further analyses.

3.2. Spatial Relationship between Gully Locations and Conditioning Factors by Applying FR Model

Analyses of the spatial relationships between gully locations and GECFs (Table 2) showed that classes of conditioning factors with FR values greater than 1 are susceptible to gully erosion [17].

For instance, among topographical factors, locations up to 1000 m. a.s.l. are the most susceptible to gully erosion—the highest value of FR is for sites with elevations from 797 to 931 m a.s.l. Locations above 1000 m a.s.l. have low susceptibility, and elevations above 1509 m a.s.l. have the lowest susceptibility and lowest FR values (FR = 0.000). All gullies in the study area occur on slopes below 15°. The highest FR values are found in slopes < 5° (1.080) and from 10 to 15° (1.119). There are no gullies on slopes > 15°. This is in accordance with the plan-curvature results. Flat areas have the highest FR value (1.391) and concave slopes have gullies (0.967). Most gullies occur on slopes exposed to the east (1.941), southeast (1.344), and northeast (1.184), whereas while northwest-, west-, and southwest aspects have more gullies (NW (0.183), W (0.429), and SW (0.536)), there are very few gullies on north-facing slopes (0.679). Based on convergence index, sites in the class of ≤38.8 (FR = 1.737) possess the most important cause of gully occurrence in the study area.

Table 1. Multi-collinearity analysis of gully erosion conditioning factors.

Factors	Collinearity Statistics	
	TOL [a]	VIF [b]
Aspect	0.904	1.123
Lithology	0.759	1.318
Slope	0.612	1.525
Normalized Difference Vegetation Index	0.596	1.674
Slope length	0.577	1.734
Convergence Index	0.559	1.780
Terrain Ruggedness Index	0.523	2.132
Distance to road	0.497	2.231
Soil type	0.431	2.312
Land use/land cover	0.423	2.383
Stream Power Index	0.419	2. 443
Elevation	0.415	2.504
Drainage density	0.411	2.561
Plan curvature	0.369	2.716
Topographic Wetness Index	0.357	2.903
Topographic Position Index	0.344	2.984
Rainfall	0.321	3.098

[a] TOL is tolerance. [b] VIF is variance inflation factor.

According to LS factor, areas with the lowest slope length have the highest susceptibility to gully occurrence, so that class of <15.2 m, with FR = 1.244, showed the strongest relationship to gullying in the study area.

Generally, TPI values > 0 indicate ridges, 0—flat areas (or constant slopes), and <0—valleys. This is confirmed with the statistical relationships between gully locations and TPI values in the study area. Most of the study area is flat and classes of TPI < 1.96 are those with the gully locations. This is in accordance with TRI values that show terrain heterogeneity. Higher TRI values show increased local relief heterogeneity. In contrast, lower TRI values indicate more level surfaces (e.g., planar surfaces or various depositional landforms). The results showed that gullies occur in areas belonging to classes of TRI values < 7.84, and the most susceptible are areas with TRI < 1.47. Despite the occurrence of gullies, the terrain is quite homogenous; most of the study area is flat. TWI reveals the areas with drainage depressions where water is likely to accumulate. Thus, the areas with high values of TWI should be more susceptible to gully formation, which is in accordance with the results that showed that higher TWI values (>11.8) have a higher occurrence of gullies in the study area. SPI values indicate potential flow-erosion at a point in the topographic surface. Most of the gullies occur in areas where SPI values are <14.9 (FR = 4.66).

Distance-to-stream and drainage density are important factors conditioning gully occurrence [17]. Gullies occur mainly in the areas close to streams (<100 m) [13]. In addition, most of the gullies occur in areas receiving 68 to 85 mm of precipitation annually [16] (Table 2).

In lithological units, class of B (phyllite, slate and meta-sandstone, and Jurassic dacite to andesite lava flows) showed the strongest correlation with gully occurrence in the study area.

According to NDVI, class of 0.043 to 0.132 had the highest FR (1.34) and therefore the strongest relationship to gully formation. Moreover, most of the gullies occur in areas of kavir and poor rangeland, which had FR values of 1.961 and 0.672, respectively. Gully erosion occurs mainly in areas with entisols/aridisols (Table 3).

Roads may intercept overland flow and promote gully formation. Most of the gullies occur near roads (<1000 m) [16]; the strongest relationship is <500 m (FR = 6.43).

Table 2. Analysis of spatial relationship between conditioning factors and gully locations using frequency ration model.

Factors	Classes	Pixels in Domain		Gullies		FR [a]
		No	%	No	%	
Elevation (m)	<797	1,050,197	43.44	30	34.88	0.803
	797–931	481,498	19.91	34	39.53	1.985
	931–1081	354,322	14.65	9	10.47	0.714
	1081–1251	334,954	13.85	9	10.47	0.755
	1251–1509	157,674	6.52	4	4.65	0.713
	>1509	39,161	1.62	0	0.00	0.000
Slope (°)	<5	2,031,134	84.01	78	90.70	1.080
	5–10	220,796	9.13	5	5.81	0.637
	10–15	75,358	3.12	3	3.49	1.119
	15–20	35,508	1.47	0	0.00	0.000
	>20	55,006	2.28	0	0.00	0.000
Aspect	F	114,082	4.72	4	4.65	0.986
	N	165,633	6.85	4	4.65	0.679
	NE	213,654	8.84	9	10.47	1.184
	E	362,097	14.98	25	29.07	1.941
	SE	460,076	19.03	22	25.58	1.344
	S	437,541	18.10	12	13.95	0.771
	SW	314,526	13.01	6	6.98	0.536
	W	196,799	8.14	3	3.49	0.429
	NW	153,398	6.34	1	1.16	0.183
Plan curvature (100/m)	Concave	755,889	31.26	26	30.23	0.967
	flat	909,452	37.61	45	52.33	1.391
	convex	752,464	31.12	15	17.44	0.560
Convergence index (100/m)	<-38.8	242,500	10.04	15	17.44	1.737
	−38.8--12.1	552,768	22.89	24	27.91	1.219
	−12.1–11.3	804,611	33.31	22	25.58	0.768
	11.3–38.8	561,527	23.25	17	19.77	0.850
	>38.8	253,921	10.51	8	9.30	0.885
LS [b] (m)	<15.2	1,423,717	58.88	63	73.26	1.244
	15.2–44.8	217,038	8.98	7	8.14	0.907
	44.8–80.1	293,604	12.14	7	8.14	0.670
	80.1–121.7	293,446	12.14	4	4.65	0.383
	>121.7	190,001	7.86	5	5.81	0.740
SPI [c]	<8.3	722,773	29.89	23	26.74	0.895
	8.3–9.9	868,697	35.93	23	26.74	0.744
	9.9–12	524,225	21.68	18	20.93	0.965
	12–14.9	223,684	9.25	9	10.47	1.131
	>14.9	78,423	3.24	13	15.12	4.660
TPI [d]	<−7.11	30,179	1.25	3	3.49	2.795
	−7.11--1.38	266,501	11.02	12	13.95	1.266
	−1.38–1.96	1,935,233	80.04	70	81.40	1.017
	1.96–9.12	159,122	6.58	1	1.16	0.177
	>9.12	26,771	1.11	0	0.00	0.000
TRI [e]	<1.47	1,774,798	73.41	67	77.91	1.061
	1.4–3.92	444,246	18.37	14	16.28	0.886
	3.92–7.84	135,731	5.61	5	5.81	1.036
	7.84–13.74	49,383	2.04	0	0.00	0.000
	>13.74	13,648	0.56	0	0.00	0.000

Table 2. *Cont.*

Factors	Classes	Pixels in Domain		Gullies		FR [a]
		No	%	No	%	
TWI [f]	<6.1	896,631	37.08	23	26.74	0.721
	6.1–8.4	964,824	39.91	28	32.56	0.816
	8.4–11.8	428,795	17.73	16	18.60	1.049
	>11.8	127,552	5.28	19	22.09	4.188
Distance to stream (m)	<100	595,385	24.63	46	53.49	2.172
	100–200	446,060	18.45	15	17.44	0.945
	200–300	395,428	16.35	9	10.47	0.640
	300–400	266,585	11.03	7	8.14	0.738
	>400	714,344	29.55	9	10.47	0.354
Drainage density (km/km^2)	<0.94	623,893	25.80	14	16.28	0.631
	0.94–1.28	966,283	39.97	20	23.26	0.582
	1.28–1.75	632,567	26.16	26	30.23	1.156
	>1.75	195,059	8.07	26	30.23	3.747
Rainfall (mm)	<68.3	490,619	20.29	6	6.98	0.344
	68.3–85.7	974,984	40.33	55	63.95	1.586
	85.7–106	830,826	34.36	25	29.07	0.846
	106–133	77,808	3.22	0	0.00	0.000
	>133	43,565	1.80	0	0.00	0.000
Distance to road (m)	<500	139,853	5.78	32	37.21	6.433
	500–1000	132,330	5.47	9	10.47	1.912
	1000–1500	127,256	5.26	4	4.65	0.884
	1500–2000	123,104	5.09	0	0.00	0.000
	>2000	1,895,259	78.39	41	47.67	0.608
NDVI [g]	<0.043	1,220,601	50.49	29	33.72	0.668
	0.043–0.132	1,196,024	49.47	57	66.28	1.340
	>0.132	860	0.04	0	0.00	0.000
Lithology	A	50,8381	21.05	10	11.63	0.552
	B	22,537	0.93	2	2.33	2.492
	C	31,795	1.32	0	0.00	0.000
	D	339,429	14.05	21	24.42	1.737
	E	183,945	7.62	13	15.12	1.985
	F	1,328,922	55.03	40	46.51	0.845
LU/LC [h]	Agriculture	2,353	0.10	0	0.00	0.000
	Bareland	20,180	0.84	0	0.00	0.000
	Kavir	629,914	26.09	44	51.16	1.961
	Poorrange	1,419,509	58.79	34	39.53	0.672
	Rock	239,538	9.92	5	5.81	0.586
	Saltlake	97,389	4.03	3	3.49	0.865
	Saltland	4,703	0.19	0	0.00	0.000
	Wetland	1,010	0.04	0	0.00	0.000
Soil type	Bad Lands	131,650	5.45	0	0.00	0.000
	Rock Outcrops/Entisols	452,055	18.72	14	16.28	0.870
	Rocky Lands	95,729	3.96	0	0.00	0.000
	Salt Flats	392,583	16.26	7	8.14	0.501
	Aridisols	387	0.02	0	0.00	0.000
	Entisols/Aridisols	1,342,192	55.59	65	75.58	1.360

[a] FR is a frequency ratio value. [b] LS is slope length. [c] SPI is Stream Power Index. [d] TPI is Topographic Position Index. [e] TRI is Terrain Ruggedness Index. [f] TWI is Topographic Wetness Index. [g] NDVI is Normalized Difference Vegetation Index. [h] LU/LC is land use/land cover.

Table 3. Area and percentage of each susceptibility classes.

Models	ADTree		Bagging-ADTree		RF-ADTree		LR	
Classes	Area (km^2)	%	Area (km^2)	%	Area (km^2)	%	Area (km^2)	%
Very Low	789.91	36.30	662.24	30.43	493.82	22.69	480.29	22.07
Low	411.20	18.90	499.29	22.95	655.34	30.12	547.92	25.18
Moderate	533.53	24.52	486.18	22.34	483.52	22.22	499.18	22.94
High	340.54	15.65	335.57	15.42	318.31	14.63	373.20	17.15
Very High	100.85	4.63	192.74	8.86	225.04	10.34	275.43	12.66

3.3. The Relative Importance of GECFs

RAF modeling revealed the importance of GECFs (Figure 6). Distance-to-road (16.95) was the most important factor in gully occurrence in the study area. The other factors, in the order of importance were drainage density (14), distance-to-stream (13.29), LU/LC (10.58), annual rainfall (9.1), TWI (6.91), NDVI (6.6), elevation (6), SPI (5.2), TPI (4.67), CI (2.87), lithology (2.76), soil type (2.57), slope (1.4), plan curvature (1.4), TRI (0.75), aspect (0.18), and LS (0.034).

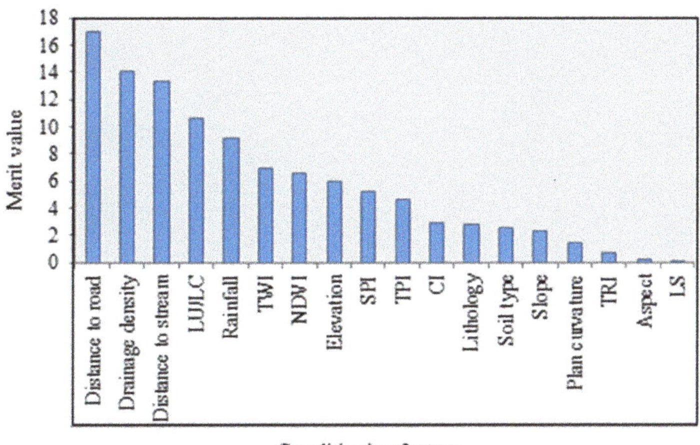

Figure 6. Relative importance of conditioning factors using a random forest model.

3.4. Gully Erosion Susceptibility Mapping Using Machine Learning Models

Gully erosion susceptibility mapping using four machine-learning models provided four predictions of gully formation zones (Table 3 and Figure 7a–d). According to all four models used in the study, most of the study area is classified as having very low and low susceptibility to gully erosion (ADTree—55.2% (1201.1 km^2), Bagging-ADTree—53.38% (1161.4 km^2), RF-ADTree—52.81% (1149.1 km^2), and LR—47.25% (1028.1 km^2)). ADTree classified the largest total area of very low susceptibility (36.30%) and the smallest total area of very high susceptibility (4.63%). The other models classified 30.43% (Bad-ADTree), 22.69% (RF-ADTree), and 22.07% (LR) as very low susceptibility, and 8.86% (Bad-ADTree), 10.34% (RF-ADTree), and 12.66% (LR) as having very high susceptibility. Among the models, LR classified the largest portion of the study area as highly susceptible (12.66%) and the smallest portion as having very low susceptibility (22.07%).

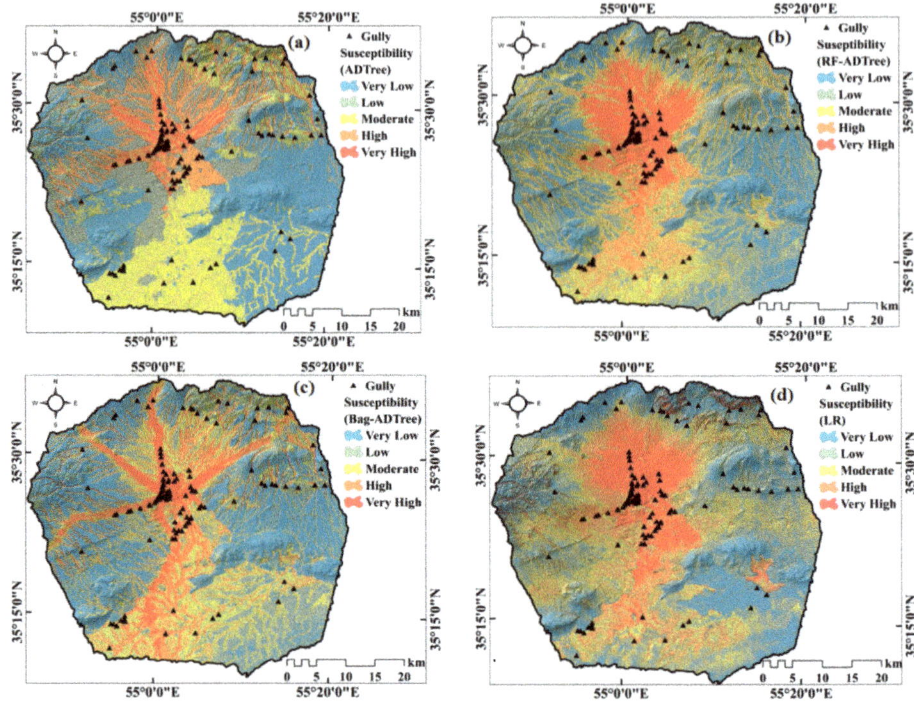

Figure 7. Gully erosion susceptibility map using (**a**) Alternating decision tree (ADTree), (**b**) Rotation Forest (RF)-ADTree, (**c**) Bagging-ADTree, (**d**) Logistic regression.

3.5. Validation of Results

The results were validated using AUC values both in SRC (success rate curve) and PRC (prediction rate curve) (Table 4, Figure 8a,b). Generally, the models tested achieved excellent accuracy. The success rate curves, a degree-of-fit measure (i.e., comparison of the susceptibility maps with training dataset), indicated that bagging-ADTree (0.964) was most accurate, and LR least accurate (0.867). The AUC values computed for prediction rate curves, indicating the predictive power of the susceptibility maps, confirmed that Bagging-ADTree was most accurate (0.978) and LR least (0.870).

Table 4. Validation of results.

Criteria	Model	AUC [a]	Standard Error	95% Confidence Interval
SRC [b]	ADTree	0.926	0.0361	0.693 to 0.822
	RF-ADTree	0.952	0.0332	0.747 to 0.867
	Bagging-ADTree	0.964	0.0318	0.763 to 0.879
	LR	0.867	0.0356	0.695 to 0.824
PRC [c]	ADTree	0.965	0.0412	0.764 to 0.929
	RF-ADTree	0.971	0.0373	0.791 to 0.945
	Bagging-ADTree	0.978	0.0334	0.818 to 0.960
	LR	0.870	0.0549	0.656 to 0.854

[a] AUC is the area under the ROC (the receiver operating characteristic) curve. [b] SRC is success rate curve. [c] PRC is prediction rate curve.

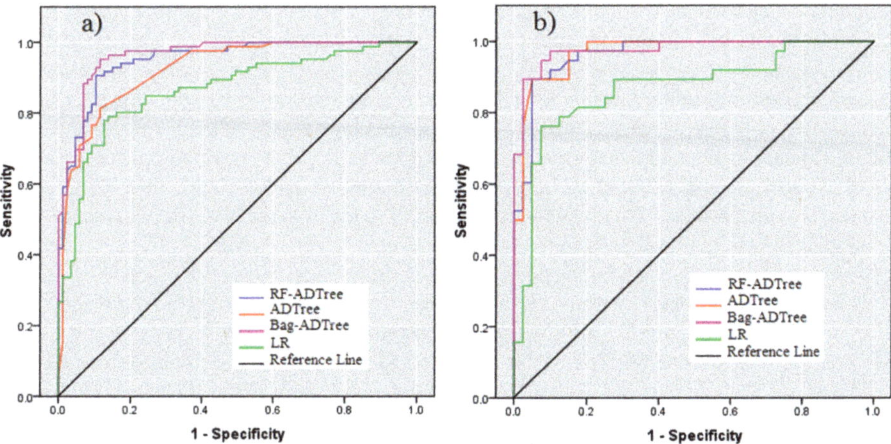

Figure 8. Validation of results using (**a**) area under the curve of success rate curve and (**b**) prediction rate curve.

4. Discussion

Different sources were used to prepare the input dataset. Because many factors used in GESM were extracted from a digital elevation model (DEM), the quality of the DEM significantly influences the accuracy of the results [96,97]. The Advanced Land Observing Satellite (ALOS) DEM with 12.5 m spatial resolution was used as it has been shown to provide better accuracy than both the Shuttle Radar Topography Mission (SRTM) and Advanced Spaceborne Thermal Emission and Reflection Radiometer (ASTER) and DEMs [98].

In this study, we developed and explored a new ensemble intelligence approach using bagging and RF as a meta-classifier and with ADTree as a base classifier, to spatially predict gully head-cut erosion in the Chah Mousi watershed. We produced GESMs based on a modeling procedure including training and validation datasets, and 18 conditioning factors (elevation, slope angle, aspect, plan curvature, CI, LS, SPI, TPI, TRI, TWI, distance to stream, drainage density, rainfall, distance to road, NDVI, lithology, land use/land cover, and soil type). These factors were checked for collinearity with statistical metrics, including TOL and VIF. The results reveal that all GECFs influenced gully erosion occurrence.

Based on FR analysis, the relationship between the factors and gully locations were assessed. Conditioning-factor classes with FR values >1 indicated areas with greater gully-erosion susceptibility [82]. Elevation plays an important role in vegetation and precipitation type and, therefore, controls the spatial distribution and gully erosion processes [99]. Elevations in the study region below 1000 m a.s.l. are more susceptible to gully erosion. Thus, the higher occurrence of gully head cut erosion in the lowland areas agrees with Dickson et al. [100]. However, Arabamiri et al. [19] determined that elevations below 829 m were most prone to gullying. In terms of slope angle and curvature, the FR analysis showed that slopes of less than 5° (including flat areas) were most likely to be sites of gully occurrence. Because lower slope angles have greater soil depth, intensive rainfall impaction and greater runoff from upslope will decrease soil strength resulting in the development and extension of the gully channel [9]. Curvature causes accumulation of runoff and enhances the velocity and volume of flow, so this variable positively correlates to locations of gully erosion. The slope aspect that controls several climate conditions, such as the intensity of precipitation, moisture, evapotranspiration, and vegetation cover [101], indirectly influences gully erosion. Among the slope-aspect classes, east- and southeast-facing slopes are the most highly correlated to gully erosion. These two slope aspect classes get more solar radiation in the northern hemisphere and, as a result, they experience more evaporation, higher soil porosity (total pore space), lower soil strength, and lower vegetation density. This is in accordance with Zabihi et al. [9], who reported that southward slope aspects are more susceptible to gully erosion. CI values below −39.6 100/m were most predictive of gully formation: the lower the CI value, the greater is the potential for gully erosion. Arabameri et al. [17] concluded, based on the WoE method, that CI values between 0 and 10 signify locations that are more susceptible to gully occurrence in their study area. LS less than 15 m indicate a more likely formation of gullies and reflects that gullies are more likely formed in flat areas with lower slope angles. This confirms the findings of Gayen et al. [102], but conflicts with the results of Zabihi et al. [9], who shows a direct relationship between LS and gully erosion locations. Zabihi et al. also implied that the higher the LS, the higher the probability of gully erosion occurrence due to increasing runoff velocity and a decreasing detachment and transport threshold of soil particles [103,104].

The most susceptible classes for the other GECFs were SPI between >14.9, TPI less than −7, TRI less than 1.4, and TWI more than 11.8. These results are confirmed by the findings of Arabameri et al. [17] who reported that, for example, the greater the TWI factor, the greater is the potential for gully occurrence. High values of TWI increase the filtration rate and provide the conditions for piping and roof collapse, resulting in the development of gully tunnels and, eventually, the appearance of gullies on the surface [105].

Moreover, the nearer sites are to a river, the higher the susceptibility to gully erosion. In this study, locations at distances less than 100 m from a stream were more likely to see gully formation. Some researchers have confirmed these results [9,13,16,42]. The sheer force of flow can overcome and decrease the strength of soil along the sides of gully forms and lead to the development of gullies of greater dimensions.

Areas with drainage densities exceeding 1.75 km/km^2 were most correlated to gully erosion. The role of this factor can be made clearer when other factors are considered. For example, a location with a lower slope angle and higher drainage density has a higher TWI, and if the soil at that location was loose and erodible, gully erosion is easier to achieve. In the study area, the lower classes of annual precipitation amounts (between 68.3 and 85.7 mm) were most susceptible to gully incidence. This suggests that though rainfall has a positive role in gully formation, it is not the most important factor. In other words, lower rainfall values are positively related to gullying.

Distances from roads are important to gully erosion and, like distances from rivers, the nearer the site, the higher the potential for gully erosion. Distances of less than 500 m from a road were positively correlated to gully locations, which underscores the importance of the roles of development and disturbance of ground surfaces in promoting landscape degradation.

Results of the NDVI factor show that vegetation plays a very important role in protecting soil against erosion, so that, with increasing vegetation, the sensitivity of an area to gully erosion decreases. Vegetation cover greatly reduces the erosion of runoff through the increase in infiltration and protection of soil through roots [106]. The findings agree with those of Arabameri et al. [13], Arabameri et al. [19], and Chaplot et al. [107] stating that low values of NDVI have a positive association with gully erosion and that it is easier for a gully to develop in areas with lower NDVI values. Generally, barren lands and sparsely vegetated areas are more susceptible to erosion than forests, where vegetation cover strongly reduces the erosive action of surface runoff.

Because gully erosion depends on the lithological properties of materials at Earth's surface, lithology is a vital factor in gullying [104]. As for lithology, Quaternary lithotypes have a high susceptibility to gully erosion. The result is in agreement with findings reported by Arabameri et al. [13], who found that Quaternary lithotypes have a strong effect on gully occurrences. In terms of land use, which plays a key role in geomorphological and hydrological processes by controlling overland flow runoff generation and sediment dynamics [108], the areas of kavir are most susceptible to gully erosion. In these regions, the complete lack of vegetation leaves the soil exposed, and it is easily eroded by precipitation. These results are in line with [13]. The entisol/aridisol soils are the most susceptible soils to gully erosion occurring in the study area, which is in accordance with [19].

In terms of the FR values, the most important GECFs in the study area were the distance to nearest road and drainage density. This is confirmed by the RAF algorithm analysis, which was used to rank the importance of the GECFs for the spatial prediction of gullies in the study area. This result is consistent with [17,109,110]. Roads are impervious surfaces, and they disrupt natural drainage systems due to improper culverts, concentration of surface runoffs, and by altering the hydrological functions of hillslopes, which significantly contribute to overland flow and allow rapid run-off, easily eroding bare soil and causing gullying [111,112]. An example of the effect of roads on gullying is shown in Figure 9. Distance to a road is the most important factor. It is followed in importance by drainage density, distance to stream, land use, rainfall, NDVI, elevation, SPI, TPI, CI, lithology, soil type, plan curvature, TRI, aspect, and LS. Though other factors affect gully erosion, the above are the most meaningful in the study area.

Figure 9. A sample of road effect on gully occurrence.

A novel ensemble intelligence approach, bagging-ADTree, and other ML algorithms—ADTree, RF-ADTree and LR—were used to create gully erosion susceptibility maps. The goodness-of-fit and the performance of the models were checked by AUROC of success and prediction rate curves. The results illustrate that bagging ADTree and RF-ADTree outperformed ADTree and LR. These results are in line with [42,113,114]. The new model accurately identified the areas that are susceptible to gully erosion based on the past patterns of formation, but it also provides excellent predictions of future development. The RF and bagging as a meta-classifier can decrease over-fitting and noise problems in the training dataset. Some researchers have confirmed the prediction power of RF in applications to some environmental problems [42,115–117].

For example, Tien Bui et al. [21] predicted gully locations in a semi-arid watershed of Iran using ADTtree and its ensembles using RF meta-classifier. They concluded that the RF model could well enhance the prediction power of ADTree as a base classifier. However, the RF-ADTree ensemble model outperformed some benchmark models, including SVM based on the polynomial and RBF kernels, LR, naïve Bayes, and ADTtree. Additionally, Shirzadi et al. [42] used four meta-classifiers, namely, multiboost, bagging, RF, and random subspace (RS), for the spatial prediction of shallow landslides in Bijar City, Kurdistan province, Iran. They used ADTree as a base classifier for the modeling process. The four ensemble models were combined with the ADTree under two scenarios of different sample sizes and raster resolutions. They reported that the RS model was more capable for sample sizes of

60%/40% and 70%/30% with a raster resolution of 10 m. According to the results, the new proposed ensemble model can spatially predict gully erosion occurrences with reasonably good accuracy.

5. Conclusions

Soil erosion is an important environmental challenge to ecosystem's condition and function. Land degradation and decreasing land productivity are a result of on-site and off-site erosion in a gully-prone area. However, detection, prediction, and management of gully-prone areas using protective measures and mitigation techniques are important efforts. Some quantitative and qualitative methods and techniques have been developed and explored for modeling and preparing the susceptibility assessments. However, due to differences in their probability distribution functions, their performances are also different. For example, some of them do not fit the data that are available. All models present advantages and disadvantages, so one of the most important aspects of the modeling strategy is selecting the appropriate model. Machine-learning models are more often used because of their ability to overcome over-fitting and noise challenges during the modeling process and because they have higher goodness-of-fits and perform better compared to other conventional models. Moreover, among the machine-learning classifiers, ensemble models are more powerful than single classifiers. They randomly divide a training dataset into subsets and perform a single classifier, which provides an output with the lowest error and the highest performance rather than the single classifier. This process overcomes the weakness of the single classifier and achieves a more powerful classifier. In response to the advantage of ensemble classifiers, a novel ensemble intelligence approach, namely bagging-ADTree, was performed and gully erosion maps were obtained. Some other machine-learning algorithms (including ADTree, Bagging-ADTree, and LR) were used for comparison and validation of the results of the new model. The random forest model is used to determine the relative importance of conditioning factors. The results indicate that distance-to-road and drainage density are very important to gully occurrence in the study area. The validation indicated that although the models achieved high goodness-of-fit scores and were powerfully predictive, the ensemble model was better than others at spatially predicting gully erosion and produced a more accurate gully-susceptibility map of the study area. Based on these results, we can recommend the new model, bagging-ADTree, for gully modeling in other zones of potential gully erosion susceptibility, but offer one caution: there may be other conditioning factors responsible for gully erosion in other areas. Finally, the results from a case study of the Chah Mousi watershed show that selecting suitable predisposing factors and combining machine-learning ensemble models with GISs can be used to efficiently predict an area's susceptibility to gully formation with high accuracy. Therefore, the gully-erosion susceptibility map generated by the method can aid decision makers, planners, and engineers in their quests to identify and develop the most effective protective measures to sustainably prevent and mitigate gully-erosion damage.

Author Contributions: Conceptualization, A.A.; data curation, A.A.; formal analysis, A.A. and W.C.; investigation, A.A. and D.T.B.; methodology, A.A., W.C., B.P., and D.T.B.; resources, A.A., software, A.A. and W.C.; supervision, A.A.; validation, A.A.; writing—original draft, A.A.; writing—review and editing, A.A., W.C., T.B., B.P., J.P.T., and D.T.B. All authors have read and agreed to the published version of the manuscript.

Funding: This research was partly funded by the Austrian Science Fund (FWF) through the Doctoral College GIScience (DK W 1237-N23) at the University of Salzburg.

References

1. Poesen, J.; Nachtergaele, J.; Verstraeten, G.; Valentin, C. Gully erosion and environmental change: Importance and research needs. *Catena* **2003**, *50*, 91–133. [CrossRef]
2. Valentin, C.; Poesen, J.; Li, Y. Gully erosion: Impacts, factors and control. *Catena* **2005**, *63*, 132–153. [CrossRef]
3. Kirkby, M.; Bracken, L. Gully processes and gully dynamics. *Earth Surf. Process. Landf. J. Br. Geomorphol. Res. Group* **2009**, *34*, 1841–1851. [CrossRef]

4. Poesen, J.; Vanwalleghem, T.; Deckers, J. Gullies and Closed Depressions in the Loess Belt: Scars of Human–Environment Interactions. In *Landscapes and Landforms of Belgium and Luxembourg*; Springer: Berlin/Heidelberg, Germany, 2018; pp. 253–267.

5. Pandey, A.; Chowdary, V.; Mal, B. Identification of critical erosion prone areas in the small agricultural watershed using USLE, GIS and remote sensing. *Water Resour. Manag.* **2007**, *21*, 729–746. [CrossRef]

6. Martınez-Casasnovas, J. A spatial information technology approach for the mapping and quantification of gully erosion. *Catena* **2003**, *50*, 293–308. [CrossRef]

7. Zinck, J.A.; López, J.; Metternicht, G.I.; Shrestha, D.P.; Vázquez-Selem, L. Mapping and modelling mass movements and gullies in mountainous areas using remote sensing and GIS techniques. *Int. J. Appl. Earth Obs. Geoinf.* **2001**, *3*, 43–53. [CrossRef]

8. Seutloali, K.E.; Beckedahl, H.R.; Dube, T.; Sibanda, M. An assessment of gully erosion along major armoured roads in south-eastern region of South Africa: A remote sensing and GIS approach. *Geocarto Int.* **2016**, *31*, 225–239. [CrossRef]

9. Zabihi, M.; Mirchooli, F.; Motevalli, A.; Darvishan, A.K.; Pourghasemi, H.R.; Zakeri, M.A.; Sadighi, F. Spatial modelling of gully erosion in Mazandaran Province, northern Iran. *Catena* **2018**, *161*, 1–13. [CrossRef]

10. Conoscenti, C.; Angileri, S.; Cappadonia, C.; Rotigliano, E.; Agnesi, V.; Märker, M. Gully erosion susceptibility assessment by means of GIS-based logistic regression: A case of Sicily (Italy). *Geomorphology* **2014**, *204*, 399–411. [CrossRef]

11. Vanwalleghem, T.; Van Den Eeckhaut, M.; Poesen, J.; Govers, G.; Deckers, J. Spatial analysis of factors controlling the presence of closed depressions and gullies under forest: Application of rare event logistic regression. *Geomorphology* **2008**, *95*, 504–517. [CrossRef]

12. Rahmati, O.; Haghizadeh, A.; Pourghasemi, H.R.; Noormohamadi, F. Gully erosion susceptibility mapping: The role of GIS-based bivariate statistical models and their comparison. *Nat. Hazards* **2016**, *82*, 1231–1258. [CrossRef]

13. Arabameri, A.; Pradhan, B.; Pourghasemi, H.; Rezaei, K.; Kerle, N. Spatial modelling of gully erosion using GIS and R programing: A comparison among three data mining algorithms. *Appl. Sci.* **2018**, *8*, 1369. [CrossRef]

14. Magliulo, P. Soil erosion susceptibility maps of the Janare Torrent Basin (southern Italy). *J. Maps* **2010**, *6*, 435–447. [CrossRef]

15. Conoscenti, C.; Agnesi, V.; Angileri, S.; Cappadonia, C.; Rotigliano, E.; Märker, M. A GIS-based approach for gully erosion susceptibility modelling: A test in Sicily, Italy. *Environ. Earth Sci.* **2013**, *70*, 1179–1195. [CrossRef]

16. Arabameri, A.; Pradhan, B.; Rezaei, K.; Lee, C.-W. Assessment of Landslide Susceptibility Using Statistical-and Artificial Intelligence-Based FR–RF Integrated Model and Multiresolution DEMs. *Remote Sens.* **2019**, *11*, 999. [CrossRef]

17. Arabameri, A.; Rezaei, K.; Pourghasemi, H.R.; Lee, S.; Yamani, M. GIS-based gully erosion susceptibility mapping: A comparison among three data-driven models and AHP knowledge-based technique. *Environ. Earth Sci.* **2018**, *77*, 628. [CrossRef]

18. Svoray, T.; Michailov, E.; Cohen, A.; Rokah, L.; Sturm, A. Predicting gully initiation: Comparing data mining techniques, analytical hierarchy processes and the topographic threshold. *Earth Surf. Process. Landf.* **2012**, *37*, 607–619. [CrossRef]

19. Arabameri, A.; Pradhan, B.; Rezaei, K.; Conoscenti, C. Gully erosion susceptibility mapping using GIS-based multi-criteria decision analysis techniques. *Catena* **2019**, *180*, 282–297. [CrossRef]

20. Amiri, M.; Pourghasemi, H.R.; Ghanbarian, G.A.; Afzali, S.F. Assessment of the importance of gully erosion effective factors using Boruta algorithm and its spatial modeling and mapping using three machine learning algorithms. *Geoderma* **2019**, *340*, 55–69. [CrossRef]

21. Tien Bui, D.; Shirzadi, A.; Shahabi, H.; Chapi, K.; Omidavr, E.; Pham, B.T.; Talebpour Asl, D.; Khaledian, H.; Pradhan, B.; Panahi, M. A Novel Ensemble Artificial Intelligence Approach for Gully Erosion Mapping in a Semi-Arid Watershed (Iran). *Sensors* **2019**, *19*, 2444. [CrossRef]

22. Jaafari, A.; Zenner, E.K.; Panahi, M.; Shahabi, H. Hybrid artificial intelligence models based on a neuro-fuzzy system and metaheuristic optimization algorithms for spatial prediction of wildfire probability. *Agric. For. Meteorol.* **2019**, *266*, 198–207. [CrossRef]

23. Taheri, K.; Shahabi, H.; Chapi, K.; Shirzadi, A.; Gutiérrez, F.; Khosravi, K. Sinkhole susceptibility mapping: A comparison between Bayes-based machine learning algorithms. *Land Degrad. Dev.* **2019**, *30*, 730–745. [CrossRef]

24. Chapi, K.; Singh, V.P.; Shirzadi, A.; Shahabi, H.; Bui, D.T.; Pham, B.T.; Khosravi, K. A novel hybrid artificial intelligence approach for flood susceptibility assessment. *Environ. Model. Softw.* **2017**, *95*, 229–245. [CrossRef]

25. Hong, H.; Panahi, M.; Shirzadi, A.; Ma, T.; Liu, J.; Zhu, A.-X.; Chen, W.; Kougias, I.; Kazakis, N. Flood susceptibility assessment in Hengfeng area coupling adaptive neuro-fuzzy inference system with genetic algorithm and differential evolution. *Sci. Total Environ.* **2018**, *621*, 1124–1141. [CrossRef] [PubMed]

26. Khosravi, K.; Pham, B.T.; Chapi, K.; Shirzadi, A.; Shahabi, H.; Revhaug, I.; Prakash, I.; Bui, D.T. A comparative assessment of decision trees algorithms for flash flood susceptibility modeling at Haraz watershed, northern Iran. *Sci. Total Environ.* **2018**, *627*, 744–755. [CrossRef] [PubMed]

27. Shafizadeh-Moghadam, H.; Valavi, R.; Shahabi, H.; Chapi, K.; Shirzadi, A. Novel forecasting approaches using combination of machine learning and statistical models for flood susceptibility mapping. *J. Environ. Manag.* **2018**, *217*, 1–11. [CrossRef]

28. Ahmadlou, M.; Karimi, M.; Alizadeh, S.; Shirzadi, A.; Parvinnejhad, D.; Shahabi, H.; Panahi, M. Flood susceptibility assessment using integration of adaptive network-based fuzzy inference system (ANFIS) and biogeography-based optimization (BBO) and BAT algorithms (BA). *Geocarto Int.* **2018**, *34*, 1–21. [CrossRef]

29. Bui, D.T.; Panahi, M.; Shahabi, H.; Singh, V.P.; Shirzadi, A.; Chapi, K.; Khosravi, K.; Chen, W.; Panahi, S.; Li, S. Novel hybrid evolutionary algorithms for spatial prediction of floods. *Sci. Rep.* **2018**, *8*, 15364. [CrossRef]

30. Miraki, S.; Zanganeh, S.H.; Chapi, K.; Singh, V.P.; Shirzadi, A.; Shahabi, H.; Pham, B.T. Mapping Groundwater Potential Using a Novel Hybrid Intelligence Approach. *Water Resour. Manag.* **2019**, *33*, 281–302. [CrossRef]

31. Rahmati, O.; Naghibi, S.A.; Shahabi, H.; Bui, D.T.; Pradhan, B.; Azareh, A.; Rafiei-Sardooi, E.; Samani, A.N.; Melesse, A.M. Groundwater spring potential modelling: Comprising the capability and robustness of three different modeling approaches. *J. Hydrol.* **2018**, *565*, 248–261. [CrossRef]

32. Tien Bui, D.; Khosravi, K.; Li, S.; Shahabi, H.; Panahi, M.; Singh, V.; Chapi, K.; Shirzadi, A.; Panahi, S.; Chen, W. New hybrids of anfis with several optimization algorithms for flood susceptibility modeling. *Water* **2018**, *10*, 1210. [CrossRef]

33. Tehrany, M.S.; Pradhan, B.; Jebur, M.N. Flood susceptibility analysis and its verification using a novel ensemble support vector machine and frequency ratio method. *Stoch. Environ. Res. Risk Assess.* **2015**, *29*, 1149–1165. [CrossRef]

34. Rahmati, O.; Samadi, M.; Shahabi, H.; Azareh, A.; Rafiei-Sardooi, E.; Alilou, H.; Melesse, A.M.; Pradhan, B.; Chapi, K.; Shirzadi, A. SWPT: An automated GIS-based tool for prioritization of sub-watersheds based on morphometric and topo-hydrological factors. *Geosci. Front.* **2019**, *10*, 2167–2175. [CrossRef]

35. Khosravi, K.; Shahabi, H.; Pham, B.T.; Adamawoski, J.; Shirzadi, A.; Pradhan, B.; Dou, J.; Ly, H.-B.; Gróf, G.; Ho, H.L.; et al. A Comparative Assessment of Flood Susceptibility Modeling Using Multi-Criteria Decision-Making Analysis and Machine Learning Methods. *J. Hydrol.* **2019**, *573*, 311–323. [CrossRef]

36. Chen, W.; Panahi, M.; Khosravi, K.; Pourghasemi, H.R.; Rezaie, F.; Parvinnezhad, D. Spatial prediction of groundwater potentiality using anfis ensembled with teaching-learning-based and biogeography-based optimization. *J. Hydrol.* **2019**, *572*, 435–448. [CrossRef]

37. Chen, W.; Pradhan, B.; Li, S.; Shahabi, H.; Rizeei, H.M.; Hou, E.; Wang, S. Novel hybrid integration approach of bagging-based fisher's linear discriminant function for groundwater potential analysis. *Nat. Resour. Res.* **2019**, *28*, 1239–1258. [CrossRef]

38. Chen, W.; Tsangaratos, P.; Ilia, I.; Duan, Z.; Chen, X. Groundwater spring potential mapping using population-based evolutionary algorithms and data mining methods. *Sci. Total Environ.* **2019**, *684*, 31–49. [CrossRef]

39. Roodposhti, M.S.; Safarrad, T.; Shahabi, H. Drought sensitivity mapping using two one-class support vector machine algorithms. *Atmos. Res.* **2017**, *193*, 73–82. [CrossRef]

40. Alizadeh, M.; Alizadeh, E.; Asadollahpour Kotenaee, S.; Shahabi, H.; Beiranvand Pour, A.; Panahi, M.; Bin Ahmad, B.; Saro, L. Social vulnerability assessment using artificial neural network (ANN) model for earthquake hazard in Tabriz city, Iran. *Sustainability* **2018**, *10*, 3376. [CrossRef]

41. Tien Bui, D.; Shahabi, H.; Shirzadi, A.; Chapi, K.; Pradhan, B.; Chen, W.; Khosravi, K.; Panahi, M.; Bin Ahmad, B.; Saro, L. Land subsidence susceptibility mapping in south korea using machine learning algorithms. *Sensors* **2018**, *18*, 2464. [CrossRef]

42. Shirzadi, A.; Soliamani, K.; Habibnejhad, M.; Kavian, A.; Chapi, K.; Shahabi, H.; Chen, W.; Khosravi, K.; Thai Pham, B.; Pradhan, B. Novel GIS based machine learning algorithms for shallow landslide susceptibility mapping. *Sensors* **2018**, *18*, 3777. [CrossRef] [PubMed]

43. Pham, B.T.; Prakash, I.; Singh, S.K.; Shirzadi, A.; Shahabi, H.; Bui, D.T. Landslide susceptibility modeling using Reduced Error Pruning Trees and different ensemble techniques: Hybrid machine learning approaches. *CATENA* **2019**, *175*, 203–218. [CrossRef]

44. Tien Bui, D.; Shahabi, H.; Shirzadi, A.; Kamran Chapi, K.; Hoang, N.-D.; Pham, B.; Bui, Q.-T.; Tran, C.-T.; Panahi, M.; Bin Ahmad, B.; et al. A Novel Integrated Approach of Relevance Vector Machine Optimized by Imperialist Competitive Algorithm for Spatial Modeling of Shallow Landslides. *Remote Sens.* **2019**, *11*, 57. [CrossRef]

45. Thai Pham, B.; Prakash, I.; Dou, J.; Singh, S.K.; Trinh, P.T.; Trung Tran, H.; Minh Le, T.; Tran, V.P.; Kim Khoi, D.; Shirzadi, A. A Novel Hybrid Approach of Landslide Susceptibility Modeling Using Rotation Forest Ensemble and Different Base Classifiers. *Geocarto Int.* **2018**. [CrossRef]

46. Chen, W.; Shahabi, H.; Zhang, S.; Khosravi, K.; Shirzadi, A.; Chapi, K.; Pham, B.T.; Zhang, T.; Zhang, L.; Chai, H.; et al. Landslide susceptibility modeling based on gis and novel bagging-based kernel logistic regression. *Appl. Sci.* **2018**, *8*, 2540. [CrossRef]

47. Chen, W.; Hong, H.; Panahi, M.; Shahabi, H.; Wang, Y.; Shirzadi, A.; Pirasteh, S.; Alesheikh, A.A.; Khosravi, K.; Panahi, S.; et al. Spatial prediction of landslide susceptibility using gis-based data mining techniques of anfis with whale optimization algorithm (woa) and grey wolf optimizer (gwo). *Appl. Sci.* **2019**, *9*, 3755. [CrossRef]

48. Chen, W.; Pourghasemi, H.R.; Naghibi, S.A. Prioritization of landslide conditioning factors and its spatial modeling in shangnan county, china using gis-based data mining algorithms. *Bull. Eng. Geol. Environ.* **2018**, *77*, 611–629. [CrossRef]

49. Kuhnert, P.M.; Henderson, A.K.; Bartley, R.; Herr, A. Incorporating uncertainty in gully erosion calculations using the random forests modelling approach. *Environmetrics* **2010**, *21*, 493–509. [CrossRef]

50. Shruthi, R.B.; Kerle, N.; Jetten, V.; Stein, A. Object-based gully system prediction from medium resolution imagery using Random Forests. *Geomorphology* **2014**, *216*, 283–294. [CrossRef]

51. Rahmati, O.; Tahmasebipour, N.; Haghizadeh, A.; Pourghasemi, H.R.; Feizizadeh, B. Evaluation of different machine learning models for predicting and mapping the susceptibility of gully erosion. *Geomorphology* **2017**, *298*, 118–137. [CrossRef]

52. Pourghasemi, H.R.; Yousefi, S.; Kornejady, A.; Cerdà, A. Performance assessment of individual and ensemble data-mining techniques for gully erosion modeling. *Sci. Total Environ.* **2017**, *609*, 764–775. [CrossRef] [PubMed]

53. Yunkai, L.; Yingjie, T.; Zhiyun, O.; Lingyan, W.; Tingwu, X.; Peiling, Y.; Huanxun, Z. Analysis of soil erosion characteristics in small watersheds with particle swarm optimization, support vector machine, and artificial neuronal networks. *Environ. Earth Sci.* **2010**, *60*, 1559–1568. [CrossRef]

54. Gutiérrez, Á.G.; Schnabel, S.; Contador, J.F.L. Using and comparing two nonparametric methods (CART and MARS) to model the potential distribution of gullies. *Ecol. Model.* **2009**, *220*, 3630–3637. [CrossRef]

55. Gutiérrez, Á.G.; Schnabel, S.; Contador, F.L. Gully erosion, land use and topographical thresholds during the last 60 years in a small rangeland catchment in SW Spain. *Land Degrad. Dev.* **2009**, *20*, 535–550. [CrossRef]

56. Angileri, S.E.; Conoscenti, C.; Hochschild, V.; Märker, M.; Rotigliano, E.; Agnesi, V. Water erosion susceptibility mapping by applying stochastic gradient treeboost to the Imera Meridionale river basin (Sicily, Italy). *Geomorphology* **2016**, *262*, 61–76. [CrossRef]

57. Azareh, A.; Rahmati, O.; Rafiei-Sardooi, E.; Sankey, J.B.; Lee, S.; Shahabi, H.; Ahmad, B.B. Modelling gully-erosion susceptibility in a semi-arid region, Iran: Investigation of applicability of certainty factor and maximum entropy models. *Sci. Total Environ.* **2019**, *655*, 684–696. [CrossRef]

58. Conoscenti, C.; Agnesi, V.; Cama, M.; Caraballo-Arias, N.A.; Rotigliano, E. Assessment of gully erosion susceptibility using multivariate adaptive regression splines and accounting for terrain connectivity. *Land Degrad. Dev.* **2018**, *29*, 724–736. [CrossRef]

59. IRIMO. Summary Reports of Iran's Extreme Climatic Events. In *Ministry of Roads and Urban Development*; Iran Meteorological Organization: Tehran, Iran, 2012. Available online: http://www.cri.ac.ir (accessed on 12 August 2018).

60. GSI. Geology Survey of Iran. 1997. Available online: http://www.gsi.ir/Main/Lang_en/index.html (accessed on 12 August 2018).

61. IUSS Working Group WRB14. *World Reference Base for Soil Resources 2014, World Soil Resources Report*; FAO: Rome, Italy, 2014.

62. Kuncheva, L.I.; Rodríguez, J.J. *An Experimental Study on Rotation Forest Ensembles*; Haindl, M., Kittler, J., Roli, F., Eds.; Springer: Berlin/Heidelberg, Germany, 2007; pp. 459–468.

63. Zhang, C.-X.; Zhang, J.-S. RotBoost: A technique for combining Rotation Forest and AdaBoost. *Pattern Recognit. Lett.* **2008**, *29*, 1524–1536. [CrossRef]

64. Xia, J.; Du, P.; He, X.; Chanussot, J. Hyperspectral Remote Sensing Image Classification Based on Rotation Forest. *IEEE Geosci. Remote Sens. Lett.* **2014**, *11*, 239–243. [CrossRef]

65. Al-Abadi, A.M. Mapping flood susceptibility in an arid region of southern Iraq using ensemble machine learning classifiers: A comparative study. *Arab. J. Geosci.* **2018**, *11*, 218. [CrossRef]

66. Tien Bui, D.; Shirzadi, A.; Shahabi, H.; Geertsema, M.; Omidvar, E.; Clague, J.J.; Thai Pham, B.; Dou, J.; Talebpour Asl, D.; Bin Ahmad, B.; et al. New Ensemble Models for Shallow Landslide Susceptibility Modeling in a Semi-Arid Watershed. *Forests* **2019**, *10*, 743. [CrossRef]

67. Kavzoglu, T.; Colkesen, I. An assessment of the effectiveness of a rotation forest ensemble for land-use and land-cover mapping. *Int. J. Remote Sens.* **2013**, *34*, 4224–4241. [CrossRef]

68. Sok, H.K.; Ooi, M.P.-L.; Kuang, Y.C. Sparse alternating decision tree. *Pattern Recognit. Lett.* **2015**, *60–61*, 57–64. [CrossRef]

69. Hong, H.; Pradhan, B.; Xu, C.; Tien Bui, D. Spatial prediction of landslide hazard at the Yihuang area (China) using two-class kernel logistic regression, alternating decision tree and support vector machines. *CATENA* **2015**, *133*, 266–281. [CrossRef]

70. Pham, B.T.; Tien Bui, D.; Prakash, I. Landslide Susceptibility Assessment Using Bagging Ensemble Based Alternating Decision Trees, Logistic Regression and J48 Decision Trees Methods: A Comparative Study. *Geotech. Geol. Eng.* **2017**, *35*, 2597–2611. [CrossRef]

71. Freund, Y.; Mason, L. *The Alternating Decision Tree Learning Algorithm*; Morgan Kaufmann Publishers Inc.: San Francisco, CA, USA, 2002; Volume 99.

72. Dietterich, T.G. An Experimental Comparison of Three Methods for Constructing Ensembles of Decision Trees: Bagging, Boosting, and Randomization. *Mach. Learn.* **2000**, *40*, 139–157. [CrossRef]

73. Breiman, L. Bagging predictors. *Mach. Learn.* **1996**, *24*, 123–140. [CrossRef]

74. Bryll, R.; Gutierrez-Osuna, R.; Quek, F. Attribute bagging: Improving accuracy of classifier ensembles by using random feature subsets. *Pattern Recognit.* **2003**, *36*, 1291–1302. [CrossRef]

75. Buhlmann, P.; Yu, B. Analyzing bagging. *Ann. Statist.* **2002**, *30*, 927–961. [CrossRef]

76. Pham, B.T.; Tien Bui, D.; Prakash, I. Bagging based Support Vector Machines for spatial prediction of landslides. *Environ. Earth Sci.* **2018**, *77*, 146. [CrossRef]

77. Tien Bui, D.; Ho, T.-C.; Pradhan, B.; Pham, B.-T.; Nhu, V.-H.; Revhaug, I. GIS-based modeling of rainfall-induced landslides using data mining-based functional trees classifier with AdaBoost, Bagging, and MultiBoost ensemble frameworks. *Environ. Earth Sci.* **2016**, *75*, 1101. [CrossRef]

78. Truongg, L.X.; Mitamura, M.; Kono, Y.; Raghavan, V.; Yonezawa, G.; Truong, Q.X.; Do, H.T.; Tien Bui, D.; Lee, S. Enhancin Prediction Performance of Landslide Susceptibility Model Using Hybrid Machine Learning Approach of Bagging Ensemble and Logistic Model Tree. *Appl. Sci.* **2018**, *8*, 71046.

79. Das, A. Logistic Regression. In *Encyclopedia of Quality of Life and Well-Being Research*; Michalos, A.C., Ed.; Springer: Dordrecht, The Netherlands, 2014; pp. 3680–3682.

80. Moon, K.-W. Logistic Regression. In *Learn ggplot2 Using Shiny App*; Moon, K.-W., Ed.; Springer International Publishing: Cham, Switzerland, 2016; pp. 51–54.

81. Raja, N.B.; Çiçek, I.; Türkoğlu, N.; Aydin, O.; Kawasaki, A. Correction to: Landslide susceptibility mapping of the Sera River Basin using logistic regression model. *Nat. Hazards* **2018**, *91*, 1423. [CrossRef]

82. Meten, M.; Bhandary, N.P.; Yatabe, R. GIS-based frequency ratio and logistic regression modelling for landslide susceptibility mapping of Debre Sina area in central Ethiopia. *J. Mt. Sci.* **2015**, *12*, 1355–1372. [CrossRef]

83. Weisburd, D.; Britt, C. Logistic Regression. In *Statistics in Criminal Justice*; Weisburd, D., Britt, C., Eds.; Springer: Boston, MA, USA, 2014; pp. 548–600.

84. Pradhan, B. Manifestation of an advanced fuzzy logic model coupled with Geo-information techniques to landslide susceptibility mapping and their comparison with logistic regression modelling. *Environ. Ecol. Stat.* **2011**, *18*, 471–493. [CrossRef]

85. Rodriguez-Galiano, V.; Sanchez-Castillo, M.; Chica-Olmo, M.; Chica-Rivas, M. Machine learning predictive models for mineral prospectivity: An evaluation of neural networks, random forest, regression trees and support vector machines. *Ore Geol. Rev.* **2015**, *71*, 804–818. [CrossRef]

86. Talaei, R. Landslide susceptibility zonation mapping using logistic regression and its validation in Hashtchin Region, northwest of Iran. *J. Geol. Soc. India* **2014**, *84*, 68–86. [CrossRef]

87. Arabameri, A.; Cerda, A.; Rodrigo-Comino, J.; Pradhan, B.; Sohrabi, M.; Blaschke, T.; Tien Bui, D. Proposing a Novel Predictive Technique for Gully Erosion Susceptibility Mapping in Arid and Semi-arid Regions (Iran). *Remote Sens.* **2019**, *11*, 2577. [CrossRef]

88. Arabameri, A.; Pradhan, B.; Rezaei, K. Spatial prediction of gully erosion using ALOS PALSAR data and ensemble bivariate and data mining models. *Geosci. J.* **2019**, *23*, 1–18. [CrossRef]

89. Arabameri, A.; Pradhan, B.; Rezaei, K.; Yamani, M.; Pourghasemi, H.R.; Lombardo, L. Spatial modeling of gully erosion using Evidential Belief Function, Logistic Regression and a new ensemble EBF–LR algorithm. *Land Degrad. Dev.* **2018**, *29*, 4035–4049. [CrossRef]

90. Arabameri, A.; Pradhan, B.; Rezaei, K. Gully erosion zonation mapping using integrated geographically weighted regression with certainty factor and random forest models in GIS. *J. Environ. Manag.* **2019**, *232*, 928–942. [CrossRef]

91. Arabameri, A.; Rezaei, K.; Cerdà, A.; Conoscenti, C.; Kalantari, Z. A comparison of statistical methods and multi-criteria decision making to map flood hazard susceptibility in Northern Iran. *Sci. Total Environ.* **2019**, *660*, 443–458. [CrossRef] [PubMed]

92. Arabameri, A.; Rezaei, K.; Cerda, A.; Lombardo, L.; Rodrigo-Comino, J. GIS-based groundwater potential mapping in Shahroud plain, Iran. A comparison among statistical (bivariate and multivariate), data mining and MCDM approaches. *Sci. Total Environ.* **2019**, *658*, 160–177. [CrossRef] [PubMed]

93. Chen, W.; Hong, H.; Li, S.; Shahabi, H.; Wang, Y.; Wang, X.; Ahmad, B.B. Flood susceptibility modelling using novel hybrid approach of reduced-error pruning trees with bagging and random subspace ensembles. *J. Hydrol.* **2019**, *575*, 864–873. [CrossRef]

94. Chen, W.; Li, Y.; Xue, W.; Shahabi, H.; Li, S.; Hong, H.; Wang, X.; Bian, H.; Zhang, S.; Pradhan, B.; et al. Modeling flood susceptibility using data-driven approaches of naïve bayes tree, alternating decision tree, and random forest methods. *Sci. Total Environ.* **2020**, *701*, 134979. [CrossRef]

95. Chen, W.; Shirzadi, A.; Shahabi, H.; Ahmad, B.B.; Zhang, S.; Hong, H.; Zhang, N. A novel hybrid artificial intelligence approach based on the rotation forest ensemble and naïve bayes tree classifiers for a landslide susceptibility assessment in langao county, china. *Geomat. Nat. Hazards Risk* **2017**, *8*, 1955–1977. [CrossRef]

96. Erasmi, S.; Rosenbauer, R.; Buchbach, R.; Busche, T.; Rutishauser, S. Evaluating the quality and accuracy of TanDEM-X digital elevation models at archaeological sites in the Cilician Plain, Turkey. *Remote Sens.* **2014**, *6*, 9475–9493. [CrossRef]

97. Pope, A.; Murray, T.; Luckman, A. DEM quality assessment for quantification of glacier surface change. *Ann. Glaciol.* **2014**, *46*, 189–194. [CrossRef]

98. Alganci, U.; Besol, B.; Sertel, E. Accuracy assessment of different digital surface models. *ISPRS Int. J. Geo-Inf.* **2018**, *7*, 114. [CrossRef]

99. Gómez-Gutiérrez, A.; Conoscenti, C.; Angileri, S.E.; Rotigliano, E.; Schnabel, S. Using topographical attributes to evaluate gully erosion proneness (susceptibility) in two mediterranean basins: Advantages and limitations. *Nat. Hazards* **2015**, *79*, 291–314. [CrossRef]

100. Dickson, J.L.; Head, J.W.; Kreslavsky, M. Martian gullies in the southern mid-latitudes of Mars: Evidence for climate-controlled formation of young fluvial features based upon local and global topography. *Icarus* **2007**, *188*, 315–323. [CrossRef]

101. Jaafari, A.; Najafi, A.; Pourghasemi, H.; Rezaeian, J.; Sattarian, A. GIS-based frequency ratio and index of entropy models for landslide susceptibility assessment in the Caspian forest, northern Iran. *Int. J. Environ. Sci. Technol.* **2014**, *11*, 909–926. [CrossRef]

102. Gayen, A.; Pourghasemi, H.R.; Saha, S.; Keesstra, S.; Bai, S. Gully erosion susceptibility assessment and management of hazard-prone areas in India using different machine learning algorithms. *Sci. Total Environ.* **2019**, *668*, 124–138. [CrossRef] [PubMed]

103. Renard, K.G.; Foster, G.R.; Weesies, G.; McCool, D.; Yoder, D. *Predicting Soil Erosion by Water: A Guide to Conservation Planning with the Revised Universal Soil Loss Equation (RUSLE)*; United States Department of Agriculture: Washington, DC, USA, 1997; Volume 703.

104. Conforti, M.; Aucelli, P.P.; Robustelli, G.; Scarciglia, F. Geomorphology and GIS analysis for mapping gully erosion susceptibility in the Turbolo stream catchment (Northern Calabria, Italy). *Nat. Hazards* **2011**, *56*, 881–898. [CrossRef]

105. Mousavi, S.M.; Golkarian, A.; Naghibi, S.A.; Kalantar, B.; Pradhan, B. GIS-based groundwater spring potential mapping using data mining boosted regression tree and probabilistic frequency ratio models in Iran. *AIMS Geosci.* **2017**, *3*, 91–115.

106. Cevik, E.; Topal, T. GIS-based landslide susceptibility mapping for a problematic segment of the natural gas pipeline, Hendek (Turkey). *Environ. Geol.* **2003**, *44*, 949–962. [CrossRef]

107. Chaplot, V.; Coadou le Brozec, E.; Silvera, N.; Valentin, C. Spatial and temporal assessment of linear erosion in catchments under sloping lands of northern Laos. *Catena* **2005**, *63*, 167–184. [CrossRef]

108. Dickie, J.A.; Parsons, A.J. Eco-geomorphological processes within grasslands, shrub lands and badlands in the semi-arid Karoo, South Africa. *Land Degrad. Dev.* **2012**, *23*, 534–547. [CrossRef]

109. Golestani, G.; Issazadeh, L.; Serajamani, R. Lithology effects on gully erosion in Ghoori chay Watershed using RS & GIS. *Int. J. Biosci.* **2014**, *4*, 71–76.

110. Arabameri, A.; Cerda, A.; Tiefenbacher, J.P. Spatial Pattern Analysis and Prediction of Gully Erosion Using Novel Hybrid Model of Entropy-Weight of Evidence. *Water* **2019**, *11*, 1129. [CrossRef]

111. Nyssen, J.; Poesen, J.; Moeyersons, J.; Luyten, E.; Veyret Picot, M.; Deckers, J.; Mitiku, H.; Govers, G. Impact of road building on gully erosion risk, a case study from the northern Ethiopian highlands. *Earth Surf. Process. Landforms* **2002**, *27*, 1267–1283. [CrossRef]

112. Svoray, T.; Markovitch, H. Catchment scale analysis of the effect of topography, tillage direction and unpaved roads on ephemeral gully incision. *Earth Surf. Process. Landf.* **2009**, *34*, 1970–1984. [CrossRef]

113. Jebur, M.N.; Pradhan, B.; Tehrany, M.S. Optimization of landslide conditioning factors using very high-resolution airborne laser scanning (lidar) data at catchment scale. *Remote Sens. Environ.* **2014**, *152*, 150–165. [CrossRef]

114. Tien Bui, D.; Pradhan, B.; Revhaug, I.; Tran, C.T. A comparative assessment between the application of fuzzy unordered rules induction algorithm and j48 decision tree models in spatial prediction of shallow landslides at lang son city, vietnam. In *Remote Sensing Applications in Environmental Research*; Springer: Berlin/Heidelberg, Germany, 2014; pp. 87–111.

115. Pham, B.T.; Bui, D.T.; Dholakia, M.; Prakash, I.; Pham, H.V.; Mehmood, K.; Le, H.Q. A novel ensemble classifier of rotation forest and Naïve Bayer for landslide susceptibility assessment at the Luc Yen district, Yen Bai Province (Viet Nam) using GIS. *Geomat. Nat. Hazards Risk* **2017**, *8*, 649–671. [CrossRef]

116. Nguyen, Q.-K.; Tien Bui, D.; Hoang, N.-D.; Trinh, P.; Nguyen, V.-H.; Yilmaz, I. A novel hybrid approach based on instance based learning classifier and rotation forest ensemble for spatial prediction of rainfall-induced shallow landslides using GIS. *Sustainability* **2017**, *9*, 813. [CrossRef]

117. Hong, H.; Liu, J.; Bui, D.T.; Pradhan, B.; Acharya, T.D.; Pham, B.T.; Zhu, A.-X.; Chen, W.; Ahmad, B.B. Landslide susceptibility mapping using J48 Decision Tree with AdaBoost, Bagging and Rotation Forest ensembles in the Guangchang area (China). *Catena* **2018**, *163*, 399–413. [CrossRef]

 *water*

Article

Models of Gully Erosion by Water

Aleksey Sidorchuk

Faculty of Geography, Lomonosov Moscow State University, Leninskiye Gory 1, 119899 Moscow, Russia; fluvial05@gmail.com

Abstract: The type of modelling of gully erosion for the projects of land management depend on the targets and degree of details of these projects, as well as on the availability of input data. The set of four models cover a broad range of possible applications. The most detailed information about predicted gullies, change of their depth, width, and volume throughout the gully lifetime is obtained with the gully erosion and thermoerosion dynamic model. The calculation requires the time series of surface runoff, catchment relief, and lithology and the complex of coefficients and parameters, some of which can be estimated only by model calibration on the measurements. The difficulty in obtaining some of these coefficients makes it necessary to use less complicated models. The stable gully model predicts final gully depths and widths and is useful for projects where only stable gully geometry is used. The modified area–slope approach is used in the two simplest models, where the position on the slopes of possible gullies is calculated without details of the gully geometry. One of these models calculates total erosion potential, taking into account all water runoff transforming a gully. The second calculates gully erosion risk, using the information about slope inclination, contributing area and maximum surface runoff.

Keywords: land management; gully geometry; dynamic erosion model; stable gully; area–slope approach

Citation: Sidorchuk, A. Models of Gully Erosion by Water. *Water* **2021**, *13*, 3293. https://doi.org/10.3390/w13223293

Academic Editor: Csaba Centeri

Received: 14 October 2021
Accepted: 19 November 2021
Published: 21 November 2021

Publisher's Note: MDPI stays neutral with regard to jurisdictional claims in published maps and institutional affiliations.

1. Introduction

The need to assess possible gully position, depth, width and volume in agricultural areas and areas of new development is well known [1–4]. This practical necessity of land management causes the emergence of a wide variety of methods for assessing gully erosion potential (see reviews [5,6]). However, in this diversity, there is an obvious bias towards using the same theoretical approach, determining the critical slope and catchment area for the formation of a gully (area–slope approach), proposed in [7,8]. Often it is enough for practical purposes to estimate the points on the initial slope, where flow achieves threshold conditions and gully erosion initiates. The area–slope approach contains a significant empirical component (empirical coefficients) and the use of such models is restricted to the region of empirical data collection ([9–11] and many others, see bibliography in [6]). Most existing models of gully erosion, more or less empirical, find such threshold points and predict a maximum length of gully. The limitation of this approach and ways to overcome them are discussed in this paper.

It is not possible to estimate other measures of the gullies, such as depth, width, area and volume within the area–slope approach. Therefore, additional empirical models are used to calculate, for example, a gully width [12,13].

Maximum morphometric measures are not achieved simultaneously during the formation of a gully. Observations and experiments show [14] that a gully formation has two main stages. The first is the stage of rapid gully erosion, when during about 5% of the gully lifetime, more than 90% of the gully length, more than 80% of its depth, and about 60% of the area is reached. The second is the stage of gully stabilisation, when the area and last of all, the volume increases. Therefore, models of gully erosion calculation naturally fall into two groups for the prediction of gully morphology at these two stages.

It is possible to use the area–slope approach to calculate threshold conditions of flow not only in the gully head points but also in all points along the gully longitudinal profile. This way led to the calculation of the longitudinal profile of the stable gully, at each point of which flow velocity (controlled by slope, contributing area, resistance to flow and runoff depth) is equal to the critical velocity of erosion initiation. The processes of gully head and wall transformation by gravitation become important in this stage. Zorina [15] proposed one of the first models of this type, and its modification is described further. The full development of the stable gully takes many hundreds of years [16].

Most sophisticated dynamic models of gully evolution are based on the solution of an equation of mass continuity [17,18]. Usually, dynamic models describe the first stage of rapid gully formation, although such models can be run up to the stage of gully stabilization.

Different types of gully erosion models are used for different purposes, depending on the targets of such calculations and on the existence of input data. We offer in this paper the set of models, which cover the range of such targets in land management projects. This set includes four models of gully erosion prediction, where all known approaches to solving this problem are applied.

Two models calculate gully depth and width along its longitudinal profile:

(1)   The dynamic model of gully formation in "real" time. Not only mechanical erosion by water on a gully bed but also erosion by thermal action of water (if frozen soil is present) and gravitation processes on the gully walls are included in this model.

(2)   The model of finite stable gully describes gully geometry when water flow and gravitation already do not deform the gully bed and walls.

In the next two models, the novel modification of the area–slope approach is used, which gives the most probable position of possible gullies:

(3)   The model of gully erosion potential, which takes into account the effect of all runoff from the catchment.

(4)   Express model of the risk of gully erosion, which takes into account only the maximum surface runoff.

The main goal of this research article is to describe these four gully erosion models of different levels of simplification, to show their similarities and differences, and to compare the results of calculations of the characteristics of the same gully using these four models. The sequential comparison of the calculation results obtained with different models for the same object is novel and allows identifying the advantages and disadvantages of the proposed models and developing recommendations for their use in practice. This comparison makes it possible to choose which of the models gives the best results for available input data and for the level of complexity of the land management project.

2. Methodology—Model Descriptions

2.1. The Models with Gully Longitudinal Profile Calculations

2.1.1. The Dynamic Model of Gully Erosion

The dynamic gully erosion and thermoerosion model GULTEM (or DYNGUL model for erosion only) was proposed for calculations at the first stage of gully evolution. At this stage, the erosion and thermoerosion are predominant at the gully bottom, and rapid mass movement occurs on the gully walls. Gully channel transformation is very intensive, and morphometric characteristics of the gully (length, depth, width, area, volume) are far from stable and changing rapidly. This model is described in detail elsewhere [17,18], therefore, only basic information is given here.

The two main processes to be described are:

(a)   Formation of a rectangular cut in the initial slope by mechanical and thermal action of the flowing water.

(b)   Transformation of the rectangular cut into a trapezoidal shape by shallow landslides during the period between adjacent runoff events.

The rate of gully incision is controlled by water flow velocity, depth, temperature, as well as by the soil mechanical pattern and the level of protection by vegetation. These characteristics are combined in the equations of mass conservation:

$$\frac{\partial Q_s}{\partial X} = C_w q_w + EW + E_b D - CU_f W \tag{1}$$

and deformation:

$$(1 - \varepsilon)\frac{\partial V}{\partial t} = \frac{\partial Q_s}{\partial X} \tag{2}$$

where $Q_s = QC$ is sediment transport rate (m^3 s^{-1}), Q—discharge (m^3 s^{-1}); X—longitudinal coordinate (m); t—time (s); C—mean volumetric sediment concentration; C_w—sediment concentration of the lateral input; q_w—specific lateral discharge (m^2 s^{-1}); E—gully bed erosion rate (m s^{-1}); E_b—channel banks erosion rate (m s^{-1}); V—gully "empty" volume (m^2); W—flow width (m); D—eroded bank height (m); U_f—particles fall velocity in the turbulent flow (m s^{-1}), ε—soil porosity.

During the episode of erosion, the accumulation of sediments on the gully bed is assumed negligible. Therefore, Equations (1) and (2) can be transformed to:

$$\frac{\Delta V}{\Delta t} = EW + E_b D \tag{3}$$

The analysis of the experiment results in the natural gullies in different environments and in experimental flumes [17,18] shows that in the conditions of steep slopes and cohesive soils, common for gullies, the mean rate of bed erosion is linearly correlated with the product of bed shear stress $\tau = g\,\rho dS$ and the mean flow velocity U (i.e., stream power):

$$E \sim \tau U = k_E H q S \tag{4}$$

where S is flow surface slope, d is flow depth (m), $q = Q/W$ is specific discharge, g is acceleration due to gravity, ρ is water density and k_E is the combined erosion coefficient, which depends on soil properties. H is Heaviside step function, equal to 0 when flow velocity U is less than the critical velocity of erosion initiation U_{cr}, and equal to 1 when $U \geq U_{cr}$. Critical velocity depends on soil and vegetation cover properties [19].

Equation (4) links the dynamic model with the methods of gully potential estimation based on the threshold slope-contributing area approach [8,9].

In the case of gully erosion in the frozen soil (thermoerosion), the water temperature becomes the main factor of erosion. Field and laboratory experiments [18] showed that as the first approximation, the soil detachment rate is linearly correlated with water temperature T °C and equal to the rate of thawing of frozen soil with the open surface:

$$E_T = k_{TE} T \tag{5}$$

where k_{TE} is the coefficient of thermoerosion. Equation (5) is used for calculation only for the case of direct contact of water with frozen soil, without any thaw layer. Therefore the inequalities

$$E < E_T \text{ or } E > E_T \tag{6}$$

are checked in the calculation algorithm. If the rate of thermal erosion is less than the rate of mechanical erosion ($E > E_T$) and a thaw layer does not form. Then Equation (5) is used for estimation of the gully bed lowering by thermoerosion. On the contrary, if the rate of the thermal front movement in soil is more than the rate of mechanical erosion ($E < E_T$) then a thaw layer is formed, and Equation (4) is used in the model for the mechanical erosion rate calculation.

The erosion rate of the gully banks can be estimated only in a first approximation as some function of the rate of gully bed erosion, controlled by the ratio of lateral velocity v

and longitudinal velocity U. Using this assumption, an approximate formula for the rate of eroded bank erosion can be proposed:

$$E_b = k_b E \tag{7}$$

where k_b is the bank erosion coefficient, the details see in [18].

Flow width W and depth d in the gullies usually is calculated via regime equations as the power functions of discharge:

$$W = pw Q^{mw} \tag{8}$$

$$d = p Q^m \tag{9}$$

Flow velocity U is

$$U = \frac{Q}{Wd} \tag{10}$$

or, according to the Chezy–Manning formula

$$U = \frac{\sqrt{S}}{n} d^{2/3} \tag{11}$$

where n is Manning's roughness coefficient.

Gully walls become practically straight after rapid sliding, following the incision. In this case, a model of straight slope stability can be used for the prediction of gully walls inclination. It is possible for the practical needs to measure the gully walls inclinations at the stable sections and use the measured angle φ for the further calculations.

When the bottom width, wall inclination φ and whole volume of the gully V are known, the shape of the gully cross-section can be represented as a trapezoid with bottom width W_b, depth

$$D_g = \left[\sqrt{W_b^2 + \frac{4V}{\tan(\varphi)}} - W_b \right] \frac{\tan(\varphi)}{2} \tag{12}$$

and top width

$$W_t = W_b + 2\frac{D_g}{\tan(\varphi)} \tag{13}$$

The result of the calculations was gully depth, bed and top width, volume for each cross-section of the gully along each flowline and for each time step. The calculations required information about initial relief of the gully catchment and about boundaries of all lithological units, including topsoil with vegetation in the form of DTM (Digital Terrain Model), the sequence of surface runoff values, and all parameters and coefficients used in Equations (4)–(13). The coefficients are empirical and local and must be estimated for a given territory. The calibration of the model parameters, especially the erosion/thermoerosion coefficients and critical velocity on measured data, is highly recommended.

The dynamic model of gully erosion is the only one with which gully evolution can be simulated in "real" time, and all details of this process (within postulated assumptions) can be adequately described. The main problem was the simplification in the model of the real process (such as straight slopes assumption) and, nevertheless, a large number of parameters and coefficients, some of which required model calibration. The level of simplification of all other models, described further, was significantly greater, and the result of the modelling was less definite. However, these models require less initial information and can be useful for preliminary investigations.

2.1.2. Stable Gully Model

The stability of the gully refers to such a stage when the gully bottom is not eroded by water, and the gully walls do not change their shape. This requires two conditions: (1) flow velocities U in the gully are less than critical velocities of erosion initiation U_{cr} for soils composing the gully bed; (2) gravitation processes become negligible and the gully walls have reached the "angle of repose" for the given soil texture, cohesion and wetness.

In the STABGUL model, the critical velocity of erosion initiation U_{cr} was calculated using the Chezy–Manning formula (Equation (11)), and the flow depth d was represented as a power-law function of discharge (Equation (9)), which was calculated as the catchment area F multiplied by the runoff depth M:

$$d = p(kMF)^m \tag{14}$$

where k is the coefficient for changing the dimension of the quantities included in the formula. Then

$$U_{cr} = \frac{\sqrt{S}}{n}d^{2/3} = \frac{\sqrt{S}}{n}p^{2/3}(kMF)^{2m/3} \tag{15}$$

Therefore, the stable gully bed inclination is:

$$S = \frac{(nU_{cr})^2}{p^{4/3}(kMF)^{4m/3}} \tag{16}$$

Equation (16) links the model of the stable gully to the methods of gully potential estimation, based on the threshold contributing area-slope approach [8,9]. It can be written in the form of an ordinary differential equation:

$$\frac{dZ}{dL} = a(L)F(L)^{-b(L)} \tag{17}$$

where the contributing area F, coefficient a and exponent b are functions of length L from the mouth of the gully with the initial altitude z_{00} to i-th point on the flowline with initial altitude z_{0i}. Equation (17) is solved numerically along all flowlines with the known altitudes of the mouth of the stable gully z_{00}, runoff depth M, critical velocity, bed roughness, parameters in Equation (16), and the required functions of length. The result of the solution is the partial altitudes z_{ij} of gully bed along flowlines for a given j-th runoff depth M_j. The partial gully depth D_{ij} at i-th point is

$$D_{ij} = z_{0i} - z_{ij}. \tag{18}$$

The partial longitudinal profiles differ substantially from each other. To calculate the most probable shape of the stable gully longitudinal profile, all discharges passing through a channel must be taken into account. The channel deformations during a given discharge are proportional to the magnitude of the sediment transport rate and duration of this discharge [20]. Following this assumption, the following sequence of calculations was proposed to calculate the most probable longitudinal profile of a stable gully:

(a) The entire range of runoff depths is divided into N equal intervals, and probability p_k of each j-th interval of runoff depth is calculated.

(b) For each j-th runoff depth M_j in the middle of the j-th interval, the partial longitudinal profiles with bottom altitudes z_{ij} are calculated at each i-th point on the flowlines with Equation (17).

(c) For each j-th runoff depth M_j the partial magnitude of the sediment transport rate E_{ij} is calculated with Equation (4), written without erosion coefficient as

$$E_{ij} \sim \frac{M_j F_i}{W_{ij}} S_{0i} \tag{19}$$

where S_0 is the inclination of the initial slope. Flow width is calculated with Equation (8).

(d) The most probable stable gully bottom altitudes Z_{pi} are calculated weighted with partial magnitudes of the sediment transport rate E_{ij} and runoff depth probability densities p_j

$$Z_{pi} = \frac{\sum_{j=1}^{N} z_{ij} E_{ij} p_j}{\sum_{j=1}^{N} E_{ij} p_j} \tag{20}$$

The result of these calculations is the most probable stable gully profile along each flowline. The calculation requires information about the initial relief of the gully catchment and about boundaries of all lithological units, including topsoil with vegetation in the form of the DTM (Digital Terrain Model); the probability density function of surface runoff depths, all parameters and coefficients used. As was already mentioned, these coefficients are empirical and local and must be estimated for a given territory.

The most probable stable gully profile can be close to one of the partial longitudinal profiles calculated with particular surface runoff. However, it did not coincide completely with any of them. It was obvious that flow velocities along the most probable profile for larger surface runoff values could be greater than the critical velocity of erosion initiation. The ultimate stable gully profile was, therefore, lower than the most probable one.

2.2. Area–Slope Approach

The product of the critical slope S and the power-law function of contributing catchment area F determine the condition of erosion initiation within this approach [7,8]:

$$a = SF^b \tag{21}$$

It is also can be written in the form:

$$S = aF^{-b} \tag{22}$$

In several works [9,21,22] Equation (21) is written in the form

$$a_m S^{-\alpha} = C \tag{23}$$

where C is critical contributing area of the catchment. It was shown in [22] that C was similar to Horton's [23] critical distance for channel initiation.

Equations (21)–(23) bear equal information, as $\alpha = 1/b$ and $a_m = a^{\frac{1}{b}}$.

Equation (21) is an empirical analogue of the formula for the critical shear stress [24] or the square of the critical flow velocity U_{cr} [25], at which erosion begins. The flow in gullies is usually of turbulent type. Therefore, flow velocity U was expressed in terms of the slope and catchment area using the Chezy–Manning formula (Equation (15)):

$$U^2 = \frac{S}{n^2} d^{4/3} = \frac{S}{n^2} p^{4/3} (kMF)^{4m/3} \tag{24}$$

The parameter a in Formula (21) will be written as

$$a = \frac{(nU)^2}{p^{4/3}(kM)^{4m/3}} \tag{25}$$

and the exponent b as $4m/3$.

According to [6], the empirical information for a set of investigated regions showed that exponent b varied across a rather narrow range and its mean was about 0.4 (0.38). Therefore, m in Equation (24) must be 0.3, and this value fits the measurements in the gullies of the Yamal peninsula [26].

The coefficient a in Equation (21) was much more variable even within one catchment area [6] since it depends on many factors. These were the shape of the gully channel, the roughness of its bottom, the critical velocity of erosion initiation and the water runoff depth.

The last characteristic was the most difficult to determine for calculation with Equation (21). The surface runoff depth varies through time (and less, in space) and is a probabilistic variable. The choice of the value (probability) of this quantity was largely subjective and at least should be justified in each specific case.

The way to avoid this uncertainty was to transform Equation (24) thus that the indefinite surface runoff depth is regarded not as a predictor but as a response:

$$M_{cr} = \frac{(nU_{cr})^{3/2m}}{S^{3/4m}p^{1/m}(kF)} \tag{26}$$

It is worth noting that Equation (26) is the version of Equation (23), where both parts are divided by the critical catchment area F (C in Equation (23)) and multiplied by the runoff depth:

$$M = (a_m M)S^{-\alpha}F^{-1} \tag{27}$$

This version is novel and was not discussed in the literature. All parameters and variables at the right side of Equation (26) can be determined unequivocally for every point on a catchment. The calculated value of surface runoff depth is also unique for each point at the catchment and has the meaning of critical. If at a given catchment point with some slope and area, the actual runoff depth is greater than or equal to the critical value calculated by the Equation (26)

$$M \geq M_{cr} \tag{28}$$

the erosion initiation is potentially possible at this point. This critical runoff depth M_{cr} and risk of erosion initiation have determinable probability (duration) P_M. Equation (26) is used further in two following models of gully erosion potential estimation.

2.2.1. Total Erosion Potential (TEP) Model

The critical runoff depth M_{cr} is the minimal runoff depth, which initiates erosion. Therefore, erosion occurs during all flows with runoff depths more than M_{cr}. The influence of all flows with runoff depths more than M_{cr} must be taken into account to calculate the gully erosion potential. In the TEP model, the similar procedure, as in the STABGUL model, is proposed:

(a) The entire range of flow depths is divided into N intervals, and probability p_j of each j-th interval is calculated.

(b) For each j-th runoff depth M_j in the middle of the j-th interval, the partial erosion rate E_{ij} is calculated at each point i-th on the catchment with Equation (4).

$$E_{ij} = \left(k_E \frac{Q_j}{W_j}S\right)_i = \left[\frac{k_E}{pw}(M_j F)^{1-mw}S\right]_i \tag{29}$$

(c) Erosion potential E_{pi} by flows with runoff depths more than critical is calculated as the sum of partial erosion rates E_{ij}, weighted with runoff depth probability p_j, beginning from the interval j_{cr} with critical runoff depth M_{cr}:

$$E_{pi} = \sum_{j=jcr}^{N} E_{ij}p_j \tag{30}$$

(d) To remove coefficients in Equation (29), relative erosion potential R_{Ei} is estimated by dividing E_{pi} by its maximum value E_{Ni}:

$$E_{TPi} = \frac{\sum_{j=jcr}^{N} E_{ki}p_j}{E_{Ni}} = \frac{\sum_{j=jcr}^{N}(M_j)^{1-mw}p_j}{\sum_{j=0}^{N}(M_j)^{1-mw}p_j} \tag{31}$$

The influence of morphometric characteristics S and F at the i-th point of the catchment is reached through the value of M_{cr} from Equation (26), which determines the number of intervals of M from jcr to N, used in the calculation of E_{TPi}.

The value of E_{TPi} varies from 1 (at $M_{cr} = 0$) to 0 (at $M_{cr} \geq M_{max}$). At $M_{cr} = 0$ ($j = 0$), erosion begins at a given point for any, the smallest values of the runoff depth and E_{TPi} is maximum. Accordingly, at $M_{cr} \geq M_{max}$ ($j = N$), the runoff depth never exceeds the critical depth, and erosion will not occur at this point.

The input information to run TEP is simpler than for the previous models, as morphometry for each flowline is not required. Instead, the digital models for flow accumulation and slope are used, which can be calculated from DEM with most GIS. Empirical coefficients and parameters in Equation (26) are used in this model. Hydrological information is represented in the form of the probability density function of runoff depths, as in the STABGUL model.

The output is the values of the relative erosion potential E_{TPi}, calculated for each pixel on the catchment area DEM. The probability density function of E_{TPi} is usually negatively skewed [26], it is better to use the logarithms of E_{TPi}.

2.2.2. Express Estimation of Gully Erosion Risk

Probability density functions of runoff depth for a small catchment can often be approximated with an exponential relationship [27]. For this case, E_{TPi} in Equation (31) is an inverse power-law function of M_{cr}. This is the base to formulate the express model of gully erosion risk GER, (the term risk is used here instead of the term potential to distinguish these two models) when power-law functions are used instead of Equation (31). Using the exponent $2m/3$ in Equation (26), the expression for such function of critical runoff depth takes the form:

$$M_{cr}^{2m/3} = \frac{n U_{cr}}{\sqrt{S}\, p^{2/3} (kF)^{2m/3}} \tag{32}$$

Therefore, Equation (31) is transformed to the expression of gully erosion risk E_{GR}:

$$E_{GR} = 1 - \left(\frac{M_{cr}}{M_{max}} \right)^{2m/3} \tag{33}$$

As in Equation (31), the value of E_{GR} varies from 1 (at $M_{cr} = 0$) to 0 (at $M_{cr} \geq M_{max}$), which corresponds to a change in the degree of potential erosion risk. At $M_{cr} = 0$, erosion begins at a given point for any, the smallest values of the runoff depth and E_{GR} is maximum. Accordingly, at $M_{cr} \geq M_{max}$, the runoff depth does not exceed the critical depth, and erosion will not occur at this point.

The input information to run GER comes from digital models for flow accumulation and slope, which can be calculated from DEM in most of GIS. Empirical coefficients and parameters p, m, U_{cr} and n are used in this model. Hydrological information is restricted to the maximum daily runoff depth. The outputs are the values of the gully erosion risk E_{GR}, calculated for each pixel on the catchment area DEM.

3. Results

Since each of the presented models of gully erosion was characterized by a certain degree of simplification of the real process, it is rational to compare calculation results with measurements of actual erosion. Of greatest interest is the comparison of calculations based on the most complicated model GULTEM with actual data. If successful, the calculation results for simpler models can be compared with calculations from GULTEM.

3.1. GULTEM Model Validation and Calibration

The common definition of validation offered by the Society for Computer Simulation Technical Committee on Model Credibility [28] is "*Substantiation that a computerized model within its domain of applicability possesses a satisfactory range of accuracy consistent with the intended application of the model"*. In our case, *"a satisfactory range of accuracy"* depends on the quality of the measurements on the natural object and on the accuracy of the input data necessary for calculations. The catchment selected for the validation and calibration procedure (Figure 1) is situated on the Yamal Peninsula, West Siberia, close to the most severely gullied part of the peninsula. The investigated gully P-1 did not exist in 1986. After construction in 1986–1987 of the exploitation camp at the top of the catchment, erosion and thermoerosion began due to the increased water supply and vegetation cover damage. In 1988, the gully P-1 length was 450 m, in 1989—740 m, in 1990–1991—940 m and in 1995—965 m. The gully head reached the buildings of the exploitation camp, and repeated filling of the gully head with heavy loam by bulldozers stopped the gully growth. The observations showed that gully P-1 was still active in 2007 [18] and further (Figure 1), increased in depth and volume, though the gully length did not increase further. Therefore, the period 1991–1995 of not regulated gully activity was used for model validation and calibration with two longitudinal profiles of the gully bed measured in July 1991 and 1995 (Figure 2).

Figure 1. The catchment of the large unnamed stable gully with tributary gully P-1 on the Yamal peninsula, Russia (the image of 27 June 2016 from Google Earth). The star on the left part of the figure shows the catchment position.

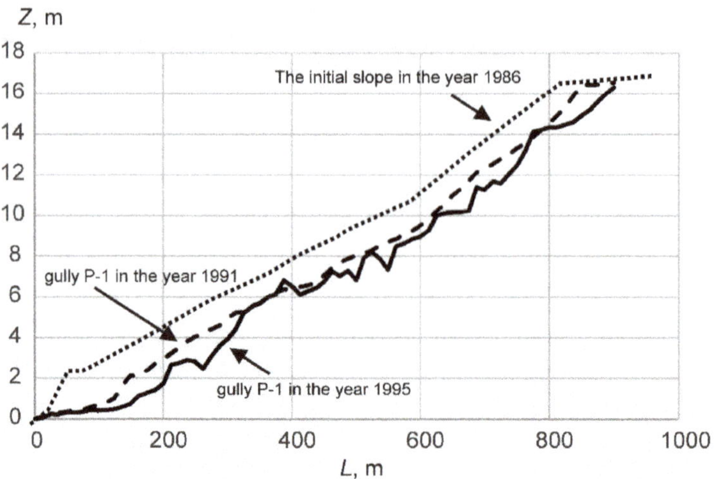

Figure 2. The longitudinal profiles of gully P-1, used for model calibration.

Empirical coefficients and parameters, required to run GULTEM, were estimated from morphometric and hydraulic measurements on selected and adjacent catchments: the coefficient of thermoerosion $k_{TE} = 2.5 \times 10^{-5}$ m s^{-1} (°C)$^{-1}$; the gully walls inclination $\varphi = 0.6$ rad; bank erosion coefficient k_b was calculated via [18]; coefficients in Equation (8) are $pw = 3.0$ m (m^3/s)$^{-mw}$ and $mw = 0.4$; coefficients in Equation (9) are $p = 0.21$ m (m^3/s)$^{-m}$ and $m = 0.3$; Manning's roughness coefficient $n = 0.06$ m^{-1} s; critical velocity of erosion initiation $U_{cr} = 0.2$ m s^{-1}. Meteorological information, in form of changing through time air temperature and precipitation, was taken from ERA5 reanalysis. The sequences of surface runoff depths were calculated with the hydrological model, validated on the hydrological measurements on the selected catchment [27].

The combined erosion coefficient k_E in Equation (4) varied in calibration procedure in the range 6.5×10^{-5}–6.5×10^{-4} m^{-1}. The difference between calculated (Z95c) and measured (Z95m) gully bed altitudes in 1995 was estimated with

$$RMSE = \frac{\sqrt{\Sigma_1^N[(Z95c - Z95m)dx]^2}}{L} \tag{34}$$

where N is the number of gully segments with the measured altitudes, dx—the distance between segments, L—gully length in the year 1995. The best fit of Z95c and Z95m was at $k_E = 3.6 \times 10^{-4}$ with $RMSE = 0.057$ m (Figure 3). This value shows a consistency between the model results and measurements and *"a satisfactory range of accuracy"*.

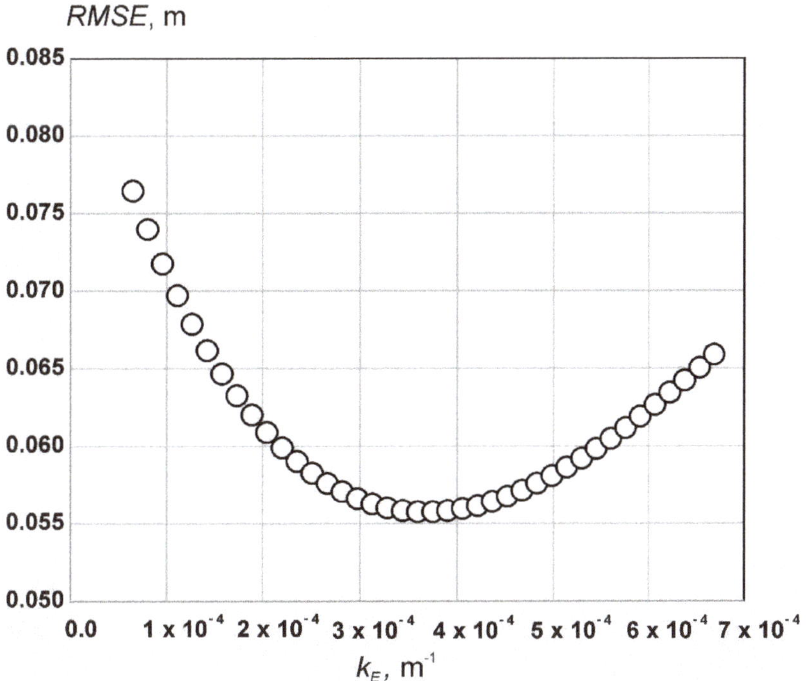

Figure 3. The result of model calibration.

3.2. The Comparison of GULTEM and STABGUL Models

The calculations with GULTEM and STABGUL models were performed for the gully P-1. The same empirical coefficients were used in both models. The probability density function for surface runoff, used in STABGUL (Figure 4), was derived from a 30-year long sequence of surface runoff depths [27], which was used in GULTEM calculations. The GULTEM model was run for a 1200-year period (40 repetitions of a 30-year long sequence) to obtain a stable longitudinal profile, suited for comparison with the stable gully calculated from the STABGUL model.

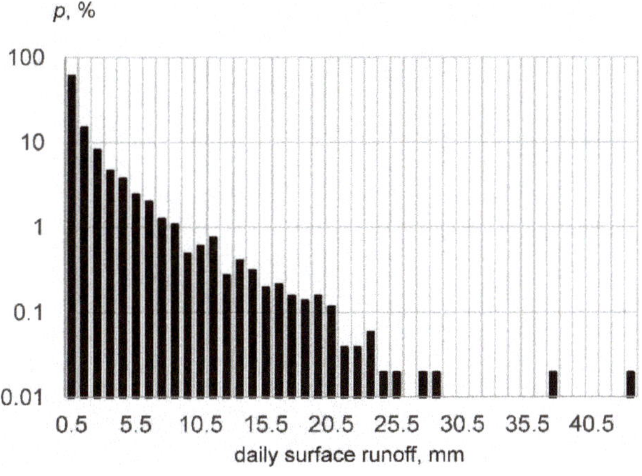

Figure 4. Histogram of daily surface runoff at the gully P-1 catchment.

The results of calculations with GULTEM in Figure 5 show a gradual increase of the gully depths and the length of the stable segment of the gully bed through time. At these segments of the gully, its depth does not change anymore. After about 1000 years in calculations, almost the entire gully bed becomes stable, and flow velocities for any discharge become less than critical at all points.

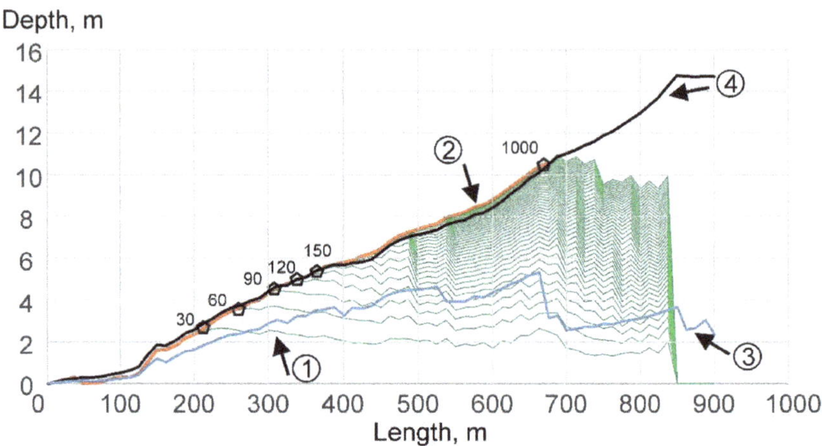

Figure 5. The sequence of gully P-1 incision depths calculated with GULTEM for the 1200-year period with 30-year step (green lines 1; pentagrams on some of the lines indicate the age). Red line 2 shows the depths of the gully after 1000 calculation years. Blue line 3 shows the depths of the most probable stable gully, black line 4—the depths of the ultimate stable gully, calculated with the STABGUL model.

Calculations with STABGUL with the above-described algorithm resulted in the most probable stable longitudinal profile with depths (line 3 in Figure 5), close to the depths from GULTEM after about 120–150 calculation years. This period is close to the period of stabilization of most human-induced gullies on the East European plain [29]. These depths are about 40% less than from GULTEM after 1000 years of calculated erosion (line 2 at Figure 5). The ultimate stable longitudinal profile calculated with STABGUL for the maximum runoff depth (line 4 at Figure 5) fits well with the stable profile after 1000 years from GULTEM. This result shows that the most probable stable longitudinal profile from the STABGUL model can accurately simulate a stable gully after 120–150 years of its lifetime, which can further slowly be transformed to its ultimate stability. It also points out the significance of high discharges on the process of gully erosion.

3.3. The Comparison of the GULTEM and TEP Models

The total erosion potential was calculated in the TEP model along the same flowline and with the same empirical coefficients and hydrological data as gully depth in GULTEM. The similarities and differences in the results (Figure 6) are explained by the similarities and differences in the models' formulations. Both models operate with the rate of erosion E (Equation (4)), which depends on local slope inclination and specific discharge.

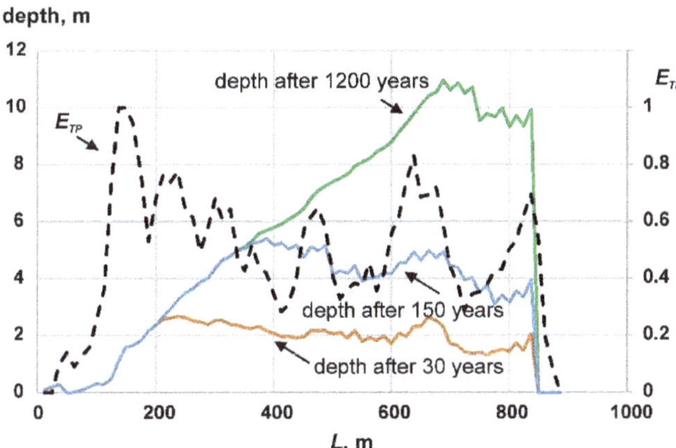

Figure 6. The gully P-1 incision depths for 30-, 150- and 1200-years calculation periods with GULTEM. Black dashed line shows total erosion potential, calculated with the TEP model.

At a given point, the gully depth in GULTEM and erosion potential in TEP are the sum of the product of the erosion rate values and their durations. In GULTEM the products are included to the sum if flow velocity is more than critical, in the TEP model, if runoff depth is more than critical. Therefore, we can expect a correlation of the calculation results. The differences shown are due to the significant simplification in the TEP model of the erosion process. The result of calculations in the TEP model shows the potential of erosion, its position at the catchment, and potential intensity, not gully geometry. In the TEP model, the influence of the altitude of the basis of erosion at the gully mouth and the processes of gully walls slumping are not taken into account. The main difference is the absence in the TEP of the gully longitudinal profile self-forming evolution. Slope inclination in the TEP model is the inclination of the initial slope. Therefore, the correlation between the results of calculation with the two models quickly decreases with time (Figure 7).

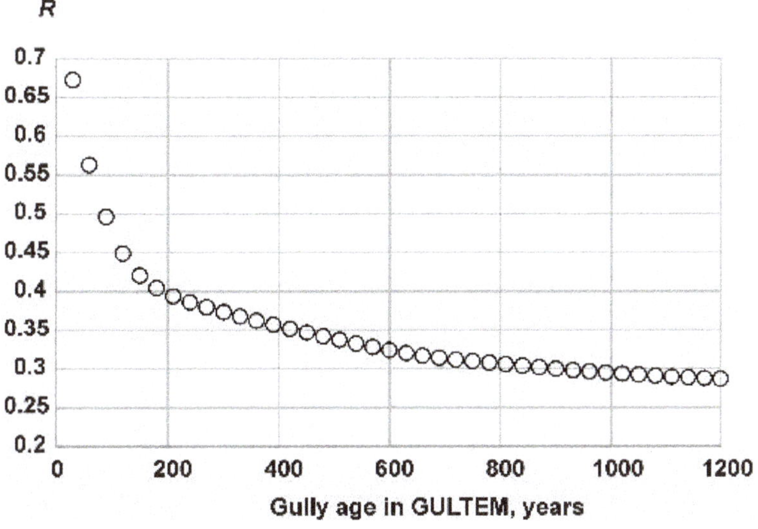

Figure 7. The decrease of correlation with the increase of prediction time between total erosion potential, calculated with the TEP model, and gully depths, calculated with GULTEM.

3.4. The Comparison of TEP with GER Model

As in previous cases, gully erosion risk was calculated in the GER model along the same flowline and with the same empirical coefficients and hydrological data. There is a close relationship (power-law or logarithmic) between calculation results from the TEP model of total erosion potential E_{TP} and of gully erosion risk E_{GR} in GER (Figure 8). Therefore, the comparison of calculation results from the GER model and GULTEM is more or less the same as in the previous paragraph.

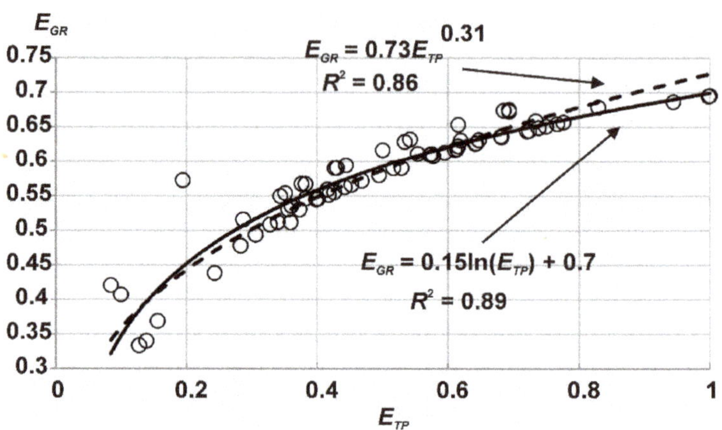

Figure 8. The relationship between total erosion potential from TEP and gully erosion risk from GER for the points along the bed profile of gully P-1, shown in Figure 2.

4. Discussion

The problems to be discussed are the availability to possible users of input information, required for calculations with detailed models, and the limitations in simplified models.

The GULTEM is the only model of the four described in this article in which gully evolution can be simulated in "real" time and gully geometry change though time can be adequately simulated. In this model, it is possible to take into account the temporal trends in hydrological information, common in the conditions of global climate change. The temporal and spatial changes in land use, for example, properties of vegetation cover, are also possible to include in the calculations. The large number of input empirical coefficients, morphometric and hydrological data required to run GULTEM is a limitation of the model. The calculations require information for all or particular flowlines in the form of digitized 2D longitudinal profiles of the altitudes of top catchment surface Z (see Figure 2) and of the surfaces Z_j of other underlying lithological units along these flowlines. The contributing catchment areas F and coordinates x, y are also collected for the same points. If a digital elevation model (DEM) is available for the catchment, most of this information can be collected using standard tools included in any geographical information system (GIS). The lithological composition of the territory may require investigations in the field.

Parameters and coefficients, used in Equations (4–13), are also better to obtain via measurements at the selected territory, as in [18]. Values of some of these coefficients can be found in special literature. The coefficients in regime Equations (8) and (9) for gullies are discussed in [6,13], the coefficient of thermoerosion in [18], the critical velocity of erosion initiation in [19,30], Manning's roughness coefficient in [31,32]. The gully walls inclination φ is assumed to equal the angle of repose for a given lithology (see discussion in [33]).

Experimental data on the values of the erosion coefficient k_E for cohesive soils, common for gullies, is limited. Experiments and model calibration were provided on soils of the Yamal peninsula [17,18], for silt loams in the gullies of the coast of George Lake in Australia [34], loess soils in the Manawatu River basin and rendzina in the Waimakariri River basin, New Zealand [35] and granodiorite saprolites at the basin of the Mbuluzi

River (Swaziland) [36]. The variability of this coefficient is very high. In the direct measurements of sediment budget in the gullies of the Yamal peninsula during the snow thaw and summer rains the value of k_E changes from 0.0008 to 0.01 for loams and silt loams with cohesion from 10 to 40 KPa, estimated with a torvane shear tester, and up to 1.3 for loose silt sand. The measurements in Australia showed a similar range of k_E: for the loams with cohesion 30–40 KPa 0.0028–0.0034, with cohesion 50–70 kPa—0.0008. The measurements in New Zealand for the loess soil with cohesion 51 kPa showed k_E = 0.004–0.005 for soil with organic carbon content 2.4% and k_E = 0.021 for the same soil with organic carbon content 0.44%. For rendzina with cohesion 22 kPa and 3.5% content of organic carbon, k_E was 2×10^{-5}. Model calibration for the gullies in Swaziland showed k_E value 0.006 for sandy saprolites with cohesion 4.5–9 kPa. The listed information was not enough to find the relationship between k_E and eroded cohesive soil properties, and the large range of its values does not permit the recommendation of values for calculations on not-investigated objects. The only way to use GULTEM for gully erosion prediction is to calibrate the model on the measurements (see Section 3.1). Usually, such calibration is possible in the framework of detailed projects of land management [36,37].

The complexity in erosion coefficient estimations is the reason to use simpler models for preliminary applications. The models STABGUL, TEP and GER do not require the erosion coefficient. Other parameters and coefficients used in these models, if not estimated in the field measurements, can be found in special literature.

Probability density functions (PDF) of surface runoff values are used in STABGUL and TEP models but for different erosion form lifetimes. The minimum prediction duration in the STABGUL model is about 150 years. Therefore, it is assumed that the time series of hydrological characteristics are stationary. This is not true in the current global climate change, and predictions with STABGUL will not include the trends in the time series of surface runoff. Keeping in mind this limitation, STABGUL is a powerful tool to find the longitudinal profiles, lengths, depths, widths, areas and volumes of the gullies close to the stage of stabilization [16]. This can be useful for the classification of the gullies in the intensively gullied territory to separate stable gullies and those that are still active [38].

The total erosion potential model is based on the modification of the area–slope approach in gully erosion investigations when critical runoff depth is calculated. The TEP model estimates possible gully position in the catchment and erosion intensity. This information can be enough in preliminary projects of land management. The absence of the need for the flowline geometry is the significant simplification: the calculations are performed directly for given points in the catchment, for which slope and contributing area are measured. Calculations of the total erosion potential with the TEP model are more flexible in the conditions of trends in hydrological time series. Much smaller prediction periods are needed in the TEP model (one or a few decades), allowing it to use step functions of PDFs to take into account temporal changes in hydrological statistics. The main limitation of the TEP model is the absence of prediction of the gully geometry.

The GER model requires very limited hydrological data—only maximum daily surface runoff depth for the prediction period. At the same time, the GER model output for a certain catchment contains nearly the same information as total erosion potential from the TEP model, which uses more hydrological data. This similarity is explained by the more or less close relationship between parameters of PDFs of surface runoff values and their maximums. Such relationships are different for different PDFs, as other parameters of PDFs must be taken into account. Therefore, transfer from TEP to GER output will also be different for different hydrological conditions, as unnamed parameters of PDFs are implicitly used in such transfer.

5. Conclusions

The four models described here encompass all now existing approaches to gully erosion prediction. Comparison of the results of calculations for the same gully using these four models makes it possible to choose the model that is most suitable for use in a land management project depending on its objectives and the available information. The set of gully erosion models discussed here can meet the requirements of land management projects with different levels of detail.

Use of the dynamic gully erosion model GULTEM is recommended for calculations of gully geometry transformation in time and space for the most detailed projects of land management. Flowline geometry, soil texture, parameters and coefficients, used in GULTEM, can be obtained via measurements at the selected territory and from DEMs. The calculations take into account the temporal trends in hydrological information, important for conditions of global climate change. The main difficulty in modelling is the estimation of the erosion coefficient, which usually requires calibrating the model.

The input data for the stable gully model STABGUL is the same as for GULTEM, except for the erosion coefficient, which is the most difficult to estimate. Therefore, STABGUL is easier to apply. The STABGUL model is useful for projects where only final gully geometry is required. Only stationary hydrological time series are used in the model.

Possible gully position in the catchment and erosion intensity are estimated with the total erosion potential (TEP) and gully erosion risk (GER) models. The novel type of area–slope approach allows effective use of hydrological information in the form of critical surface runoff for erosion initiation. These models can be useful for preliminary estimates of the possible effects of different land uses.

The presented models, with proper selection of initial data and necessary calibration, can be used for territories with different climates, soils, and land use. They are most effective for areas with possible linear disturbance of the natural vegetation cover, which is typical for the use of land for pastures, for the initial stages of urbanization and the industrial development of new territories.

Funding: This research was funded by Russian State Task 0110—1.13, N 121051100166-4 "Hydrology, morphodynamics, and geoecology of erosion-channel systems".

Conflicts of Interest: The author declares no conflict of interest.

References

1. Poesen, J.; Nachtergaele, J.; Verstraeten, G.; Valentin, C. Gully erosion and environmental change: Importance and research needs. *Catena* **2003**, *50*, 91–133. [CrossRef]
2. Valentin, C.; Poesen, J.; Li, Y. Gully erosion: Impacts, factors and control. *Catena* **2005**, *63*, 132–153. [CrossRef]
3. Bennett, S.; Wells, R. Gully erosion processes, disciplinary fragmentation, and technological innovation. *Earth Surf. Process. Landf.* **2019**, *44*, 46–53. [CrossRef]
4. Sidorchuk, A.Y. Assessing the gully potential of a territory (a case study of central Yamal. *Geogr. Nat. Resour.* **2020**, *41*, 178–186. [CrossRef]
5. Poesen, J.W.A.; Torri, D.; Vanwalleghem, T. Gully erosion: Procedures to adopt when modelling soil erosion in landscapes affected by gullying. In *Handbook of Erosion Modelling*; Morgan, R.P.C., Nearing, M.A., Eds.; Wiley-Blackwell: Chichester, UK, 2011; pp. 360–386. [CrossRef]
6. Torri, D.; Poesen, J. A review of topographic threshold conditions for gully head development in different environments. *Earth Sci. Rev.* **2014**, *130*, 73–85. [CrossRef]
7. Schumm, S.A. Geomorphic thresholds: The concept and its applications. *Trans. Inst. Brit. Geogr.* **1979**, *NS4*, 485–515. [CrossRef]
8. Patton, P.C.; Schumm, S.A. Gully erosion, north western Colorado: A threshold phenomenon. *Geology* **1975**, *3*, 83–90. [CrossRef]
9. Desmet, P.J.; Poesen, J.; Govers, G.; Vandaele, K. Importance of slope gradient and contributing area for optimal prediction of the initiation and trajectory of ephemeral gullies. *Catena* **1999**, *37*, 377–392. [CrossRef]
10. Vandekerckhove, L.; Poesen, J.; Oostwoudwijdenes, D.; Nachtergaele, J.; Kosmas, C.; Roxo, M.J.; de Figueiredo, T. Thresholds for gully initiation and sedimentation in Mediterranean Europe. *Earth Surf. Process. Landf.* **2000**, *25*, 1201–1220. [CrossRef]
11. Garrett, K.K.; Wohl, E.E. Climate-invariant area–slope relations in channel heads initiated by surface runoff. *Earth Surf. Process. Landf.* **2017**, *42*, 1745–1751. [CrossRef]
12. Woodward, D.E. Method to predict cropland ephemeral gully erosion. *Catena* **1999**, *37*, 393–399. [CrossRef]
13. Nachtergaele, J.; Poesen, J.; Sidorchuk, A.; Torri, D. Prediction of concentrated flow width in ephemeral gully channels. *Hydrol. Process.* **2002**, *16*, 1935–1953. [CrossRef]
14. Kosov, B.F.; Nikolskaya, I.I.; Zorina, Y.F. Experimental research of gullies formation. *Exp. Geomorphol.* **1978**, *3*, 113–140. (In Russian)
15. Zorina, E.F. Quantitative methods of estimating of gully erosion potential. In *Soil Erosion and Channel Processes (Eroziya Pochv i Ruslovye Protsessy)*; Chalov, R.S., Ed.; Moscow Univ. Press: Moscow, Russia, 1979; Volume 7, pp. 81–90. (In Russian)

16. Sidorchuk, A.; Märker, M.; Moretti, S.; Rodolfi, G. Gully erosion modelling and landscape response in the Mbuluzi River catchment of Swaziland. *Catena* **2003**, *50*, 507–525. [CrossRef]
17. Sidorchuk, A. Dynamic and static models of gully erosion. *Catena* **1999**, *37*, 401–414. [CrossRef]
18. Sidorchuk, A. Gully erosion in the cold environment: Risks and hazards. *Adv. Environ. Res.* **2015**, *44*, 139–192.
19. Sidorchuk, A.; Grigor'ev, V. Soil erosion on the Yamal peninsula (Russian Arctic) due to gas field exploitation. *Adv. GeoEcol.* **1998**, *31*, 805–811.
20. Wolman, M.G.; Miller, J.P. Magnitude and frequency of forces in geomorphic processes. *J. Geol.* **1960**, *68*, 54–74. [CrossRef]
21. Montgomery, D.R.; Dietrich, W.E. Landscape dissection and drainage area-slope thresholds. In *Process Models and Theoretical Geomorphology*; Kirby, M.J., Ed.; Wiley: New York, NY, USA, 1994; pp. 221–246.
22. Istanbulluoglu, E.; Tarboton, D.G.; Pack, R.T.; Luce, C. A sediment transport model for incision of gullies on steep topography. *Water Resour. Res.* **2003**, *39*, 1103. [CrossRef]
23. Horton, R.E. Erosional development of streams and their drainage basins: Hydro-physical approach to quantitative morphology. *Geol. Soc. Am. Bull.* **1945**, *56*, 275–370. [CrossRef]
24. Harvey, M.D.; Watson, C.C.; Schumm, S.A. *Gully Erosion*; Tech. Note 366; U.S. Dept. of the Inter., Bur. of Land Management: Washington, DC, USA, 1985; p. 181.
25. Mirtskhulava, Ts.E. *Principles of Physics and Mechanics of Channel Erosion (Osnovy Fiziki i Mekhaniki Erozii Rusel)*; Gidrometeoizdat: Leningrad, Russia, 1988; p. 303. (In Russian)
26. Sidorchuk, A. The potential of gully erosion on the Yamal peninsula, West Siberia. *Sustainability* **2020**, *12*, 260. [CrossRef]
27. Matveeva, T.; Sidorchuk, A. Modelling of surface runoff on the Yamal peninsula, Russia, using ERA5 reanalysis. *Water* **2020**, *12*, 2099. [CrossRef]
28. Schlesinger, S.; Crosbie, R.E.; Gagne, R.E.; Innis, G.S.; Lalwani, C.S.; Loch, J.; Sylvester, R.J.; Wright, R.D.; Kheir, N.; Bartos, D. Terminology for model credibility. *Simulation* **1979**, *32*, 103–104.
29. Sidorchuk, A.Y.; Golosov, V.N. Erosion and sedimentation on the Russian plain, II: The history of erosion and sedimentation during the period of intensive agriculture. *Hydrol. Process.* **2003**, *17*, 3347–3358. [CrossRef]
30. Van Rijn, L.C. Erodibility of mud-sand bed mixtures. *J. Hydraul. Eng.-ASCE* **2020**, *146*, 04019050. [CrossRef]
31. Chow, V.T. *Open-Channel Hydraulics*; McGraw-Hill Book Co.: New York, NY, USA, 1959; p. 680.
32. Engman, E.T. Roughness coefficients for routing surface runoff. *J. Irrig. Drain. Eng.-ASCE* **1986**, *112*, 39–53. [CrossRef]
33. Statham, I. Some limitations to the application of the concept of angle of repose to natural hillslopes. *Area* **1975**, *7*, 264–268.
34. Sidorchuk, A. Dynamic model of gully erosion. In *Modelling Soil Erosion by Water*; Boardman, J., Favis-Mortlock, D., Eds.; Springer: Berlin/Heidelberg, Germany, 1998; pp. 451–460. [CrossRef]
35. Sidorchuk, A.; Schmidt, J.; Cooper, G. Variability of shallow overland flow velocity and soil aggregate transport observed with digital videography. *Hydrol. Process.* **2008**, *22*, 4035–4048. [CrossRef]
36. Sidorchuk, A.; Märker, M.; Moretti, S.; Rodolfi, G. Soil erosion modelling in the Mbuluzi River catchment (Swaziland, South Africa). Part I: Modelling the dynamic evolution of gullies. *Geogr. Fis. E Din. Quat.* **2001**, *24*, 177–187.
37. Kayode-Ojo, N.; Ehiorobo, J.O.; Uzoukwu, N. Modelling and prediction of gully initiation in the University of Benin using the GULTEM dynamic model. *J. Energy Technol. Policy* **2016**, *6*, 22–32.
38. Wasson, R.J. Annual and decadal variation of sediment yield in Australia, and some global comparisons. In *Variability in Stream Erosion and Sediment Transport*; Olive, L.J., Loughran, R.J., Kesby, J.A., Eds.; IAHS Publ. N 224: Wallinford, UK, 1994; pp. 269–279.

Article

Controlling Factors of Badland Morphological Changes in the Emilia Apennines (Northern Italy)

Paola Coratza * and **Carlotta Parenti**

Department of Chemical and Geological Sciences, University of Modena and Reggio Emilia, Via Campi 103, 41125 Modena, Italy; carlotta.parenti@unimore.it
* Correspondence: paola.coratza@unimore.it

Abstract: Badlands are typical erosional landforms of the Apennines (Northern Italy) that form on Plio-Pleistocene clayey bedrock and rapidly evolve. The present study aimed at identification and assessment of the areal and temporal changes of badlands within a pilot area of the Modena Province (Emilia Apennines), where no previous detailed investigation has been carried out. For this purpose, a diachronic investigation was carried out to map the drainage basin and the drainage networks of the linear erosion features in the study area during the last 40 years, and to evaluate changes in badlands drainage basins morphometry and surface, land use and pluviometry. The investigation carried out indicated a general stabilisation trend of the badlands in the study area. In fact, a reduction in the bare surface area from 6187.1 m^2 in 1973 to 4214.1 m^2 in 2014 (31%), due to an intensified revegetation process around the badland areas, has been recorded. This trend, in line with the results of research carried out in other sector of the Northern Apennines, is mainly due to intensive land use changes, mostly the increase in forest cover and the reduction of agricultural land, that occurred in the study area from the 1970s onwards.

Keywords: badlands; morphological changes; land use change; Emilia Apennines (Northern Italy)

Citation: Coratza, P.; Parenti, C. Controlling Factors of Badland Morphological Changes in the Emilia Apennines (Northern Italy). *Water* **2021**, *13*, 539. https://doi.org/10.3390/w13040539

Academic Editor: Csaba Centeri

Received: 29 December 2020
Accepted: 16 February 2021
Published: 19 February 2021

Publisher's Note: MDPI stays neutral with regard to jurisdictional claims in published maps and institutional affiliations.

1. Introduction

Soil erosion is one of the most significant land degradation processes worldwide and has produced diverse geomorphological effects in different environments according to anthropogenetic and climatic forcing.

Badlands are deeply and densely dissected accelerated erosional landforms often developed on unconsolidated or poorly cemented materials [1]. They are generally characterised by steep, unvegetated slopes, a high drainage density and high erosion rates. Badlands are similar to miniature fluvial systems, and it is possible to observe interconnections between hillslope processes and landforms [2]. The morphogenesis and evolution of badlands are complex and still under debate. Among the controlling factors suggested in the literature, those indicated as fundamental are as follows: (i) The lithology [3,4] and physical, chemical and mineralogical properties (e.g., sediment size, clay mineralogy, Atterberg's limits, porosity and pore water chemistry) [5,6]; (ii) climatic factors such as strong seasonal humid/arid contrasts favoured by south-facing slopes [6–11]; (iii) the landscape morphology/topography, particularly the slope gradient, orientation and exposure [12]; (iv) anthropic activities such as deforestation for land reclamation, land levelling, cropland abandonment, extensive agriculture and pasture [13,14]. Moreover, badland areas are often characterised by scarce or absent vegetation cover [15–19], related to unfavourable climate conditions, and by the occurrence of mass wasting processes such as slides, earthflows, mudflows and creeping [8,14,20,21].

Badlands are typically frequent in drylands [20,22–29], but they also occur in wetter areas (from semiarid to humid), with annual rain ranging from 400 to 1200 mm/year, often concentrated in the autumn after marked semiarid conditions during the summer period, as in the Mediterranean Basin [30–35].

Hillslope degradation is an important process that can have negative impacts on agriculture and terrestrial ecosystems as a whole [36], causing the loss and depletion of soil, decreases land productivity and accessibility, economic damage, risky conditions and environmental changes in the landscape [37–45]. On the other hand, it can also increase the geodiversity of a territory, contributing to its tourism development [14,46–48]. Moreover, spectacular erosion landforms (i.e., badlands) have important ecological functions and are considered hotspot areas of biodiversity [49]. The different phases of historical gully erosion can reflect significant environmental changes such as changes in climate and human–environment interactions [48]. During times of global change, it is of remarkable interest to trace the environmental changes in different terrains, including badlands, and to properly analyse and study these processes in order to plan effective strategies for landscape conservation and enhancement.

For the last 30 years there has been a growing scientific interest in badlands and, more in general, in gully erosion. This reflects the concern to increase knowledge of factors, controls, dynamics and impacts of badlands in the context of global change [39,44,50–52]. The study of Martinez-Murillo and Nadal-Romero [53] reviewed studies during recent decades on badlands, presenting a summary of the main specific research topics: origin, lithology, human activities and land use, vegetation, hydrology, piping, erosion and erosion rates, modelling and use of emerging technologies, reclamation and restoration, geoheritage and geotourism. Many studies have examined the effects of human activities and land use changes on badlands [54–61]. Farming systems and practices, such as development of livestock and croplands, rural abandonment and land levelling, can strongly affect the persistence and the evolution of badlands. In particular, land levelling is responsible for badland shrink in several parts of the world, as reported by Poesen [62]. Clarke and Rendell [56] highlight how the remodelling and levelling activities to increase agricultural productivity can cause the loss of badland landforms and an increase in soil erosion. Hillslope degradation in response of land use change, also promoted by agricultural policy, has been extensively documented in semi-arid regions, particularly in many areas of the Mediterranean Basin [35,56,63–69]. Reforestation is also blamed to play an active role on landscape transformation and particularly in decreasing erosion rates in badland areas [35,70–73]. The relationship between changes in rainfall patterns (annual precipitation, intensity and inter-annual variability) and badland inter-rill erosion has received attention as well [56,66,67,74–80].

In Italy, badlands are spread widely and discontinuously in areas of the Apennines mountain chain and Sicily, where Plio-Pleistocene marine clayey and marly terrains outcrop [81]. In the Italian literature, the investigation of badland landforms and landscapes—their identification, characterisation, and temporal and spatial variation—dates back to the beginning of the last century and is mainly concentrated on central and southern parts of Italy [2,6,11–14,17,23,28,49,55,67,80–102].

Badlands consisting of Pliocene and Pleistocene claystone are widespread in the pre-Apennine hills in the Emilia Romagna region, but, apart from the studies by Bucciante [6] and Farabegoli and Agostini [87], no recent and detailed work has been conducted. The local erosion rates have been very high in this area, causing the development of widespread badland landforms, accompanied by hazardous processes, such as the retrogression of badland scarps and rapid soil depletion. The Modena Apennines are widely affected by intense hillslope degradation, and typical badland examples are observed in the hilly area between Fiorano and Marano sul Panaro. According to the Modena Province-Coordinated Territorial Plan (PTCP), badlands are divided and mapped according to their landscape value, as:

- Type A or peculiar badlands (type A sensu [17,83]), characterised by sharp and dissected landforms with knife-shape slopes, sparse vegetation and high drainage. This type of badland has a high intrinsic landscape value, and all interventions and activities that could alter or compromise the conditions of the places, the morphogenetic or biological processes, or the perception of the landscapes are forbidden [103];

- Type B or typical badlands (type B sensu [17,83]), characterised by gentle slopes affected by recurrent surficial slides and mudflows and a less-dense drainage system. For this type of badland, with a medium landscape value, some anthropogenic activities (e.g., the installation of telecommunication lines and systems, and systems for water supply) in the surrounding areas are allowed;
- Type C or pseudo-badlands, characterised by a gentle morphology, with little-to-no landscape value. Anthropogenic activities are allowed in these areas, but measures to mitigate the impacts on the landscape should always be taken;

In Modena Province, 8% of the badlands are Type A and cover an area of 3.6 km^2, 48.2% belong to Type B and cover 10 km^2, and 43.8% belong to Type C and cover 10.1 km^2.

Starting from these premises, the present study focuses on badlands of a pilot area in the Modena Province (Emilia Apennines, Northern Italy), representative of the morpho-evolution and degradation processes that affect hilly areas of the Emilia Apennines. The study area is part of the industrial district of Sassuolo-Fiorano-Maranello, the largest tile making district in the world, that faced significant land use changes in the past century. Moreover, the area represents a growing tourist attraction due to the high aesthetic value of hilly landscapes and to the presence of the best developed and largest mud volcano field of Italy. As mentioned above, no previous systematic studies on badlands in Emilia Apennines have been conducted. Thus, the main aims of this study are the: (i) Identification, morphometrical characterisation and mapping of spatial and temporal changes of badlands; (ii) identification and characterization of the main geo-environmental features, such as topographic aspect, climate, and human induced land use changes, to detect the controlling factors in the morphodynamics of badlands in the study area.

For this purpose, a diachronic investigation is carried out, mapping the drainage basin and the drainage networks of each badlands area over the last 40 years, and evaluating changes in badland drainage basins morphometry and surface, land use and pluviometry. The obtained results are compared with studies on badlands in Apennines and in the central Mediterranean regions.

2. Study Area

The study area, located on the foothills of the Northern Apennines (Municipality of Fiorano Modenese, Modena Province, Italy) has an extent of about 10 km^2 and an elevation ranging from 143 to 308 m a.s.l. (Figure 1). The area is; therefore, a low-hill territory, belonging to the River Secchia catchment. The Torrente Fossa, the main stream channel in the study area, is a right tributary of the River Secchia and flows in a SW–NE direction, collecting the water of its left tributaries, Rio Chianca and Rio delle Salse.

Figure 1. Location of the study area (white dashed line) and localities cited in this paper (©2017 Google).

2.1. Climatic Setting

The climate of the area is defined as temperate, a mild temperate climate (Cfa) according to the Köppen classification [104]. An arithmetic mean of climate data for the period 1954–2018, related to three significant stations in the surrounding regions of the study area (San Valentino, Castellarano and Sassuolo cf. Figure 1), are summarised in Table 1. The mean annual rainfall is around 800 mm. Rainfall is concentrated in the spring and autumn months, with a maximum precipitation in November at about 90 mm; conversely, the summer months are hot and dry, with a minimum precipitation in July at only 39 mm (Figure 2a). The mean annual temperature is about 13 °C. In July, the hottest month, the temperature reaches 29 °C in July, the hottest month, and 2 °C in January, the coldest (Figure 2b).

Table 1. Climate data of the study area.

	1954–2018
Mean annual rainfall (mm)	840
Minimum annual rainfall (mm)	449
Maximum annual rainfall (mm)	1160
Absolute minimum monthly rainfall (mm)	39 (July)
Absolute maximum monthly rainfall (mm)	93 (November)
Mean annual temperature (°C)	13
Minimum monthly temperature (°C)	2 (January)
Maximum monthly temperature (°C)	29 (July)

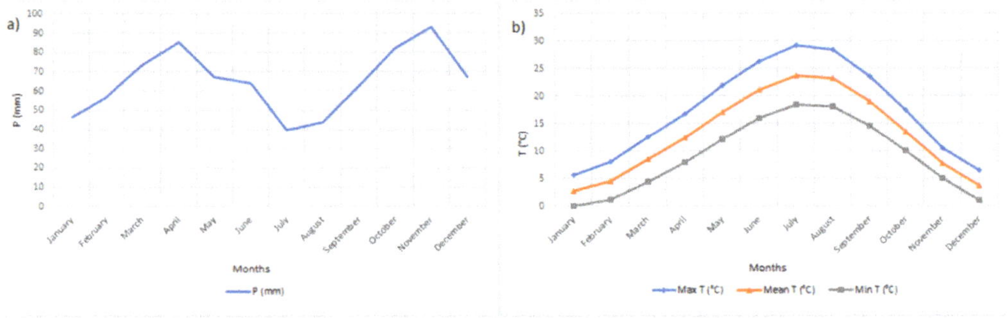

Figure 2. (**a**) Average monthly precipitation for the study area; (**b**) average monthly maximum, minimum and mean temperature of 1954–2018.

2.2. Geological Setting

From a geological viewpoint, the area is located in the Modena Apennine margin, where the outcropping units reflect the most recent (since the Upper Eocene) geological history of the rising Apennine chain (Figure 3). In particular, marine silty-clayey rock types belonging to the Argille Azzurre Formation (Lower Pliocene–Lower Pleistocene) extensively crop out [105]. They are marly clays, grey and blueish-grey clayey and silty marls, cropping out in medium-to-thin layers, with joints hardly visible owing to bioturbation. Discontinuous, thin, laminated layers of fine biocalcarenites and yellowish siltites are found locally [105,106]. The carbonate content is quite low (25–30%) [107], and the main components are kaolin and montmorillonite [108]. The general setting of the strata shows a NW–SE strike, with an 8°-to-40° NE dip. In the SW part of the study area, the Termina Formation (Tortonian–Lower Messian) crops out. The Termina Formation is composed of greyish clayey marl characterised by undistinguished stratification and by the presence of bioclasts, biosoma and glauconite, intercalated by thin layers of fine sandstone. Continental

Quaternary deposits, which belong to the Modena Unit sediments dating from the sixth century CE, crop out along the Torrente Fossa riverbed. This unit comprises gravel deposits turning to alluvial terrace sand and silt deposits [105,106]. Corresponding to the Salse di Nirano Natural Reserve, there are sediments belonging to the emissions of Salse. They are considered Quaternary deposits and are composed of "mud flow" locally complemented by debris related to the upwelling of water and hydrocarbons.

Figure 3. Litho-structural sketch of the study area.

From a tectonic viewpoint, the study area is located approximately 2 km southwest of the active Pedo-Apennine thrust. An anticline, folding the Pliocene–Early Pleistocene claystone, crosses the area, with its major axis trending NW–SE. Two main families of steep, nearly perpendicular, tectonic discontinuities (high angle faults and/or fractures) are oriented with NW-SW direction and, orthogonal to the latter, with SW-NE and ENE-WSW direction.

2.3. Geomorphological Outlines

The geomorphological features of the study area are determined by the widespread presence of clay. The main geomorphological processes and related landforms are associated to mass movements and surface runoff. Landslides are mainly ascribable to shallow earth slides and earth flows, in most cases affecting cultivated fields. Many landslide deposits along the Rio Serra and Rio delle Salse valleys have been colonised by spontaneous vegetation, which has contributed to their stabilisation. Solifluction is a secondary process particularly widespread in the northern part of the area, particularly in slopes with fine grain size lithologies and bare of vegetation, where it has been favoured by creep due to grazing. Landforms and deposits due to running waters are particularly widespread. The alluvial deposits are limited in the central region of the study area in correspondence with Rio del Petrolio and Rio Chianca. Due to the mutual interaction between gravitational processes and running water action, a remarkable badland morphology, developed on slopes constituted by the Argille Azzurre Formation, characterises this area (Figure 4). In fact, according to PTCT, Type A badlands make up more than 40%, covering an area of 1.4 km^2, while Type B and C badlands represent 29% each, covering 0.5 and 0.4 km^2, respectively. Landslides and earth flows can contribute to the regression of the badlands' slopes and can fill the drainage network.

Figure 4. Main geological and geomorphological features of the study area: (**a**) Active landslide; (**b**) solifluction characterising the Argille Azzurre Formation; (**c**) badland morphology characterising the Argille Azzurre Formation.

3. Materials and Methods

Initially badlands were mapped on 1:5000 topographical maps by aerial photo and satellite image interpretation and field visits, allowing an inventory map of badlands to be constructed. In our study, we consider only Type A badland [17,83] and every little horseshoe type hydrographic unit characterised by bare soil or signs of intense denudations were mapped.

The morphological evolutionary trend of the badlands in the study area was evaluated according to the different steps described below.

3.1. Multitemporal Analysis of Aerial Photos and Satellite Images and Mapping of Badlands' Drainage Basin Changes

The multitemporal mapping of the drainage basins of each badland area, with particular reference to the badlands' head retreats and sidewall failures, was performed in GIS (ESRI software ArcGis 10.2.1, ArcMap) through manual delineation method. Gully walls and headcuts were digitalized on screen and on orthophotographs and satellite images with appropriate and comparable scales (Table 2). Therefore, images dated to 1954 and 1981, with non-comparable scales, were excluded from the mapping. Considering this source data, with a resolution of 0.5 m, the expected error is of ±2% [11,103]. Subsequently, the area (A) of each drainage basin and its bare surface was calculated and uploaded into a geodatabase, and the percentages of surface variations throughout the investigated periods were calculated with respect to the initial surface area size in order to compare their evolution and perform the succeeding morphometric analysis.

Table 2. Cartographic documents examined. GAI: Aerial Italian Group; RER: Emilia-Romagna Region; GN: Geoportale Nazionale website; AGEA: Agenzia per le Erogazioni in Agricoltura; TeA: Consorzio Telerilevamento Agricoltura; B/W: Black and white; C: Colour.

Type	Year	Scale	Flight	Color
Aerial photos	1954	1:66,000	Volo GAI	B/W
Aerial photos	1973	1:15,000	Volo RER	B/W
Aerial photos	1981	1:33,000	Volo Romagna	B/W
Orthophotos	1988	1:10,000	National Web Map Service	B/W
Orthophotos	1996	1:10,000	National Web Map Service	B/W
Orthophotos	2006	1:10,000	National Web Map Service	C
Orthophotos AGEA	2008	1:10,000	National Web Map Service	C
Orthophotos AGEA	2011	1:10,000	National Web Map Service	C
Orthophotos TeA	2014	1:10,000	Regione Emilia-Romagna Web Map Service	C

3.2. Morphometric and Multiparametric Analysis

The drainage network in the digitalised basins was reconstructed through the aerial photointerpretation of two pairs of orthophotos from the years 1973 and 2014 that show the most significant changes in the badland area and morphology. The stream network was digitalised in GIS, through manual delineation method; the lengths of the channels (L), the number, and the stream hierarchy were identified; and the following morphometric indices were calculated: The drainage density (D) and the direct bifurcation ratio (Rdb). These indices are related to the characteristics of the hydrographic system and depend on the erosion rate. The D parameter is the sum of the lengths of all the streams and rivers in a drainage basin (ΣL) divided by the total area of the drainage basin (A) [109,110]. The intensification of the precipitation and slope acclivity cause the D to increase. The higher the D, the higher the erosion rate for the ground. The direct bifurcation ratio is the ratio of the number of stream branches of a given order (Ndu) to the number of stream branches of the next higher order (Nu + 1) [109–111].

In order to correlate different factors, a multiparametric analysis was performed, taking into account the following parameters: The direct bifurcation ratio, drainage density, slope exposure, and basin evolution typology.

3.3. Pluviometric Data and Land Use

Although the pluviometric data were fragmented for the study area, three measuring stations adjacent to the area, and located at San Valentino (314 m a.s.l.), Sassuolo (121 m) and Castellarano (135 m), were taken into consideration and an arithmetic mean of data from these stations was used for analysis. The dataset covers the period 1954–2018. In order to better understand the role played by precipitation variability in badland development, total annual and monthly precipitation data over a period of 60 years were computed. Moreover, the rainfall intensity, maximum number of consecutive dry days and maximum number of consecutive wet days were calculated. According to several authors [45,66,76,112], daily rainfall >10.0 mm can be considered the threshold at which runoff commences in semiarid environments. Accordingly, pluviometric events >2.0, >10.0 and <30.0, and >30.0 mm were considered, and the trend from 1954 to 2018 was analysed. Finally, the average annual precipitation over 10-year periods was then correlated with the results obtained by the analysis of the evolutionary trend of the badland area and the multitemporal photointerpretation analysis.

Because land use changes and variation in vegetation cover are possible driving factors for badland evolution, land use changes were examined using two datasets for 1994 and 2014 provided by the Geoportale della Regione Emilia-Romagna, and the percentages of the major land use classes, according to the CORINE Land Cover, were calculated.

4. Results and Discussion

4.1. Multitemporal Analysis and Mapping of Badland Areas

The multitemporal photointerpretation allowed the identification of the surface-area changes caused by water runoff over the last 50 years. In 1954, the badlands were characterised by a dense network of deep incisions, and steep and bare slopes with sharp, knife-edge ridges. In 1973 and 1981, the vegetation cover increased, and rills and gullies were less evident. Significant changes cannot be recognised in the aerial photos of 1988/89 and 1994 due to their small scales and black and white colours. From 2003 to the present, the vegetation cover has colonised the foot of the slopes, and a general trend of decreasing bare surface areas and a progressive increase in vegetation growing upward can be observed. Moreover, the bottom of the gullies has frequently become filled with flow deposits over the last 50 years.

The analysis of the orthophotos spanning 1973 to 2014 allowed the mapping and digitalisation of 55 drainage basins (Figure 5). Inside each basin, the segment of furrows was traced for the years 1973 and 2014, showing the most significant changes in badland surface area (Figure 5). The multitemporal landform mapping provided information about badland changes and evolution in time and space.

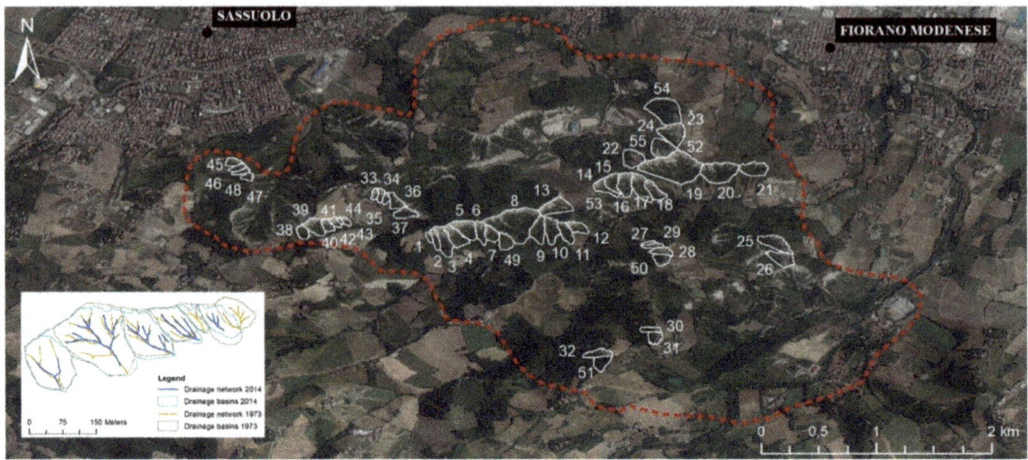

Figure 5. Badlands inventory map (white line) in the study area (red dashed line). The box shows an example of the area variation over a 40-year period and the segment of furrows traced.

Observing the spatial distribution of badlands, it can be noticed that 60% of badlands is aligned preferentially along an ENE-WSW direction, perpendicular to the anticline which cross the area, while the remaining 40% is aligned preferentially parallel to the anticline with the major axis trending NW-SE, documented in local geological maps.

On average, the recorded trend highlighted a moderate reduction in the bare surface area from 6187.1 m^2 in 1973 to 4214.1 m^2 in 2014 (31%), due to an intensified revegetation process around the badland areas, while the total badland area did not show a significant modification, from 575,778 m^2 in 1973 to 583,565 m^2. In fact, the increase of about 1.3% in the total badland area is in line with expected error related to orthophotos resolution and digitalization process.

In particular, for a 40-year period in the study area, it can be observed that:

- In 26 basins, there is no evidence of significant evolution of the upper margin of the badland; the upper margin is indeed apparently stable;
- In 10 basins, the upper margin is moving downward because of the vegetation preventing the erosion;
- In 11 basins, the ridge of the badlands is affected by a regression of the margin;
- Four basins are visible only in the aerial photos of 1973 and not in the orthophotos of 2014: This is due to intensified revegetation processes;
- Four basins are visible only in the orthophotos of 2014 and not in the aerial photos of 1973.

The changes in the bare surface area may suggest that, in the study area, the revegetation process is more intense than the erosion one.

4.2. Morphometric and Multiparametric Analysis

Regarding the morphometric analysis, the data relating to the geometric parameters of the badland basins in 1973 and 2014 can be summarised as follows:

- D fluctuated between 0.018 and 0.064 m/m^2 in 1973 and 0.006 and 0.044 m/m^2 in 2014. The average values were 0.04 m/m^2 (in 1973) and 0.02 m/m^2 (in 2014) (Figure 6). In 1973, the maximum value was 0.064 m/m^2, and the minimum, 0.018 m/m^2; in 2014, the maximum was 0.044 m/m^2, and the minimum, 0.006 m/m^2. The averages were 0.04 m/m^2 (in 1973) and 0.02 m/m^2 (in 2014);
- Rdb fluctuated between 2 and 6 in 1973 and between 2 and 5 in 2014, showing that the hydrographic system is not very articulated. The maximum value in 1973 was 6,

and the maximum in 2014 was 5; the minimum value was 2 in both 1973 and 2014 (Figure 7).

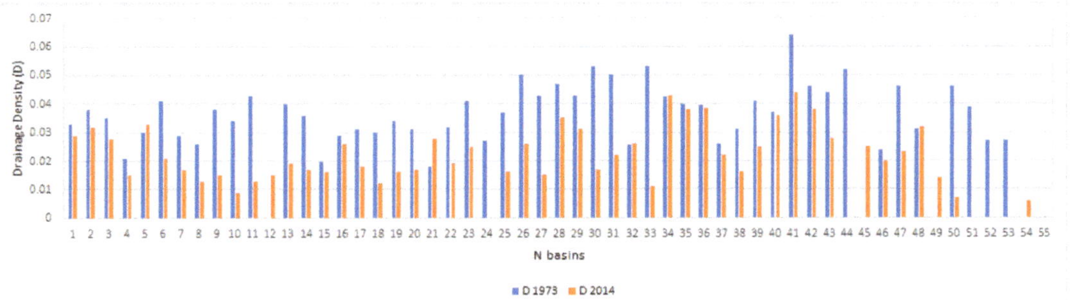

Figure 6. Variation of drainage density (D) values in 1973 (in blue) and in 2014 (in orange).

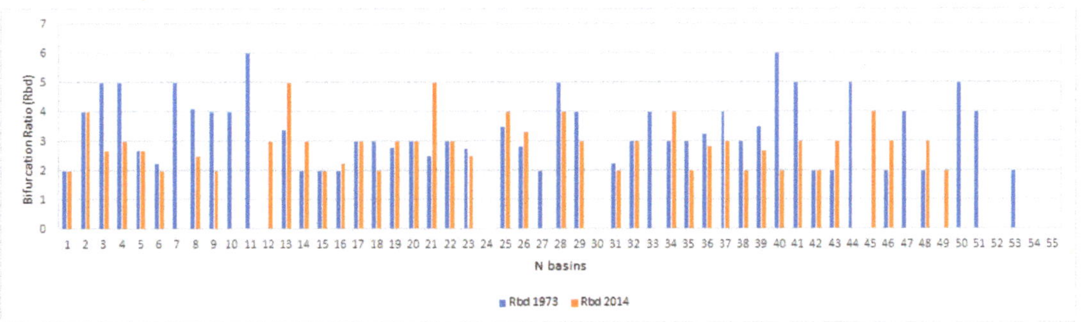

Figure 7. Variation of Bifurcation Radio (Rdb) values in 1973 (in blue) and in 2014 (in orange).

Comparing the values of D in 1973 and in 2014, it can be noticed that D is generally decreasing through time, indicating a decreasing erosion rate for the ground. This could be mainly due to the decreasing total length of the channels of the hydrographic system (ΣL) and not to the variation of the badland area extent. There was an increase in D in only five basins—5, 21, 32, 34, and 48.

Comparing the values of Rdb in 1973 and in 2014, it can be observed that the Rdb values are generally decreasing, which means that the hydrographic systems are not so articulated and the erosion rate is decreasing. However, in 11 basins (13, 14, 16, 19, 21, 25, 26, 34, 43, 46, and 48), Rdb is increasing, and the values did not change in nine basins (1, 2, 5, 15, 17, 20, 22, 32, and 42).

Finally, we evaluated the correlation between the Rbd parameter and the D parameter for both years (the D values increased a hundred times) (Figure 8). Both D and Rdb tended to decrease over time, although with differently distributed values. In particular, only in nine basins (numbers 8, 9, 18, 23, 27, 28, 29, 33, and 41) there was a significant decrease in both D and Rdb; in four basins (numbers 3, 4, 40, and 47), Rdb decreased significantly, unlike D, which showed a limited decrease; in seven basins (numbers 1, 2, 15, 17, 20, 22, and 42), D decreased significantly, but the corresponding Rdb values do not show variations between the two years considered; finally, the increase in Rdb was also related to an increase in D only in three basins, namely, numbers 21, 34, and 48.

The data obtained were then correlated with the type of evolution of badlands and with slope aspect. Therefore, only basins characterised by significant evidence of evolution through time were considered. In 11 basins, the scarp of the badlands was affected by a regression, with the Rbd values showing a random trend and D values decreasing (showing a possible correlation). In 10 basins, the upper margin of the badlands is moving downward because of the presence of the vegetation, and the Rbd and D values have random trends, showing no correlation between the evolution typologies and the variation of parameters over time.

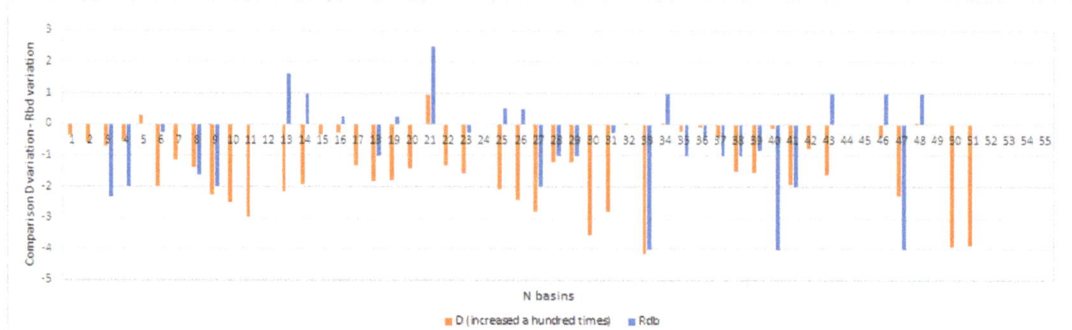

Figure 8. Comparison between D values in 1973 and D values in 2014 (increased a hundred times) (in orange) and differences between Rbd values in 1973 and Rbd values in 2014 (in blue) (y axis) and number of basins (x axis).

The slope exposure does not seem to have particularly influenced the evolution of the badlands. In fact, 69% of badlands form preferentially on SW and S-facing flanks (135°–225°); of the 11 badlands that show a regression of the upper margin, five basins are exposed to western directions, five to a southern direction and only one to a northern direction (Figure 9). On the other hand, all 10 basins that show a revegetation process are exposed to southern directions.

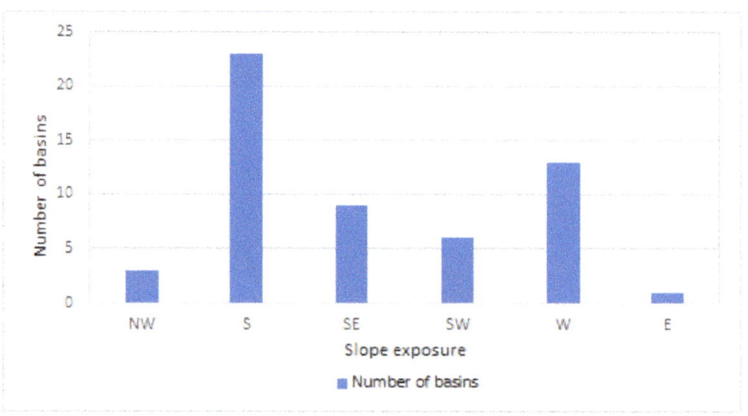

Figure 9. Relationship between the slope aspect direction and mapped badlands in the study area.

4.3. Analysis of Pluviometric Data and Land Use Changes

The analysis of the annual precipitation during the period 1954–2018 showed a general decreasing tendency for the average annual precipitation, although the decreases were very low and characterised by a sinusoidal trend (Figure 10a); this is consistent with the results reported by several authors on long-term climate variability in Italy [113–118]. A general decrease in rainfall intensity (mm/rain day) also occurred during the period (Figure 10b). Nevertheless, the decrease in intensity was less than that in total annual precipitation. From the trend analysis of events >2.0, >10.0 and <30.0, and >30.0 mm, it emerges that a significant decrease in pluviometric events >2.0, >10.0, and >30.0 mm has occurred

(Figure 11), reflecting a general decrease in rainfall intensity and total annual precipitation. No significant trend was recorded for >30.0 mm pluviometric events.

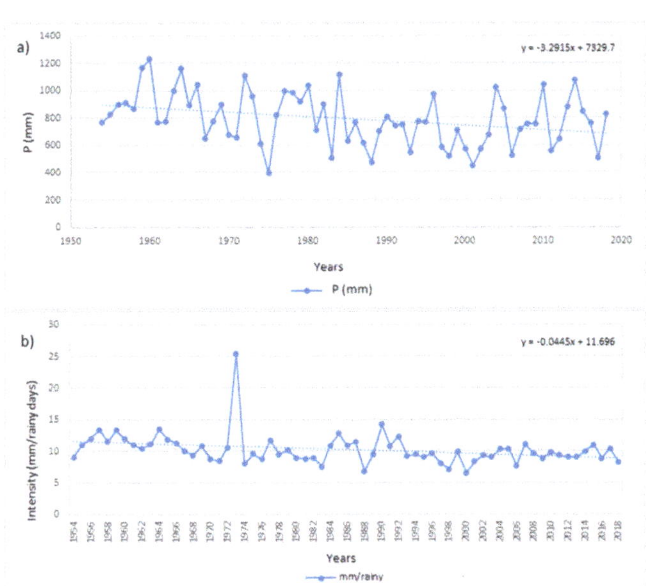

Figure 10. Pluviometric pattern for the study area: (**a**) Distribution of average annual precipitation during the observation period; (**b**) annual intensity (mm/rain day).

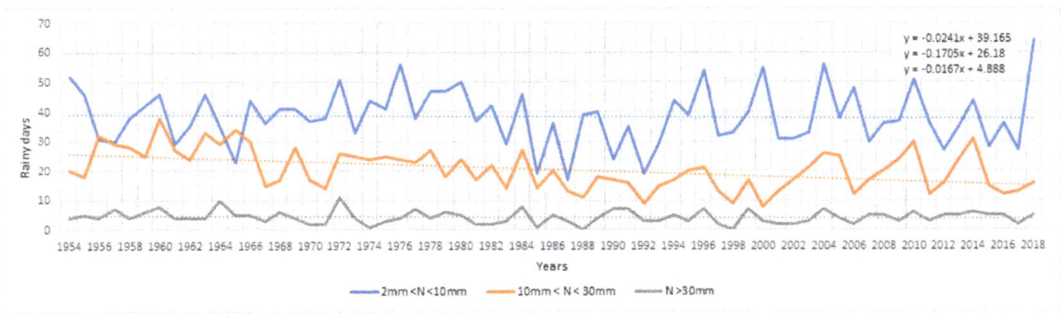

Figure 11. Total annual number of rainy days > 2 mm, > 10 mm and <30 mm, and <30 during 1954–2018.

In order to evaluate the role of dry/wet cycles, the maximum numbers of consecutive dry and wet daily events were calculated. The analysis shows a slight increase in consecutive wet days and decrease in consecutive dry days. This trend is in line with the bare surface area reduction recorded for the study area. In fact, a substantial decrease in the amount and frequency of rainfall tend to reduce the effectiveness of rilling, leading to less-intensive dynamics for the soil-erosion process.

The analysis of the average annual precipitation over 10-year periods (Figure 12) and of the results obtained by multitemporal photointerpretation analysis showed that, in 1954–1973, the variation of the average annual precipitation is represented by a decrease of about 15%, accompanied by an increment in vegetation, as revealed by the multitemporal photointerpretation analysis. In 1973–1983, the average annual precipitation decreased by about 9%, and the analysis of the evolutionary trend of the badland areas revealed that in 7% there was a retrogression of badland scarps by about 9 m. No significant differences were detected in the period 1983–1993, while an increase of 15% in average annual precipitation was recorded in the subinterval 1994–2013. In this time period, 15% of the badlands were characterised by active retrogressive movement, which may be related to the precipitation increase recorded.

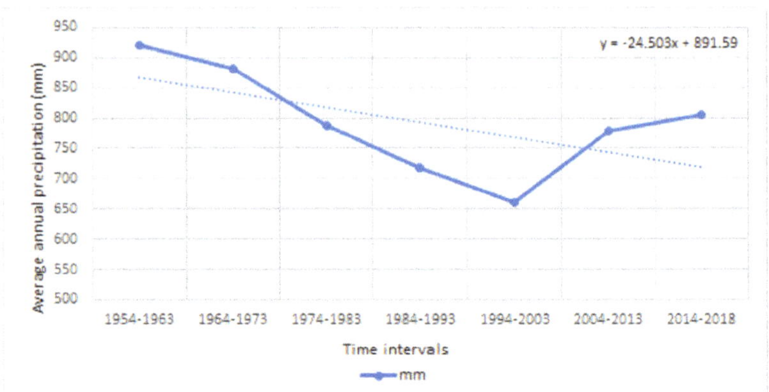

Figure 12. Average annual precipitation over 10-year periods. The average annual precipitation decreases constantly from 921.82 mm/year to 661.56 mm/year (28%) in the period 1954–2003, while a slight increase is recorded in the period 2004–2014.

Land use changes were deduced using two datasets for 1994 and 2014 provided by the Geoportale della Regione Emilia-Romagna, and the percentages of the major land use classes, according to the CORINE Land Cover, were calculated (Table 3). Arable lands and shrub and/or herbaceous vegetation associations were significantly reduced in the study area, by >30%, compared with the situation elucidated by the 1994 land use map, favouring a revegetation process. In fact, forest areas increased by 34% (Table 3), while areas with little or no vegetation slightly increased, by 5% (Figure 13).

Table 3. Land use change in the period 1994–2014 in the study area.

CODE	Land Use	1994 (m^2)	2014 (m^2)	Δ (%)
11	Urban fabric	118,361	301,359	155%
12	Industrial, commercial, and transport units	99,964	100,606	1%
13	Mine, dump, and construction sites	225,865	224,716	−1%
14	Artificial, non-agricultural vegetated areas	66,297	74,995	13%
21	Arable land	3,182,080	2,098,310	−34%
22	Permanent crops	159,478	365,372	129%
23	Pastures	14,033	517,243	3586%
31	Forest	1,861,610	2,497,780	34%
32	Shrubs and/or herbaceous vegetation associations	1,862,140	1,195,760	−36%
33	Open spaces with little or no vegetation	1,254,490	1,320,920	5%
51	Inland waters	71,782	119,043	66%

The increase in the vegetation cover confirms a gradual stabilisation of the badland slopes over time and is in line with results of the analysis of the detailed maps of the vegetation cover and land use of the territory of the Salse di Nirano Natural Reserve—with a surface area of approximately 75,000 m^2—of 1973 and 2006. In 1973, areas with no vegetation and sparse and discontinuous herbaceous cover accounted for approximately 15% of the total area, while in 2006 they were reduced to less than 5%. Arable lands were significantly reduced, by 62%, while the forest cover increased significantly from 7.5% in 1973 to 37% in 2006, tending to encompass the surrounding areas adjacent to badlands. In general, the forests are in a juvenile state, mostly originated by the abonnement of agriculture or iterated cutting in the past century. In 2006, different typologies of grassy plants covered the ground, such as *Agropyron pungens*, *Podospermum canum* and *Aster linosyris*, stabilising the slope and limiting the water erosion. Subsequently, bushes and shrubs have taken root (e.g., *Spartium junceum*, *Rosa canina*, *Prunus spinosa*, and *Cornus sanguinea*), which present strong root systems and make the slope more stable. Arboreal plants have also colonised the slopes, such as *Quercus pubescens*, *Ulmus minor* and *Fraxinus ornus* [119]. This trend is coincident with the institution in 1982 of the Regional Natural Reserve of Salse di Nirano, and with the resulting land use restrictions and regulations implemented for the area (Figure 14).

Figure 13. Land use change occurred in a sector of the study area over a 50-year period. (**a**) IGMI, GAI 1954 aerial photo and (**b**) orthophotos TeA of 2014. Despite the images' different resolution, the increase in forest cover and the reduction of agricultural lands are evident.

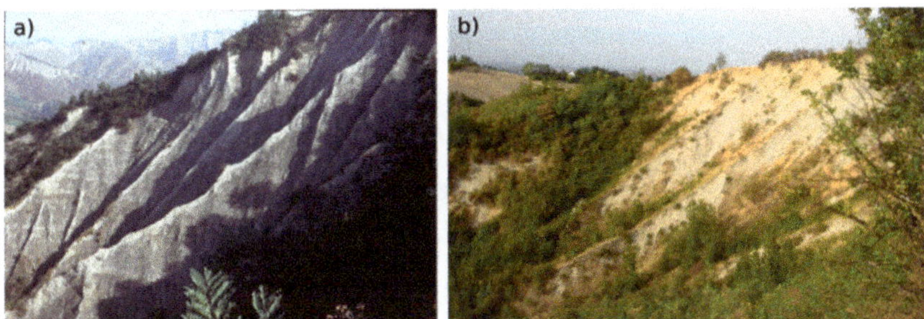

Figure 14. Vegetation change occurred in badland n. 26 between the late 1980s (**a**) and 2006 (**b**) (photos: (**a**) M. Panizza; (**b**) M. Bedetti).

5. Discussion and Conclusions

The investigation conducted in the study area allowed the identification and assessment of the areal and temporal changes of badlands over the last 40 years. In total, 55 Type A badlands were identified and digitalised, together with their stream network. Badlands are more extensively developed on SW and S-facing slopes, representing 69% of the total badlands. The presence of two main families of nearly perpendicular tectonic discontinuities, roughly parallel and orthogonal to the anticline axis, is likely to have influenced the distribution of the badland slopes. In fact, the 60% of badlands is aligned along an ENE-WSW direction, while the remaining 40% is aligned along a NW-SE direction. The drainage density and the direct bifurcation ratio for the years 1973 and 2014 were calculated; both decreased over time, showing a decreasing erosion rate for the ground. This trend was confirmed by the results of the multitemporal analysis of the orthophotos, which reveal a reduction of the bare surface area from 6187.1 m^2 in 1973 to 4214.1 m^2 in 2014 due to an intensified revegetation process. The area reduction of the badlands observed for the study area is related to the land-cover changes, mainly due to agricultural activities, that occurred from the 1970s onwards. These consist essentially of a significant increase in more-protective land use types such as forests, permanent crops and pastures. This is particularly significant in the badlands located within the Regional Natural Reserve of Salse di Nirano, where the land use restrictions and regulations implemented along with its institution in 1982 has played a predominant role in badland evolution. In fact, the lack of typical management practices, such as regular mowing activities, and the reduction of arable lands due to their progressive abandonment have contributed to favouring revegetation on the badland slopes. The pluviometric variation during the period 1954–2018 likely favoured increase in revegetation processes, even if further analysis of pluviometric data (e.g., annual number of consecutive dry and wet days and annual spell length of dry and wet days spells) as well as temperature variation are needed.

The stabilisation trend of the badland area in the study area is in line with the results of research carried out in the Northern Apennines by Bosino et al. [101], who observed that 73% of badlands show a decrease in average area of ca. 34% over a 40-year period, caused by a combination of natural and anthropic processes. The authors state that the decrease of average annual precipitation and of agricultural activity (especially vineyards) and the increase of the forest cover caused a decrease in surface runoff, favouring revegetation process and a reduction of badland areas. Piccarreta et al. [66] show, as well, a decrease in degraded areas between 1955 to 2002 of about 28% for the areas around Pisticci, Aliano, and Craco, in southern Italy. This reduction is mainly due to the widespread badland re-modelling for the durum wheat cultivation. The levelling of gully heads for the production of cereals and orchards is reported to be the contributing factor for a land degradation decrease from 1949 to 1986 in the Basilicata region (southern Italy) [66] as well. However, several studies carried out in areas of central and southern Apennines, characterised by comparable climatic condition, reported an intensification of erosional processes and an increase of degraded areas, both on short and long term. Within the Radicofani badlands of southern Tuscany (central Italy), Ciccacci et al. [120] describe an increase of the total area occupied by badlands between 1955 and 2006 accompanied by an intensification of erosion rate. These changes appear to be due to the markedly increase of intense agricultural practices by mechanical device through time. The anthropic pressure related to the mismanagement of agricultural practices for durum wheat cultivation, which has led to the reclamation of scrub lands and badlands, appears to have played a significant role in the increasing of soil erosion in some areas in Basilicata (Southern Italy) [67] as well. In fact, abandoned areas, previously devoted to sown ground cultivation, are affected by intense erosional processes especially when extreme rainy event occurs, and deep gullies and new badland channels can evolve rapidly [121]. Human activities, particularly pastures or arable lands, played an important role in land degradation in Sicily (Southern Italy) as well [14,122]. The human impact on land use changes has been credited as main triggering drivers causing in increased soil erosion in several areas of the central Mediterranean region [123–126].

Results from this study provide a first insight into badlands morphological changes through time in a sector of Apennines that has been poorly investigated, highlighting the evolutionary trend. Detailed field surveys to supplement and verify results obtained are ongoing. Moreover, future research for the quantification of the morpho-evolution rate through UAV-derived high-resolution digital elevation model comparison are planned.

Author Contributions: Conceptualization, P.C.; methodology, P.C.; software, C.P.; validation, P.C. and C.P.; formal analysis, P.C. and C.P.; investigation, P.C.; writing—original draft preparation, P.C. and C.P.; writing—review and editing, P.C.; supervision, P.C. All authors have read and agreed to the published version of the manuscript.

Funding: This research was conducted with financial support of FAR 2020 program of the Chemical and Geological Science Department of University of Modena and Reggio Emilia.

Institutional Review Board Statement: Not applicable.

Informed Consent Statement: Not applicable.

Data Availability Statement: Not applicable.

Acknowledgments: The authors are grateful to anonymous referees for their useful comments and suggestions to improve the original draft.

Conflicts of Interest: The authors declare no conflict of interest.

References

1. Harvey, A. Badlands. In *Encyclopedia of Geomorphology*; Goudie, A., Ed.; Routledge: London, UK, 2004; pp. 45–47.
2. Cappadonia, C.; Coco, L.; Buccolini, M.; Rotigliano, E. From slope morphometry to morphogenetic processes: An integrated approach of field survey, geographic information system morphometric analysis and statistics in Italian badlands. *Land Degrad. Dev.* **2016**, *27*, 851–862. [CrossRef]
3. Bouma, N.A.; Imeson, A. Investigation of relationships between measured field indicators and erosion processes on badland surfaces at Petrer, Spain. *Catena* **2000**, *40*, 147–171. [CrossRef]
4. Kasanin-Grubin, M.; Bryan, R. Lithological properties and weathering response on badland hillslopes. *Catena* **2007**, *70*, 68–78. [CrossRef]
5. Battaglia, S.; Leoni, L.; Sartori, F. Mineralogical and grain size composition of clays developing calanchi and biancane erosional landforms. *Geomorphology* **2002**, *49*, 153–170. [CrossRef]
6. Bucciante, M. Sulla Distribuzione Geografica dei Calanchi in Italia. *L'Universo* **1922**, *3*, 585–605.
7. Rapetti, F.; Vittorini, S. La temperatura del suolo in due versanti contrapposti del preappennino argilloso toscano. *Boll. Soc. Ital. Sci. Suolo* **1975**, *9*, 9–25.
8. Alexander, D.E. I calanchi, accelerated erosion in Italy. *Geography* **1980**, *65*, 95–100.
9. Torri, D.; Bryan, R.B. Micropiping processes and biancana evolution in southest Tuscany, Italy. *Geomorphology* **1997**, *20*, 219–235. [CrossRef]
10. Buccolini, M.; Coco, L.; Cappadonia, C.; Rotigliano, E. Relationships between a new slope morphometric index and calanchi erosion in northern Sicily, Italy. *Geomorphology* **2012**, *149–150*, 41–48. [CrossRef]
11. Bollati, I.; Masseroli, A.; Mortara, G.; Pelfini, M.; Trombino, L. Alpine gullies system evolution: Erosion drivers and control factors. Two examples from the western Italian Alps. *Geomorphology* **2019**, *327*, 248–263. [CrossRef]
12. Buccolini, M.; Coco, L. The role of the hillside in determining the morphometric characteristics of "calanchi": The example of Adriatic central Italy. *Geomorphology* **2010**, *191*, 142–149. [CrossRef]
13. Caraballo-Arias, N.A.; Conoscenti, C.; Di Stefano, C.; Ferro, V. Testing GIS-morphometric analysis of some Sicilian badlands. *Catena* **2014**, *113*, 370–376. [CrossRef]
14. Brandolini, P.; Pepe, G.; Capolongo, D.; Cappadonia, C.; Cevasco, A.; Conoscenti, C.; Marsico, A.; Vergari, F.; Del Monte, M. Hillslope degradation in representative Italian areas: Just soil erosion risk or opportunity for development? *Land Degrad. Dev.* **2018**, *29*, 3050–3068. [CrossRef]
15. Vittorini, S. Osservazioni sull'origine e sul ruolo di due forme di erosione nelle argille: Calanchi e biancane. *Boll. Soc. Geogr. Ital.* **1977**, *10*, 25–54.
16. Faulkner, H. Vegetation cover density variations and infiltration pattern on piped alkali sodic soils; implications for the modelling of overland flow in semi-arid areas. In *Vegetation and Erosion*; Thornes, J.B., Ed.; John Wiley and Sons Ltd.: Chichester, UK, 1990; pp. 317–346.
17. Moretti, S.; Rodolfi, G. A typical "calanchi" landscape on the Eastern Apennine margin (Atri, Central Italy): Geomorphological features and evolution. *Catena* **2000**, *40*, 217–228. [CrossRef]
18. Biondi, E.; Pesaresi, S. The badland vegetation of the northern-central Apennines (Italy). *Fitosociologia* **2004**, *41*, 155–170.
19. Regüés, D.; Guàrdia, R.; Gallart, F. Geomorphic agents versus vegetation spreading as causes of badland occurrence in a Mediterranean subhumid mountainous area. *Catena* **2000**, *40*, 173–187. [CrossRef]
20. Howard, A.D. Badlands. In *Geomorphology of Desert Environments*; Abrahams, A.D., Parsons, A.J., Eds.; Chapman and Hall: London, UK, 1994; pp. 213–242.
21. Gallart, F.; Pérez-Gallego, N.; Latron, J.; Catari, G.; Martínez-Carreras, N.; Nord, G. Short- and long-term studies of sediment dynamics in a small humid mountain Mediterranean basin with Badlands. *Geomorphology* **2013**, *196*, 242–251. [CrossRef]
22. Yair, A.; Lavee, H.; Bryan, R.B.; Adar, E. Runoff and erosion processes and rates in the Zion Valley Badlands, northern Negev, Israel. *Earth Surf. Process.* **1980**, *5*, 205–225. [CrossRef]
23. Alexander, D.E. Difference between «calanchi» and «biancane» badlands in Italy. In *Badland Geomorphology and Piping*; Bryan, R., Yair, A., Eds.; Geo Books: Norwich, UK, 1982; pp. 71–87.

24. Harvey, A. The role of piping in the development of badlands and gully systems in south-east Spain. In *Badland Geomorphology and Piping*; Bryan, R., Yair, A., Eds.; Geo Books: Norwich, UK, 1982; pp. 317–335.
25. Imeson, A.C. Studies of erosion thresholds in semiarid areas: Field measurement of soil loss and infiltration in northern Morocco. *Catena Suppl.* **1983**, *4*, 79–89.
26. Sdao, G.; Simone, A.; Vittorini, S. Osservazioni geomorfologiche su calanchi e biancane in Calabria (Geomorphology observation of calanchi and biancane in Calabria). *Geografia Fisica e Dinamica Quaternaria* **1984**, *7*, 10–16.
27. Berndtsson, R. *Spatial Hydrological Processes in a Water Resources Planning Perspective. An Investigation of Rainfall and Infiltration in Tunisia*; Report No.1009; Department of Water Resources Engineering, Institute of Technology, University of Lund: Lund, Sweden, 1988; p. 315.
28. Alexander, D.E.; Calvo, A. The influence of lichens on slope processes in some Spanish Badlands. In *Vegetation and Erosion: Process and Environments*; Thornes, J.B., Ed.; John Wiley and Sons Ltd.: Chichester, UK, 1990; pp. 385–398.
29. McCloskey, G.L.; Wasson, R.J.; Boggs, G.; Douglas, M.M. Timing and causes of gully erosion in the riparian zone of the semi-arid tropical Victoria River, Australia: Management implications. *Geomorphology* **2016**, *266*, 96–104. [CrossRef]
30. Bryan, R.B.; Yair, A. *Badland Geomorphology and Piping*; Geo Books: Norwich, UK, 1982.
31. Campbell, I.A. Badlands and badlands gullies. In *Arid Zone Geomorphology*; Thomas, D.S.G., Ed.; Belhaven: London, UK, 1989; pp. 159–183.
32. Regüés, D.; Pardini, G.; Gallart, F. Regolith behaviour and physical weathering of clayey mudrock as dependent on seasonal weather conditions in a badland area at Vallcebre, Eastern Pyrenees. *Catena* **1995**, *25*, 199–212. [CrossRef]
33. Pardini, G.; Vigna Guidi, G.; Pini, R.; Regues, D.; Gallart, F. Structure and porosity of smectitic mudrocks as affected by experimental wetting-drying cycles and freezing-thawing cycles. *Catena* **1996**, *27*, 149–165. [CrossRef]
34. Torri, D.; Calzolari, C.; Rodolfi, G. Badlands in changing environments: An introduction. *Catena* **2000**, *40*, 119–125. [CrossRef]
35. Nadal-Romero, E.; Regüés, D. Geomorphological dynamics of subhumid mountain badland areas—Weathering, hydrological and suspended sediment transport processes: A case study in the Araguás catchment (Central Pyrenees) and implications for altered hydroclimatic regimes. *Prog. Phys. Geogr.* **2010**, *34*, 123–150. [CrossRef]
36. Dube, H.B.; Mutema, M.; Muchaonyerwa, P.; Poesen, J.; Chaplot, V. A Global Analysis of the Morphology of Linear Erosion Features. *Catena* **2020**, *190*, 104542. [CrossRef]
37. Vandekerckhove, L.; Poesen, J.; Wijdenes, D.O.; Nachtergaele, J.; Kosmas, C.; Roxo, M.; De Figueiredo, T. Thresholds for gully initiation and sedimentation in Mediterranean Europe. *Earth Surf. Process.* **2000**, *25*, 1201–1220. [CrossRef]
38. Martínez-Casasnovas, J.A.; Antón-Fernández, C.; Ramos, M.C. Sediment production in large gullies of the Mediterranean area (NE Spain) from high-resolution digital elevation models and geographical information systems analysis. *Earth Surf. Process.* **2003**, *28*, 443–456. [CrossRef]
39. Poesen, J.; Nachtergaele, J.; Verstraeten, G.; Valentin, C. Gully Erosion and Environmental Change: Importance and Research Needs. *Catena* **2003**, *50*, 91–133. [CrossRef]
40. Müller-Nedebock, D.; Chaplot, V. Soil carbon losses by sheet erosion: A potentially critical contribution to the global carbon cycle: Soil Carbon Erosion by Sheet Erosion. *Earth Surf. Process. Landf.* **2015**, *40*. [CrossRef]
41. Vanwalleghem, T.; Poesen, J.; Nachtergaele, J.; Verstraeten, G. Characteristics, controlling factors and importance of deep gullies under cropland on loess-derived soils. *Geomorphology* **2005**, *69*, 76–91. [CrossRef]
42. De Santisteban, L.M.; Casalí, J.; Lopez, J.J. Assessing soil erosion rates in cultivated areas of Navarre (Spain). *Earth Surf. Process. Landf.* **2006**, *31*, 487–506. [CrossRef]
43. Nadal-Romero, E.; Martínez-Murillo, J.F.; Vanmaercke, M.; Poesen, J. Scale-dependency of sediment yield from badlands areas in Mediterranean environments. *Prog. Phys. Geogr. Earth Environ.* **2011**, *35*, 297–332. [CrossRef]
44. Gallart, F.; Marignani, M.; Pérez-Gallego, N.; Santi, E.; Maccherini, S. Thirty years of studies on badlands, from physical to vegetational approaches. A succinct review. *Catena* **2013**, *106*, 4–11. [CrossRef]
45. Boardman, J.; Favis-Mortlock, D.; Foster, I. A 13-year record of erosion on badland sites in the Karoo, South Africa. *Earth Surf. Process. Landf.* **2015**, *40*, 1964–1981. [CrossRef]
46. Migoń, P. *Geoturystyka*; Wydawnictwo Naukowe PWN: Warsaw, Poland, 2012.
47. Palacio-Prieto, J.L.; Rosado-González, E.; Ramírez-Miguel, X.; Oropeza-Orozco, O.; CramHeydrich, S.; Ortiz-Pérez, M.A.; Figueroa-Mah-Eng, J.M.; Fernández de Castro-Martínez, G. Erosion, culture and geoheritage; the case of Santo Domingo Yanhuitlán, Oaxaca, México. *Geoheritage* **2016**, *8*, 359–369. [CrossRef]
48. Zgłobicki, W.; Poesen, J.; Cohen, M.; Del Monte, M.; García-Ruiz, J.M.; Ionita, I.; Niacsu, L.; Machová, Z.; Martín-Duque, J.F.; Nadal-Romero, E.; et al. The potential of permanent gullies in Europe as geomorphosites. *Geoheritage* **2017**, *11*, 217–239. [CrossRef]
49. Bollati, I.; Vergari, F.; Del Monte, M.; Pelfini, M. Multitemporal dendrogeomorphological analysis of slope instability in upper Orcia valley (Southern Tuscany, Italy). *Geogr. Fis. Dinam. Quat.* **2016**, *39*, 105–120.
50. Poesen, J.; Valentin, C. Gully erosion and global change—Preface. *Catena* **2003**, *50*, 87–89. [CrossRef]
51. Li, Y.; Poesen, J.; Valentin, C. *Gully Erosion under Global Change*; Sichuan Science Technology Press: Chengu, China, 2004.
52. Valentin, C.; Poesen, J.; Li, Y. Gully erosion: Impacts, factors and control. *Catena* **2005**, *63*, 132–153. [CrossRef]
53. Martínez-Murillo, J.F.; Nadal-Romero, E. Perspectives on Badland Studies in the Context of Global Change. In *Badlands Dynamics in a Context of Global Change*; Nadal-Romero, E., Martínez-Murillo, J.F., Kuhn, N.J., Eds.; Elsevier: Amsterdam, The Netherlands, 2018; pp. 1–25.

54. Dramis, F.; Gentili, B.; Coltorti, M.; Cherubini, C. Geological observations on badlands in the Marche region. *Geogr. Fis. Dinam. Quat.* **1982**, *5*, 38–45.
55. Phillips, C.P. The Crete Senesi, Tuscany. A vanishing landscape? *Landscape Urban Plann.* **1998**, *41*, 19–26. [CrossRef]
56. Clarke, M.L.; Rendell, H.M. The impact of the farming practice of remodelling hillslope topography on badlands morphology and soil erosion processes. *Catena* **2000**, *40*, 229–250. [CrossRef]
57. Buccolini, M.; Gentili, B.; Materazzi, M.; Aringoli, D.; Pambianchi, G.; Piacentini, T. Human impact and slope dynamics evolutionary trends in the monoclinal relief of Adriatic area of central Italy. *Catena* **2007**, *71*, 96–109. [CrossRef]
58. Gallart, F. Algunos criterios topográficos para identificar el origen antrópico decárcavas. *Cuadernos Investig. Geogr.* **2009**, *35*, 215–221. [CrossRef]
59. Keay-Bright, J.; Boardman, J. The influence of land management on soil erosion in the Sneeuberg Mountains, Central Karoo, South Africa. *Land Degrad. Dev.* **2007**, *18*, 423–439. [CrossRef]
60. Torri, D.; Santi, E.; Marignani, M.; Rossi, M.; Borselli, L.; Maccherini, S. The recurring cycles of biancana badlands: Erosion, vegetation and human impact. *Catena* **2013**, *106*, 22–30. [CrossRef]
61. Vergari, F.; Della Seta, M.; Del Monte, M.; Barbieri, M. Badlands denudation "hot spots": The role of parental material properties on geomorphic processes in 20 years monitored sites of Southern Tuscany (Italy). *Catena* **2013**, *106*, 31–41. [CrossRef]
62. Poesen, J. Soil erosion in the Anthropocene: Research needs. *Earth Surf. Process.* **2017**, *43*, 64–84. [CrossRef]
63. Vandaele, K.; Poesen, J.; Govers, G.; Van Wesemael, B. Geomorphic threshold conditions for ephemeral gully incision. *Geomorphology* **1996**, *16*, 161–173. [CrossRef]
64. Garcia-Ruiz, J.M.; White, S.M.; Làsanta, T.; Marti, C.; Gonzalez, C.; Paz Errea, M.; Valero, B. Assessing the effects of land-use changes on sediment yield and channel dynamics in the central Spanish Pyrenees. *IAHS Publ.* **1997**, *245*, 151–158.
65. Vandekerckhove, L.; Poesen, J.; Oostwoud Wijidenes, D.; Figueiredo, T. Topographical threshold for ephemeral gully initiation in intensively cultivated areas of the Mediterranean. *Catena* **1998**, *33*, 271–292. [CrossRef]
66. Piccarreta, M.; Capolongo, D.; Boenzi, F.; Bentivenga, M. Implications of decadal changes in precipitation and land use policy to soil erosion in Basilicata, Italy. *Catena* **2006**, *65*, 138–151. [CrossRef]
67. Capolongo, D.; Pennetta, L.; Piccarreta, M.; Fallacara, G.; Boenzi, F. Spatial and temporal variations in soil erosion and deposition due to land-levelling in a semi-arid area of Basilicata (southern Italy). *Earth Surf. Process. Landf.* **2008**, *33*, 364–379. [CrossRef]
68. García-Ruiz, J.M.; Lana-Renault, N. Hydrological and Erosive Consequences of Farmland Abandonment in Europe, with Special Reference to the Mediterranean Region—A Review. *Agric. Ecosyst. Environ.* **2011**, *140*, 317–338. [CrossRef]
69. Arnáez, J.; Lasanta, T.; Errea, M.P.; Ortigosa, L. Land abandonment, landscape evolution, and soil erosion in a Spanish Mediterranean mountain region: The case of Camero Viejo. *Land. Degrad. Dev.* **2011**, *22*, 537–550. [CrossRef]
70. Lukey, B.; Sheffield, J.; Bathurst, J.; Hiley, R.A.; Mathys, N. Test of the SHETRAN technology for modelling the impact of reforestation on badlands runoff and sediment yield at Draix, France. *J. Hydrol.* **2000**, *235*, 44–62. [CrossRef]
71. Vallauri, D.; Aronson, J.; Barbero, M. An analysis of forest restoration 120 years after reforestation on badlands in the Southwestern Alps. *Restor. Ecol.* **2002**, *10*, 16–26. [CrossRef]
72. Castaldi, F.; Chiocchini, U. Effects of land use changes on badland erosion in clayey drainage basins, Radicofani, Central Italy. *Geomorphology* **2012**, *169–170*, 98–108. [CrossRef]
73. Ballesteros-Canovas, J.A.; Stoffel, M.; Martín-Duque, J.F.; Corona, C.; Lucía, A.; Bodoque, J.M.; Montgomery, D.R. Gully evolution and geomorphic adjustments of badlands to reforestation. *Sci. Rep.* **2017**, *7*, 45027. [CrossRef] [PubMed]
74. Renschler, C.S.; Mannaerts, C.; Diekkrüger, B. Evaluating spatial and temporal variability in soil erosion risk—Rainfall erosivity and soil loss ratios in Andalusia, Spain. *Catena* **1999**, *34*, 209–225. [CrossRef]
75. Torri, D.; Regües-Muñoz, D.; Pellegrini, S.; Bazzoffi, P. Within-storm soil surface dynamics and erosive effects of rainstorms. *Catena* **1999**, *38*, 131–150. [CrossRef]
76. Mannaerts, C.M.; Gabriels, D. A probabilistic approach for predicting rainfall soil erosion losses in semiarid areas. *Catena* **2000**, *40*, 403–420. [CrossRef]
77. Archibold, O.W.; Lévesque, L.M.J.; Boer, D.H.; Aitken, A.E.; Delanoy, L. Gully retreat in a semi-urban catchment in Saskatoon, Saskatchewan. *Appl. Geogr.* **2003**, *23*, 261–279. [CrossRef]
78. Boardman, J.; Parsons, A.J.; Holland, R.; Holmes, P.J. Development of badlands and gullies in the Sneeuberg, Great Karoo, South Africa. *Catena* **2003**, *50*, 165–184. [CrossRef]
79. Øygarden, L. Rill and gully development during extreme winter runoff event in Norway. *Catena* **2003**, *50*, 217–242. [CrossRef]
80. Vergari, F.; Della Seta, M.; Del Monte, M.; Fredi, P.; Palmieri, E.L. Long-and short-term evolution of several Mediterranean denudation hot spots: The role of rainfall variations and human impact. *Geomorphology* **2013**, *183*, 14–27. [CrossRef]
81. Del Monte, M. The typical Badlands Landscape between the Tyrrhenian Sea and Tiber River. In *Landscape and Landforms of Italy*; World Geomorphological Landscapes; Soldati, M., Marchetti, M., Eds.; Springer: Cham, Switzerland, 2017; pp. 281–291.
82. Castiglioni, B. Osservazioni sui calanchi appenninici. *Boll. Soc. Geol. Ital.* **1933**, *52*, 357–360.
83. Rodolfi, G.; Frascati, F. Cartografia di base per la programmazione in aree marginali (area rappresentativa dell'alta Vadera): Memorie illustrative della carta geomorfologica. *Annali dell'Istituto Sperimentale per lo Studi e la Difesa del Suolo* **1979**, *10*, 37–80.
84. Biancotti, A.; Cortemiglia, G.C. Morphogenetic evolution of the river system of southern Piedmont (Italy). *Geogr. Fis. Dinam. Quat.* **1982**, *5*, 10–13.

85. Del Prete, M.; Bentivenga, M.; Coppola, L.; Rendell, H. Aspetti evolutivi dei reticoli calanchivi a sud di Pisticci. *Geol. Romana* **1994**, *30*, 295–306.

86. Calzolari, C.; Ungaro, F. Geomorphic features of a badland (biancane) area (Central Italy): Characterisation, distribution and quantitative spatial analysis. *Catena* **1998**, *31*, 237–256. [CrossRef]

87. Farabegoli, E.; Agostini, C. Identification of Calanco, a badland landform in the northern Apennines, Italy. *Earth Surf. Process. Landf.* **2000**, *25*, 307–318. [CrossRef]

88. Battaglia, S.; Leoni, L.; Rapetti, F.; Spagnolo, M. Dynamic evolution of badlands in the Roglio basin (Tuscany, Italy). *Catena* **2011**, *86*, 14–23. [CrossRef]

89. Ciccacci, S.; Galiano, M.; Roma, M.A.; Salvatore, M.C. Morphological analysis and erosion rate evaluation in badlands of Radicofani area (Southern Tuscany-Italy). *Catena* **2008**, *74*, 87–97. [CrossRef]

90. Della Seta, M.; Del Monte, M.; Fredi, P.; Palmieri, E.L. Space–time variability of denudation rates at the catchment and hillslope scales on the Tyrrhenian side of Central Italy. *Geomorphology* **2009**, *107*, 161–177. [CrossRef]

91. Clarke, M.L.; Rendell, H.M. Climate-driven decrease in erosion in extant Mediterranean badlands. *Earth Surf. Process. Landf.* **2010**, *35*, 1281–1288. [CrossRef]

92. Piccarreta, M.; Caldara, M.; Capolongo, D.; Boenzi, F. Holocene geomorphic activity related to climatic change and human impact in Basilicata, Southern Italy. *Geomorphology* **2011**, *128*, 137–147. [CrossRef]

93. Buccolini, M.; Coco, L. MSI (morphometric slope index) for analyzing activation and evolution of calanchi in Italy. *Geomorphology* **2013**, *191*, 142–149. [CrossRef]

94. Bollati, I.; Della Seta, M.; Pelfini, M.; Del Monte, M.; Fredi, P.; Lupia Palmieri, E. Dendrochronological and geomorphological investigations to assess water erosion and mass wasting processes in the Apennines of Southern Tuscany (Italy). *Catena* **2012**, *90*, 1–17. [CrossRef]

95. Caraballo-Arias, N.A.; Conoscenti, C.; Di Stefano, C.; Ferro, V. A new empirical model for estimating calanchi Erosion in Sicily, Italy. *Geomorphology* **2015**, *231*, 292–300. [CrossRef]

96. Coco, L.; Cestrone, V.; Buccolini, M. Geomorphometry for studying the evolution of small basins: An example in the Italian Adriatic foredeep. In *Geomorphometry for Geosciences*; Jasiewicz, J., Zwoliński, Z., Mitasova, H., Hengl, T., Eds.; Bogucki Wydawnictwo Naukowe, Adam Mickiewicz University in Poznań—Institute of Geoecology and Geoinformation, International Society for Geomorphometry: Poznań, Poland, 2015; pp. 107–110.

97. Aucelli, P.P.C.; Conforti, M.; Della Seta, M.; Del Monte, M.; D'uva, L.; Rosskopf, C.M.; Vergari, F. Multi-temporal digital photogrammetric analysis for quantitative assessment of soil erosion rates in the Landola catchment of the Upper Orcia Valley (Tuscany, Italy). *Land Degrad. Dev.* **2016**, *27*, 1075–1092. [CrossRef]

98. Bianchini, S.; Del Soldato, M.; Solari, L.; Nolesini, T.; Pratesi, F.; Moretti, S. Badland susceptibility assessment in Volterra municipality (Tuscany, Italy) by means of GIS and statistical analysis. *Environ. Earth Sci.* **2016**, *75*, 889. [CrossRef]

99. Caraballo-Arias, N.A.; Ferro, V. Are calanco landforms similar to river basins? *Sci. Total Environ.* **2017**, *603–604*, 244–255. [CrossRef] [PubMed]

100. Caraballo-Arias, N.A.; Ferro, V. Assessing, measuring and modelling erosion in calanchi areas: A review. *J. Agric. Eng.* **2016**, *47*, 181–190. [CrossRef]

101. Bosino, A.; Omran, A.; Maerker, M. Identification, characterization and analysis of the Oltrepo Pavese Calanchi in the northern Apennines, Italy. *Geomorphology* **2019**, *340*, 53–66. [CrossRef]

102. Provincia di Modena. *PTCP. Piano Territoriale di Coordinamento Provinciale. Variante di Adeguamento in Materia di Dissesto Idrogeologico ai Piani di Bacino dei Fiumi Po e Reno*; Area Programmazione e Pianificazione Territoriale: Modena, Italy, 2009.

103. Smiraglia, C.; Azzoni, R.S.; D'Agata, C.; Maragno, D.; Fugazza, D.; Diolaiuti, G.A. The evolution of the Italian glaciers from the previous data base to the New Italian Inventory. Preliminary considerations and results. *Geogr. Fis. Dinam. Quat.* **2015**, *38*, 79–87.

104. Köppen, W. *Grundriß der Klimakunde*, 2nd ed.; Walter de Gruyter and Co.: Berlin, Germany; Leipzig, Germany, 1931.

105. Gasperi, G.; Bettelli, G.; Panini, F.; Pizziolo, M. *Note Illustrative della Carta Geologica d'Italia alla Scala 1: 50.000, Foglio 219 Sassuolo*; S.EL.CA (Società Elaborazioni Cartografiche): Florence, Italy, 2005.

106. Regione Emilia-Romagna. Cartografia Geologica Online in Scala 1:10.000 della Regione Emilia-Romagna. 2017. Available online: https://geo.regione.emilia-romagna.it/cartografia_sgss/user/viewer.jsp?service=geologia (accessed on 28 January 2021).

107. Boni, A.; Casnedi, R. *Note Illustrative della Carta Geologica d'Italia alla Scala 1:100.000. Fogli 69 e 70 "Asti" e "Alessandria"*; Servizio Geologico d'Italia: Rome, Italy, 1970; p. 83.

108. Tomadin, L. Ricerche sui Sedimenti Argillosi Fluviali dal Brenta al Reno. *Giorn. Geol.* **1969**, *36*, 1.

109. Horton, R.E. Erosional development of streams and their drainage basins: Hydrophysical approach to quantitative morphology. *Geol. Soc. Am. Bull.* **1945**, *56*, 275–370. [CrossRef]

110. Strahler, A. Dynamic Basis of Geomorphology. *Geol. Soc. Am. Bull* **1952**, *63*, 923–938. [CrossRef]

111. Avena, G.C.; Giuliano, G.; Palmieri, E.L. Sulla valutazione quantitativa della gerarchizzazione ed evoluzione dei reticoli fluviali. *Boll. Soc. Geol. Ital.* **1967**, *86*, 781–796.

112. Caloiero, D.; Mercuri, T. *Le alluvioni in Basilicata dal 1921 al 1980*; CNR-IRPI, Geodata 16: Cosenza, Italy, 1982.

113. Buffoni, L.; Maugeri, M.; Nanni, T. Precipitation in Italy from 1833–1996. *Theor. Appl. Climatol.* **1999**, *63*, 33–40. [CrossRef]

114. Brunetti, M.; Maugeri, M. Variations of Temperature and Precipitation in Italy from 1866 to 1995. *Theor. Appl. Clim.* **2000**, *65*, 165–174. [CrossRef]

115. Brunetti, M.; Maugeri, M.; Monti, F.; Nanni, T. Temperature and precipitation variability in Italy in the last two centuries from homogenised instrumental time series. *Int. J. Clim.* **2006**, *26*, 345–381. [CrossRef]
116. Giorgi, F.; Lionello, P. Climate change projections for the Mediterranean region. *Glob. Planet Chang.* **2008**, *63*, 90–104. [CrossRef]
117. Pavan, S.; Zheng, Z.; Berg, P.; Lotti, C.; Giovanni, C.; Borisova, M.; Lindhout, P.; Jong, H.; Ricciardi, L.; Visser, R.; et al. Map- vs homology-based cloning for the recessive gene ol-2 conferring resistance to tomato powdery mildew. *Euphytica* **2008**, *162*, 91–98. [CrossRef]
118. Lionello, P.; Abrantes, F.F.; Gacic, M. The climate of the Mediterranean region: Research progress and climate change impacts. *Reg. Environ. Chang.* **2014**, *14*, 1679–1684. [CrossRef]
119. Dallai, D.; Del Prete, C. Gli Ambienti delle Salse: Problemi di Tutela della Biodiversità Vegetale. In Proceedings of the Convegno Internazionale "I Vulcani di Fango", Modena, Italy, 9–10 June 2007.
120. Ciccacci, S.; Galiano, M.; Roma, M.A.; Salvatore, M.C. Morphodynamics and morphological changes of the last 50 years in a badland sample area of Southern Tuscany (Italy). *Z. Geomorphol.* **2009**, *53*, 273–297. [CrossRef]
121. Piccarreta, M.; Capolongo, D.; Miccoli, M.N.; Bentivenga, M. Global change and long-term gully sediment production dynamics in Basilicata, southern Italy. *Environ. Earth Sci.* **2012**, *67*, 1619–1630. [CrossRef]
122. Cappadonia, C.; Conoscenti, C.; Rotigliano, E. Monitoring of erosion on two calanchi fronts—Northern Sicily (Italy). *Landf. Anal.* **2011**, *17*, 21–25.
123. García-Ruiz, J.M.; Lasanta, T.; Ruiz-Flaño, P.; Ortigosa, L.; White, S.; González, C.; Martí, C. Land-use changes and sustainable development in mountain areas: A case study in the Spanish Pyrenees. *Landsc. Ecol.* **1996**, *11*, 267–277. [CrossRef]
124. Molinillo, M.; Lasanta, T.; García-Ruiz, J.M. Managing mountainous degraded landscapes after farmland abandonment in the Central Spanish Pyrenees. *Environ. Manag.* **1997**, *21*, 587–598. [CrossRef] [PubMed]
125. Gómez-Gutiérrez, A.; Schnabel, S.; Lavado-Contador, F. Gully erosion, land use and topographical thresholds during the last 60 years in a small rangeland catchment in SW Spain. *Land. Degrad. Dev.* **2009**, *20*, 535–550. [CrossRef]
126. Lesschen, J.P.; Cammeraat, E.L.H.; Nieman, T. Erosion and terrace failure due to agricultural land abandonment in semi-arid environment. *Earth Surf. Process. Landf.* **2008**, *33*, 1574–1584. [CrossRef]

Article

Structural Connectivity of Sediment Affected by Check Dams in Loess Hilly-Gully Region, China

Leichao Bai [1,2], Juying Jiao [1,2,3,*], Nan Wang [3] and Yulan Chen [3]

[1] State Key Laboratory of Soil Erosion and Dryland Farming on the Loess Plateau, Institute of Soil and Water Conservation, Northwest A&F University, Xianyang 712100, China; beleit@nwafu.edu.cn
[2] Key Laboratory of the Loess Plateau Soil Erosion and Water Loss Process and Control, Ministry of Water Resources, Zhengzhou 450003, China
[3] Institute of Soil and Water Conservation, CAS and Ministry of Water Resources, Xianyang 712100, China; wangnan163@mails.ucas.ac.cn (N.W.); chenyulan@nwafu.edu.cn (Y.C.)
* Correspondence: jyjiao@ms.iswc.ac.cn

Citation: Bai, L.; Jiao, J.; Wang, N.; Chen, Y. Structural Connectivity of Sediment Affected by Check Dams in Loess Hilly-Gully Region, China. *Water* **2021**, *13*, 2644. https://doi.org/10.3390/w13192644

Academic Editor: Csaba Centeri

Received: 6 September 2021
Accepted: 23 September 2021
Published: 25 September 2021

Publisher's Note: MDPI stays neutral with regard to jurisdictional claims in published maps and institutional affiliations.

Abstract: Check dams play an irreplaceable role in soil and water conservation in the Chinese Loess Plateau region. However, there are few analyses on the connection between check dams and the downstream channel and the impact on structural connectivity and sediment interception efficiency. Based on a field survey, this study classified the connection mode between check dams and the downstream channel, and the actual control area percentage by discharge canal in dam land was used to quantitatively evaluate the degree of the structural connectivity of sediment between the check dam and the downstream channel. The analysis results show that the connection mode can be divided into eleven categories with different structural connectivity. The different connection modes and its combination mode of check dams and downstream channels in dam systems have a large difference, and the structural connectivity of the dam system is less than or equal to that of the sum of single check dams in a watershed. The degree of structural connectivity of a dam system will be greatly reduced if there is a main control check dam with no discharge canal in the lower reaches of the watershed. Compared with a single check dam, the structural connectivity of a dam system is reduced by 0–42.38%, with an average of 11.18%. According to the difference in connection mode and structural connectivity of check dams and dam systems in the four typical small watersheds, the optimization methods for connection mode in series, parallel and hybrid dam systems were proposed. The research results can provide a reference for the impact of a check dam on the sediment connectivity and the sediment interception efficiency in a watershed and can also guide the layout of a dam system and the arrangement of drainage facilities.

Keywords: watershed; sediment connectivity; connection mode; connection degree; Loess Plateau

1. Introduction

The Loess Plateau is the area most seriously affected by soil and water loss in the world, and its soil and water loss control has been the focus of attention and research by scholars. Since the 1950s, with the gradual implementation of a series of restoration measures, such as returning farmland to forests (grass) and the construction of check dams, the amount of sediment discharged into the Yellow River has decreased significantly. The average annual sediment discharge at the Tongguan hydrological station has decreased from 1.6 billion tons before the 1970s to 179 million tons in 2010–2020 [1], which shows that the water and soil control and management of the Loess Plateau have achieved remarkable results. However, the reduction of the sediment put into the Yellow River does not fully explain that the soil erosion of the Loess Plateau has been effectively controlled because the whole watershed has been treated as a "black box" when the sediment discharge was monitored at the outlet, and it is difficult to explain the degree and process of soil erosion on multiple scales in the watershed [2,3]. The low sediment load does not only not indicate

Water **2021**, *13*, 2644. https://doi.org/10.3390/w13192644 https://www.mdpi.com/journal/water

that the watershed is in a healthy state [4], it may also ignore the hot spots of erosion in the watershed and the potential harm caused by sedimentation [5].

Among the many soil and water conservation measures in the Chinese Loess Plateau region, check dams are one of the most effective ways to intercept sediment and control erosion in both the short-term and the long-term [6]. There were 56,422 check dams in the area controlled by Tongguan hydrological station on the Loess Plateau, of which 41,008 check dams had been filled up [7,8], and there are still many uncounted small check dams [9]. As soon as a check dam is built, it begins to play a role in flood control and sediment retention in the dam-controlled watershed. Sediment deposition in check dams leads to a decrease in bed slope, which makes significant alterations in the cross-sectional geometry and characteristics of bed sediments [10–12]. However, the influence of check dams on longitudinal and transverse sediment connectivity has not been analyzed in depth. Some scholars considered that check dams disrupted the longitudinal sediment connectivity in a catchment [13–15], while some others say the opposite [12].

Usually, most of the previous studies have focused on the sediment interception efficiency of check dams under the ideal conditions that the check dam could completely intercept the upstream sediment [9], but less attention has been paid to the connection mode of check dams and its effect on the sediment connectivity [12,16], and the analysis about the connection mode and degree between check dam and downstream channel based on the different drainage buildings and spillways (DBS) [16]. In fact, as the barrier node of sediment transport in the channel, the connection modes and degrees of check dam to the downstream channel are different due to the different types of drainage measures, the sediment interception efficiency is also quite different. Due to the influence of extreme rainfall and human activities, many check dams have been damaged to varying degrees [17]; the gaps formed in damaged dam bodies act as spillways when check dams were damaged under extreme rainfall [18,19]. Whether it is the discharge channel, drainage building and spillway to meet the design standards, the discharge channel excavated in the course of agricultural production or the dam body damaged by natural or human activities, all of these will lead to the connection between dam land and the downstream channel. The discharge canal was built at the foot of the slope on one side of dam land and caused the originally disconnected check dam to be connected to the downstream channel. Therefore, many check dams have not completely intercepted the sediment from the dam land from the dam-controlled watershed; part of the sediment is transported to the downstream channel by the drainage building, discharge canal, spillway or dam body damaged gap. The connection mode between the check dam and the downstream channel is much more complicated, especially in a dam system; this has led to a huge difference in the degree of sediment connectivity. It is necessary to find new methods to evaluate the sediment interception efficiency of check dams based on the new understanding of sediment connectivity.

Therefore, based on the field investigation, the objectives of this study were to (1) analyze the connection mode between a single check dam and its downstream channel, (2) select typical small watersheds to discuss the connection mode of dam system composed of different connection modes of check dams, (3) evaluate the structural connectivity of single check dams and dam systems and (4) propose the optimization methods of the connection mode between check dams and downstream channels in the Chinese Loess Plateau region. It is expected to provide some references for evaluating the effect of check dams on the sediment connectivity and calculating the sediment interception efficiency of check dams in the regions with check dams for soil and water conservation.

2. Materials and Methods

2.1. Study Area

In this study, four small watersheds with different watershed scales, including Hejiapan small watershed (3.27 km^2) and Lujiabian small watershed (15.43 km^2) of Chabagou watershed, the Yangjuangou small watershed (54.00 km^2) and the Majiagou small watershed (74.70 km^2) of Yanhe River basin were selected as the study areas (Figure 1). The Chabagou watershed is the second tributary of the Wuding River basin, and the total area is 205 km^2 with an altitude of 870–1286 m. The total area of the Yanhe River basin is 7687 km^2, with an altitude of 454–1765 m. The study areas belong to the Loess hilly and gully region of the Loess Plateau, with a broken and complicated topography. The gully density in the Chabagou watershed and Yanhe River basin are 5–6 km/km^2 and 2.1–4.6 km/km^2, respectively. As a result of the dry continental climate, the average annual rainfall of the Chabagou watershed and Yanhe River basin are about 450 and 500 mm, respectively, with 70% concentrated from July to September, and most of them are short heavy rainfalls [20], the average annual temperature is about 9.2 °C and 8.8–10.2 °C, respectively [21,22]. Although returning farmland to a forest (grass) has been implemented, more or less sloping cropland still exists [23], especially in the Chabagou watershed. The soil types of Chabagou watershed are mainly loessial soil, the soil particles are mainly composed of silt particles, with soil particles larger than 0.05 mm accounting for 25.8%, soil particles between 0.01 and 0.05 mm accounting for 57.7%, and soil particles smaller than 0.01 mm accounting for 16.5% of the total [24]. The soil types of the Yanhe River basin are mainly loessial soil and heilusoil, of which loessial soil is widely distributed in the Loess-hilly region, hillside and gully, and heilusoil is only distributed on the top of Loess-hilly region, watershed and large gully terrace [25], which has poor soil erosion resistance, so it is easy to cause water and soil loss.

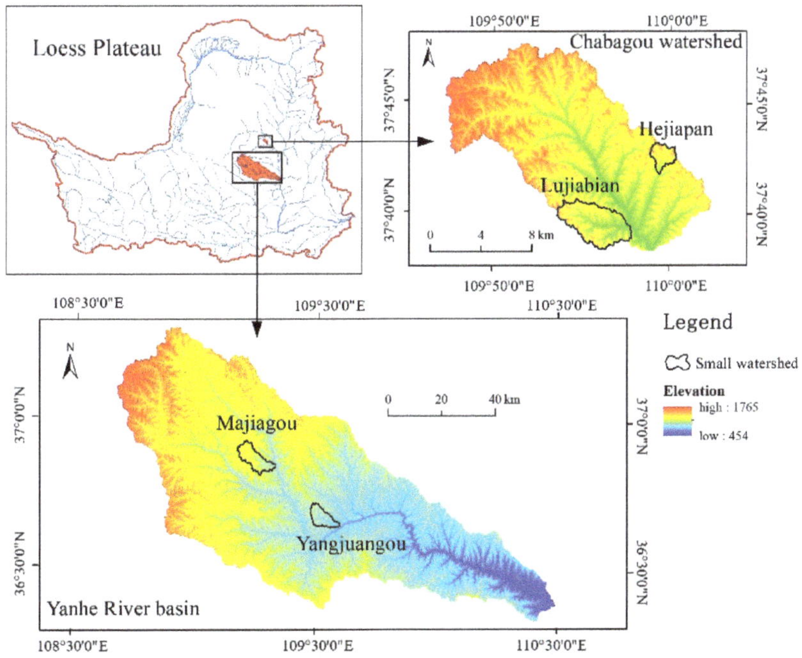

Figure 1. Location of small watersheds.

2.2. Field Investigation and Classification of Connection Mode

To better understand the different connection modes of check dams with downstream channels, we first carried out a detailed survey of check dams in the above four small watersheds in the field, including the location, DBS, whether the dam body was damaged, the discharge canals and its distribution in the dam land and the interconnection of all the check dams in the dam system of the small watershed. Then the check dams were marked in detail on remote sensing of Google Earth, for facilitating the calculation of later related parameters.

According to the survey results, the connection modes between check dams and downstream channels can be classified into eleven categories: (1) Disconnected (Figure 2a); (2) Connected through spillway (Figure 2b); (3) Connected through shaft (Figure 2c) or horizontal pipe (Figure 2d); (4) Connected through shaft and spillway (Figure 2e), or horizontal pipe and spillway (Figure 2f); (5) Connected through shaft, horizontal pipe and spillway (Figure 2g); (6) Connected through dam body damaged gap (Figure 2h); (7) Connected through discharge canal (DC) (Figure 2i); (8) Connected through discharge canal to shaft (DS$_1$) (Figure 2j); (9) Connected through discharge canal to spillway (DS$_2$) (Figure 2k); (10) Connected through discharge canal to shaft and spillway (DSS) (Figure 2l); (11) Connected through discharge canal to dam body damaged gap (DD) (Figure 2m), and this kind of check dam originally did not have a drainage building, the dam body was damaged and formed a gap during its operation period or a gap was dug in the dam body artificially for the convenience of drainage. Due to the differences in the degree of ecological restoration and check dam management, the number of check dams and connection modes in different watersheds varied greatly.

2.3. Distribution and Number of Check Dams in four Watersheds

Table 1 and Figure 3 show the number and distribution of check dams in four typical watersheds. There are 8, 29, 49 and 50 check dams in the Hejiapan, Lujiabian, Yangjuangou and Majiagou watersheds, respectively. The number of check dams that were disconnected and connected through discharge canals are the highest, with 38 and 31, respectively, while connected through shaft or horizontal pipe, spillway, dam body damaged gap, DD and DS$_2$ are relatively large, and the remaining are very small.

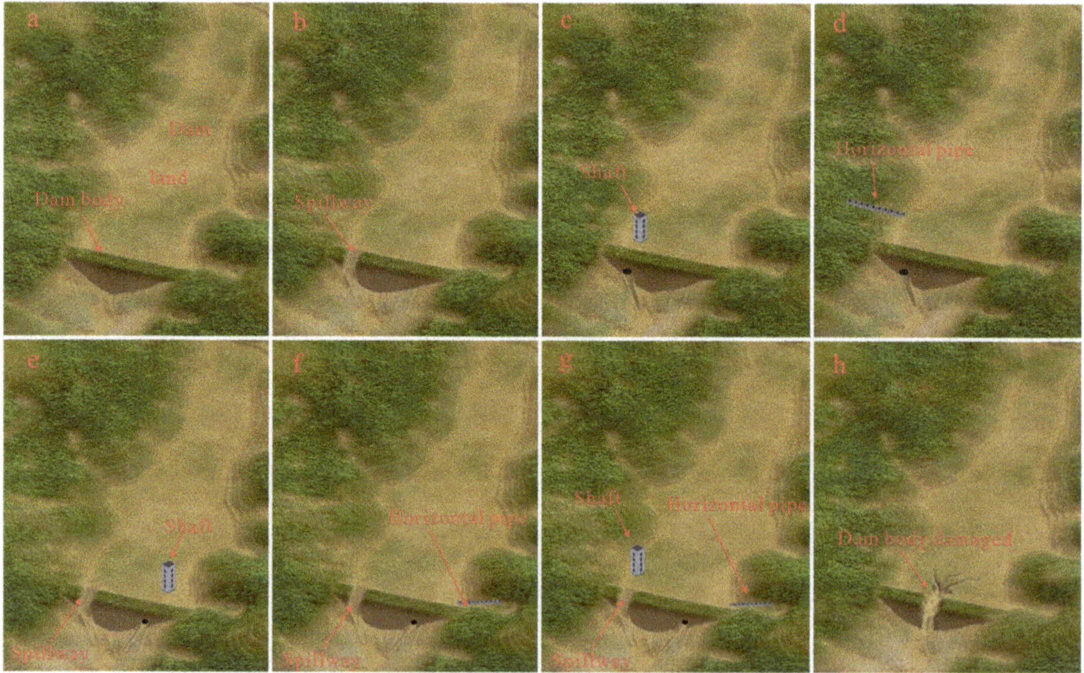

Figure 2. *Cont.*

Figure 2. Different connection modes of check dams with a downstream channel.

Table 1. Number of check dams with different connection modes in the four typical watersheds.

Connection Modes	Number of Check Dams				
	Hejiapan	Lujiabian	Yangjuangou	Majiagou	Total
Disconnected	5	5	20	8	38
Shaft or horizontal pipe	1	3	/	10	14
Spillway	/	2	2	6	10
Shaft and spillway or horizontal pipe and spillway	/	/	/	2	2
Shaft, horizontal pipe and spillway	/	/	/	1	1
Dam body damaged gap	/	2	2	4	8
Discharge canal to shaft (DS_1)	/	/	/	1	1
Discharge canal to dam body damaged gap (DD)	2	9	1	2	14
Discharge canal (DC)	/	4	23	4	31
Discharge canal to spillway (DS_2)	/	4	1	8	13
Discharge canal to shaft and spillway (DSS)	/	/	/	4	4
Total	8	29	49	50	136

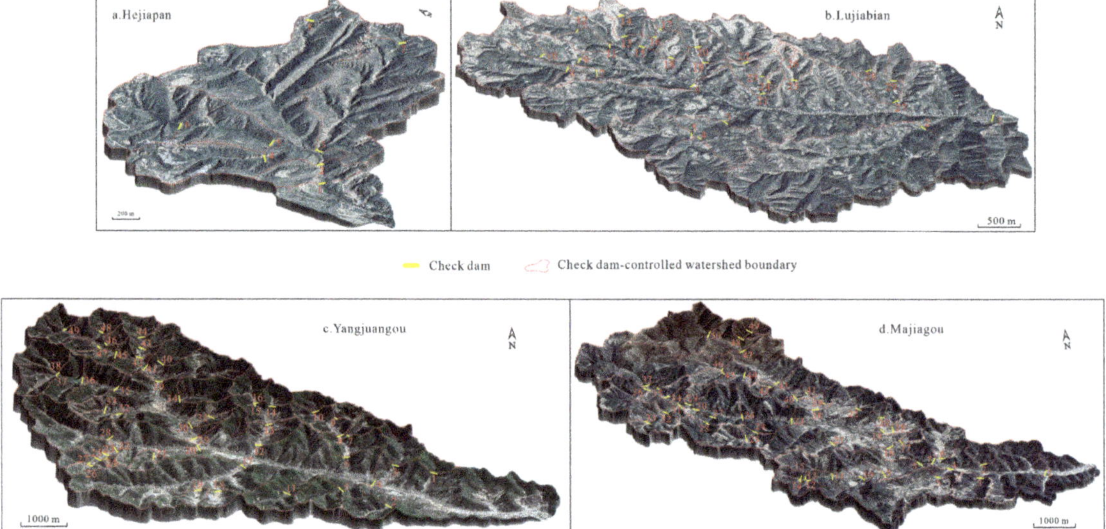

Figure 3. Distribution of check dams in four typical watersheds.

There are only three connection modes in the Hejiapan watershed, which are disconnected, connected through shaft and DD; the number of disconnected is the highest. There are seven types of connection modes in the Lujiabian watershed; the largest number of connection modes is DD, while the other types are fewer. There are six types of connection modes in the Yangjuangou watershed; the number of disconnected and connected through discharge canal is the largest, with 20 and 23, respectively, while the other types are very few. There are eleven types of connection modes in the Majiagou watershed; the largest number of connection modes are disconnected, connected through shaft or horizontal pipe, spillway and DS_2, while the other types are few.

2.4. Evaluation of Structural Connectivity Degree

In general, the area controlled by check dams is the entire area above the dam body in the catchment. In fact, due to the different connection modes of check dams to downstream channels, the actual area controlled by check dams is not the entire area above the dam body. The actual control area of the 1–6 types of connection modes is the entire dam-controlled area, while the 7–11 types of check dam with discharge canal are not the entire dam-controlled area, which is divided into two parts: the dam land-controlled area and the discharge canal-controlled area.

Figure 4 is a schematic diagram of check dams with discharge canals, in which part of the sediment from the slope directly flows into the discharge canal and then is transported to the downstream channel. Therefore, the sediment actually intercepted by the check dam is only from the slope that the dam land-controlled area. Considering that the characteristics of topography and vegetation of the check dam-controlled watershed are not very different, the degree of structural connectivity of the sediment of check dams to downstream channels (SCCD) can be evaluated by the ratio of the actual control area controlled by the discharge canal to the area of the entire area above check dam body in the watershed.

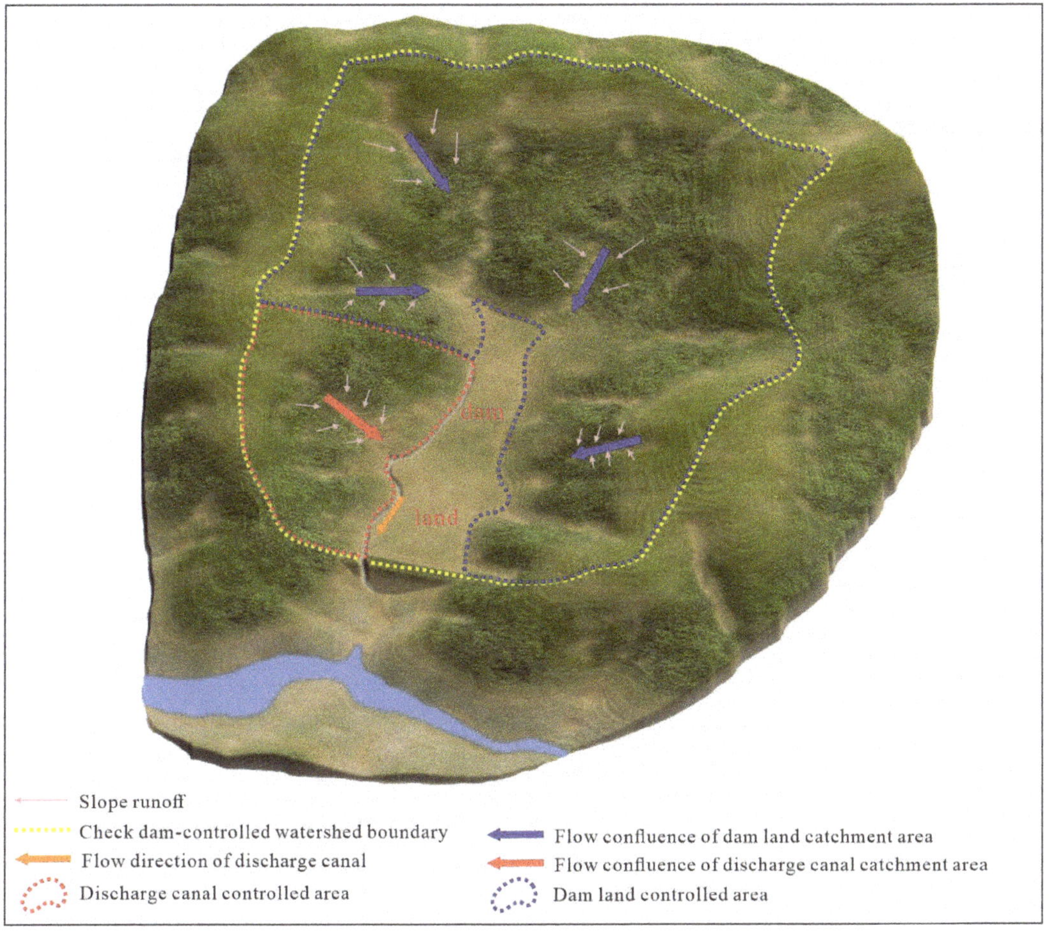

Figure 4. Check dam connected to the downstream channel through the discharge canal.

The different connection mode types of check dams with downstream channels are identified on the remote sensing image. If there is a discharge canal in the dam land, the check dam-controlled watershed is divided into discharge canal-controlled area and dam land-controlled area. Based on the ArcGIS software platform using Formula (1), the *SCCD* was calculated.

$$SCCD = \frac{A_{dcc}}{A_{cdc}} \tag{1}$$

where *SCCD* is the degree of structural connectivity of sediment for different connection mode check dams (%), the value of *SCCD* is between 0% and 100%, the smaller the value, the weaker the degree of connectivity of sediment. A_{dcc} is the discharge canal-controlled area (km^2), A_{cdc} is the total area of the check dam-controlled watershed (km^2).

According to the monitoring of the sediment in the Chabagou watershed and the Yanhe River basin under the conditions of a single rainfall, the sediment in the dam land has basically been unable to be transported to the dam body under non-rainstorm conditions in recent years [24]. With the restoration of ecology and less human disturbance, there is no sediment transport into the dam land. Therefore, it can be considered that under non-rainstorm conditions, although the check dams with connection modes of types 1–6 are connected to the downstream channel through DBS, their structural connectivity is disconnected. However, for the check dams with connection mode of types 7–11, sediment from the slope will be transported through the discharge canal to the downstream channel even under non-rainstorm conditions.

3. Results

3.1. Connection Mode and Degree

Table 2 shows the average SCCD of different connection modes of check dams in the four watersheds. From the perspective of different connection modes, check dams without drainage canals are disconnected, while check dams with discharge canals have different combinations of DBS and dam body damaged gap, resulting in a difference of the SCCD. The order of the average SCCD of check dams with discharge canal is: $DSS > DS_2 > DC > DD > DS_1$. The average SCCD of check dam with DS1 is the smallest, only 5.32%. While the average SCCD of check dam with DD and DC is 49.83% and 50.06%, respectively. The average SCCD of check dams with DSS and DSS_2 is relatively high, 61.63% and 53.36%, respectively. It is for this reason that check dams of these two types have a longer operation time, which makes the discharge canal longer, and leads to a lower actual control area of dam land.

Table 2. Average SCCD of different connection modes in the four watersheds.

Connection Modes	Number	Average SCCD (%)				
		Hejiapan	Lujiabian	Yangjuangou	Majiagou	Average
Disconnected	38	0	0	0	0	0
Shaft or horizontal pipe	14	0	0	/	0	0
Spillway	10	/	0	0	0	0
Shaft and spillway or horizontal pipe and spillway	2	/	/	/	0	0
Shaft, horizontal pipe and spillway	1	/	/	/	0	0
Dam body damaged gap	8	/	0	0	0	0
Discharge canal to shaft (DS_1)	1	/	/	/	5.32 ± 0	5.32 ± 0
Discharge canal to dam body damaged gap (DD)	14	3.39 ± 0.03	58.20 ± 0.36	40.17 ± 0	63.42 ± 0.27	49.83 ± 0.37
Discharge canal (DC)	31	/	82.36 ± 0.09	42.99 ± 0.29	58.36 ± 0.06	50.06 ± 0.29
Discharge canal to spillway (DS_2)	13	/	54.69 ± 0.34	31.50 ± 0	55.43 ± 0.16	53.36 ± 0.23
Discharge canal to shaft and spillway (DSS)	4	/	/	/	61.63 ± 0.23	61.63 ± 0.23
Sum/Average	136	1.13 ± 0.02	27.89 ± 0.33	19.11 ± 0.19	22.20 ± 0.28	

From the perspective of the four watersheds, the average SCCD of check dams in the Hejiapan watershed accounted for the lowest percentage, only 1.13%. The reason is that the average actual control area percentage of the discharge canal is very low. The average SCCD of check dams in the Yangjuangou watershed and the Majiagou watershed is 19.11% and 22.20%, respectively. The average SCCD of the Lujiabian watershed is the largest, at 27.89%. Although the average SCCD of check dams connected through DD and DS_2 ate both less than 60%, the average SCCD of check dams with only a discharge canal is 82.36%, resulting in the actual control area of discharge canal accounting for the largest proportion among the four watersheds. Therefore, if the structural connectivity of the sediment of the four watersheds is ranked by the average SCCD of check dams, the order is Lujiabian > Majiagou > Yangjuangou > Hejiapan. The SCCD is mainly affected by the length of discharge canal, and the length of the discharge canal is ultimately affected by the management degree of check dams. Due to the long operating life of the check dams in the Yangjuangou watershed, coupled with inadequate management, the check dams are generally damaged, many shafts, horizontal pipes and spillways have been completely destroyed or buried, resulting in a small number of check dams with drainage building or spillway. Although the number of check dams with discharge canals is relatively large, accounting for 55.10%, the length of the discharge canal is short, and the actual control area percentage of the discharge canal is relatively low. On the contrary, check dams of the Lujiabian watershed and the Majiagou watershed have better management, the number of check dams with a discharge canal is also large, accounting for 65.52% and 46.00%, respectively, and the length of the discharge canal is longer, the actual control area percentage of discharge canal is higher. Therefore, in order to reduce SCCD, the management of check dams should not only pay attention to the safe operation of check dams but also avoid the existence of discharge in the dam land or ensure the discharge canal is as short as possible.

3.2. Degree of Structural Connectivity for Dam System

The dam systems have been formed on the Loess Plateau [26], and the dam systems are composed of check dams with different connection modes. Whether they are series, parallel or hybrid dam systems, the connection mode is more complicated and variable (Figure 5).

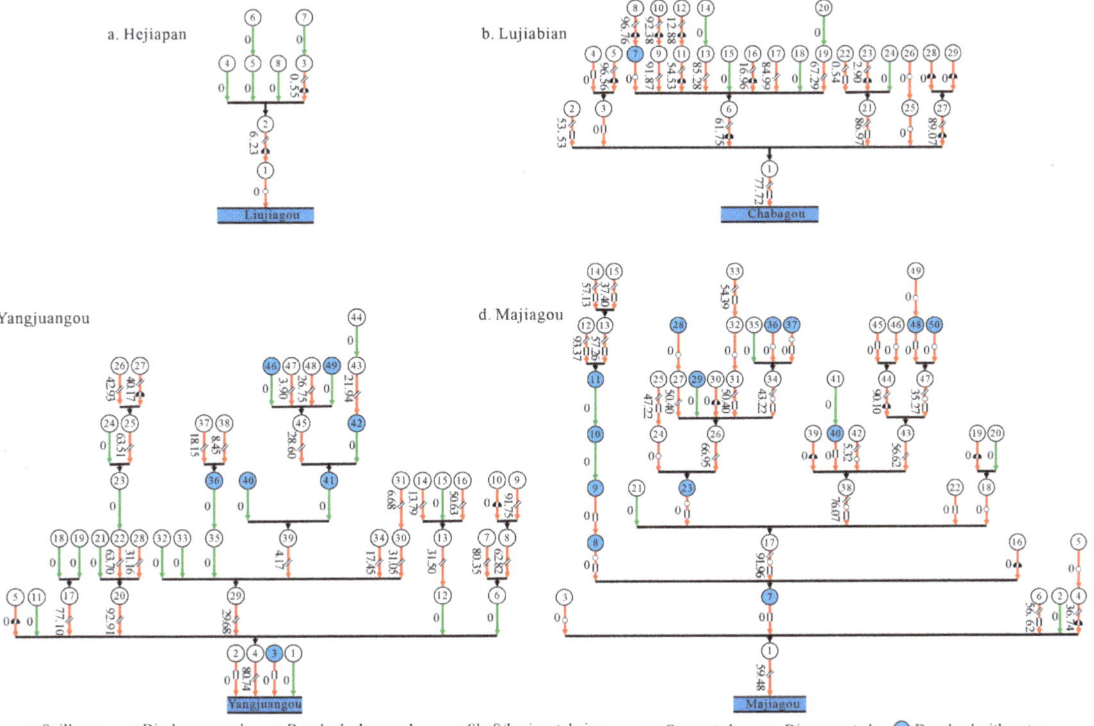

Figure 5. Position and connection modes of check dams in the four watersheds (the number is the SCCD of each check dam).

According to the different cascade modes of each dam system, the dam system of the four watersheds can be divided into 27 small dam systems, and the area of each dam system ranges from 0.82 to 71.50 km^2, with an average value of 14.25 km^2. The SCCD of a single check dam is between 2.33% and 66.41%, with an average of 27.53%. While the SCCD of each small dam system is between 0% and 42.38%, with an average of 11.18% (Table 3). According to the different connection modes of each check dam in small dam systems, the connection modes of dam systems can be divided into four categories: (1) Check dam at the bottom of the dam system has a discharge canal, and if the upstream check dam has a discharge canal, it must be connected to the check dam at the bottom of the dam system (dam systems #4, #11, #23 and #24). (2) Check dam at the bottom of the dam system has a discharge canal, but other check dams upstream have no discharge canal (dam systems #5, #7 and #9). (3) Check dams upstream and at the bottom of the dam system have discharge canals, but the check dams between the upstream and the bottom of the dam system have no discharge canals, which leads to the upstream check dam not being connected to the check dam at the bottom of the dam system (dam systems #3, #6, #8, #10, #13, #17, #20, #21, #25 and #27). (4) Check dam at the bottom of the dam system has no discharge canal, while some other check dams have discharge canals (dam systems #1, #2, #9, #12, #14, #15, #16, #18, #22 and #26). Compared with the SCCD of a single dam, the SCCD of the first two types of dam systems have not changed, the SCCD of the third type of dam system decreases 1.76–22.56%, with an average of 10.03%, and the last type of dam system decreases 2.33–42.38%, with an average of 20.15%.

Table 3. SCCD of the sum of a single check dam and dam system in the four watersheds.

Watershed	Dam System Number	Dam System	Dam System Area (km^2)	Sum of Single Check Dam		Dam System		Difference (%)
				A_{dcc} (km^2)	SCCD (%)	A_{dcc} (km^2)	SCCD (%)	
Hejiapan	1	1–8	3.02	0.07	2.33	0.00	0	2.33
	2	3–5	1.54	0.47	30.25	0.00	0	30.25
	3	6–20	3.76	2.31	61.49	1.54	40.86	20.63
Lujiabian	4	21–24	0.82	0.18	21.65	0.18	21.65	0
	5	27–29	0.83	0.22	26.36	0.22	26.36	0
	6	1–29	14.99	8.59	57.32	7.35	49.03	8.29
	7	6–10	3.03	4.10	66.41	4.10	66.41	0
	8	12–16	2.78	3.46	54.02	3.08	48.12	5.90
	9	17–19	6.17	0.25	6.23	0.00	0	6.23
	10	20–28	6.41	0.85	11.28	0.08	1.08	10.20
	11	29–49	17.42	0.67	21.69	0.67	21.69	0
Yangjuangou	12	23–27	1.08	0.10	8.18	0.00	0.00	8.18
	13	39–49	7.56	4.11	23.61	3.09	17.74	5.87
	14	35–38	4.04	0.38	35.14	0.00	0	35.14
	15	45–49	3.08	0.58	20.74	0.00	0	20.74
	16	42–44	1.26	1.28	42.38	0.00	0	42.38
	17	1–49	47.03	19.02	40.45	15.76	33.51	6.94
	18	8–15	8.02	0.92	11.47	0.00	0	11.47
	19	34–37	8.46	0.37	4.41	0.37	4.41	0
	20	31–33	2.52	0.50	20.00	0.23	8.97	11.03
	21	26–37	15.77	2.88	18.25	2.60	16.49	1.76
	22	23–37	20.39	3.38	16.60	0.00	0	16.60
Majiagou	23	43–50	12.63	3.14	24.86	3.14	24.86	0
	24	38–50	16.19	4.75	29.31	4.75	29.31	0
	25	17–50	47.38	15.21	32.10	11.82	24.96	7.14
	26	7–50	57.18	16.13	28.21	0.00	0.00	28.21
	27	1–50	71.50	20.48	28.65	4.35	6.09	22.56

Therefore, if the four types of dam systems are ranked according to the reduction degree of SCCD, the order is (4) > (3) > (2) = (1). If the check dam at the bottom of the dam system has no discharge canal, the SCCD of the dam system has decreased the most, and it is possible to intercept all the sediment from the whole catchment above under non-rainstorm conditions so as to maximize the sediment interception efficiency of the dam system. The SCCD of the other three types of connection modes with a discharge canal have not changed or are little changed. Therefore, the SCCD of a dam system is less than or equal to that of the sum of a single check dam; and the SCCD depends on the connection mode between check dams and downstream channels in the watershed.

4. Discussion

4.1. The Effect of Connection Mode on Structural Connectivity of Sediment

It is convenient to quantitatively evaluate the SCCD by using the actual control area percentage of the discharge canal. The higher the SCCD, the lower the sediment interception efficiency. At the same time, the influence of the height difference between dam land and the outlet of DBS was considered, which can also reflect the functional connectivity of sediment between the check dam and the downstream channel to a certain extent. Gao et al. [16] defined the degree of check dam connected to the downstream channel through a spillway is strong, but through shaft or horizontal pipe is weak. Hassanli et al. [27] also reported that the sediment interception efficiency of a check dam built with broken and angular rocks is higher than that built with rounded rocks, which is actually a combination of structural connectivity and functional connectivity to measure the degree of connectivity between a check dam and downstream channel. It was considered that the degree of connectivity depends on the outlet size of drainage facilities, which has certain limitations, and the influence of a discharge canal is not considered.

From the number of different connection modes between check dams and the downstream channel in the four small watersheds, it can be seen that the number of disconnected and connected (connected through the spillway, shaft, horizontal pipe and dam body damaged gap) are 38 and 35, respectively, accounting for 27.94% and 25.74%, respectively. Among them, the sediment interception efficiency of the check dam that is disconnected is 100%, and the sediment interception efficiency of check dam with a shaft, horizontal pipe, spillway and dam body damaged gap is also 100% if the sediment is not transported to the dam body. However, even if the sediment is not transported to the dam body, the sediment on the slope controlled by the discharge canal will not be intercepted in the dam land but will be directly transported to the downstream channel through the discharge canal. While there are 63 check dams with a discharge canal, accounting for 46.32% of the total, the average SCCD of the 63 check dams with a discharge canal is 50.71%, which shows that the existence of a discharge canal in the dam land has a great influence on the structural connectivity of check dams, and also affects the functional connectivity.

In fact, the connection mode between check dams and a downstream channel is not fixed; natural factors, such as rainfall, and human factors, such as lack of management, will cause changes in the connection mode, especially in the Loess Plateau, due to frequently occurring short heavy rainfall, which provides more possibilities for the changes [28–30]. If the check dam is disconnected from the downstream channel is damaged in extreme rainfall, it would be connected to the downstream channel through the dam body damaged gap; the check dams connected through DBS would also become the check dams connected through the dam body damaged gap due to the drainage buildings or spillway being damaged [31].

If the headrace is formed in dam land after a long period of erosion, the connection mode will be changed: the headrace becomes the discharge canal, and a check dam without a discharge canal is transformed into a check dam with a discharge canal. According to the field surveys, it is found that the discharge canal in small and medium-sized check dams only plays the role of sediment discharge, which has little influence on the dam land area (Figure 6). While drainage channels of some check dams will encroach and shrink the dam land area (Figure 7) and will not intercept sediment, some will become the sediment source with aggravating erosion [27] and promoting sediment connectivity [12].

Figure 6. Check dam with a discharge canal.

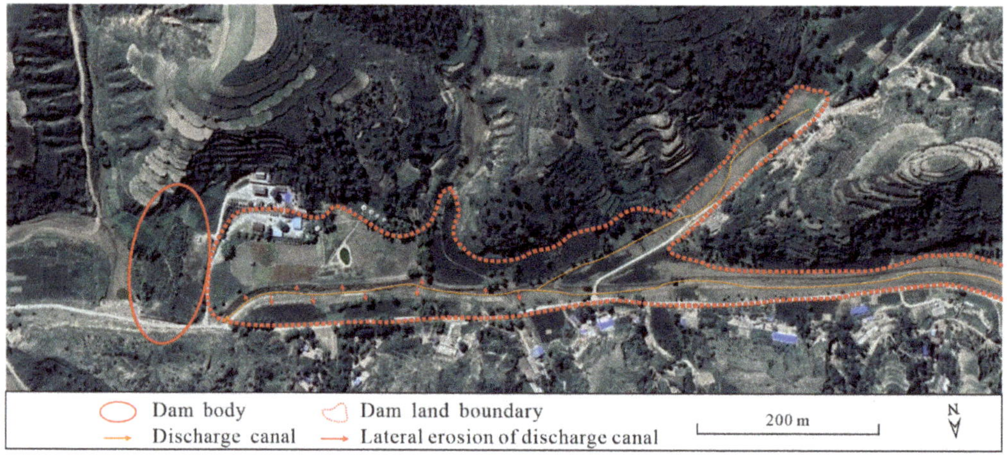

Figure 7. Discharge canals can encroach and shrink dam land area.

4.2. Optimization of Connection Mode in Dam System

From the analysis of structural connectivity of the small watersheds in this study, it can be seen that the degree of structural connectivity of dam systems will be less than the sum of single check dams. The larger dam systems in a watershed are composed of different small dam systems, and the cascade modes of the small dam systems include series, parallel and hybrid. The layout of a dam system can adjust the water and sediment process by changing the degree of connectivity of channels. The simulation results with the MIKE model show that different cascade modes of dam systems have different degrees of attenuation on flood peak and total flood discharge: hybrid > parallel > series [32]. Zhang et al. [33] also reported that hybrid and parallel systems, as well as check dams with DBS, had better flood control capacity. In fact, this mainly focuses on the functional connectivity of dam systems and does not consider the influence of the structural connectivity of check dams and downstream channels separately. If the structural connectivity is considered, the degree of sediment connectivity and sediment interception efficiency of check dams are irregular, which is mainly affected by the connection mode.

The Hejiapan and Lujiabian watersheds are mainly parallel dam systems due to the low level of gully, while the Yangjuangou and Majiagou watersheds area mainly hybrid dam systems due to the high level of gully. It is necessary to evaluate the connection mode of each small dam system and consider the influence of different cascade modes on the structural connectivity in the whole watershed when optimizing the connection mode of check dams. The difference in connection modes and their combinations will lead to great differences in the degree of connectivity and sediment interception efficiency of the whole dam system.

Ideally, if all the check dams are not connected with the downstream channels, the sediment interception efficiency of the whole dam system will be 100%. In fact, spillway, shaft and horizontal pipes should be configured according to the actual situation and the construction requirements of check dams when building check dams, and with the extension of the operation time of check dams, the formation of gap due to the dam body was damaged, and the formation of a discharge canal in the dam land is inevitable [31], which eventually leads to various connection modes between the check dam and the downstream channel, and the degree of structural connectivity is mainly affected by the discharge canal. The existence of a discharge canal leads to the reduction of the actual control area of the check dam, thus increasing the degree of structural connectivity.

The design of connection modes between a check dam and downstream channel should follow the principles: the check dam must not be at the risk of dam-break under extreme rainstorm conditions but should ensure the connectivity between the check dam and downstream channel, and the excess water and sediment can be transported to the downstream channel in time, thus ensuring the safe operation of a single check dam and the dam system, minimizing the structural connectivity between check dam and downstream channel, and retaining water and sediment in the dam land as much as possible under extreme rainstorm conditions. Therefore, two preconditions should be followed: (1) all check dams in the dam system should have DBS and ensure the outlet of the drainage building is not blocked. Otherwise, check dams would be easily damaged under extreme rainfall conditions. For example, in the "7·26" extreme rainstorm, the dam body was intact when the discharge outlet of the shaft was opened in time, while some check dams were damaged due to the discharge outlets of the shaft were blocked [19,34]. (2) The discharge canal should be avoided in the dam land, and its length should be kept as short as possible, especially in a large check dam-controlled watershed.

For example, in Figure 8: Check dams (#6 and 9) at the top of the series of dam systems (#6→#5→#4 and 9→#8) and check dam (# 5) between the top and the bottom of the series dam system can be the check dam with DBS or a discharge canal, and check dams (#4 and #8) at the bottom of the series dam system can be a check dam with only DBS. The advantage of this design is that, because the check dam at the bottom of the series dam system has no discharge canal, the actual control area of the check dam is still the area of the whole check dam-controlled watershed; that is, the SCCD is minimized.

Parallel dam systems can be divided into two situations: (1) The parallel check dams are all single check dams, and there are no other check dams upstream (#3 and #7), these single check dams are connected to a downstream channel through DBS or discharge canal. (2) Series dam systems and single check dams together form a parallel dam system (#3, #6→#5→#4, #7 and 9→#8). The series dam system is configured with the principle of a series dam system, while the single check dam is configured with the connection mode according to the principle of (1).

The series and parallel dam systems are combined to form smaller hybrid dam systems, and the smaller hybrid dam systems further constitute the larger hybrid dam systems, in which the connected mode should be optimized with reference to the principle of the series and parallel dam systems. However, it should be noted that no matter whether it is the small or the large hybrid dam system, the bottom key check dam (#1) is very important for the structural connectivity of sediment in the whole dam system [27]. The key check dam should not have a discharge canal and only be connected to the downstream channel through DBS, and if there have other check dams connected to the key check dam in series upstream (#2), these check dams can only have a discharge canal or DBS.

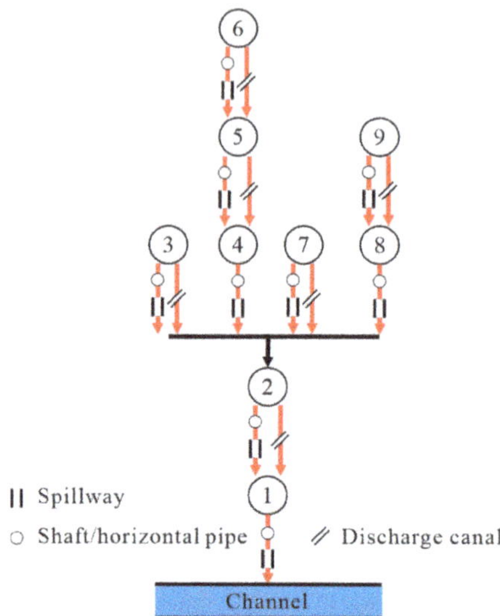

Figure 8. Optimal connection mode in a dam system (the shaft, horizontal pipe and spillway can be in separate or combined forms).

5. Conclusions

Based on the field survey, the connection modes between check dams and downstream channels are divided into eleven categories. The degree of structural connectivity of sediment between the check dam and downstream channel was quantitatively evaluated according to the actual control area percentage of the discharge canal. When the sediment cannot be transported to the dam body under non-rainstorm conditions, the order of the structural connectivity of sediment of the 11 connection modes is: Disconnected = shaft or horizontal pipe = spillway = shaft and spillway, or horizontal pipe and spillway = shaft, horizontal pipe and spillway = dam body damaged gap < discharge canal to shaft < discharge canal to dam damage gap < discharge canal < discharge canal to spillway < discharge canal to shaft and spillway. The different connection modes and combination forms of check dams lead to a great difference in structural connectivity of dam systems, the structural connectivity of the dam system is less than or equal to that of the sum of single check dams. The degree of structural connectivity of a dam system will be greatly reduced if the main control check dam of the dam system has no discharge canal. The optimization method of the connection mode of the series, parallel and hybrid dam systems is proposed according to the different cascade modes of dam systems. This study only discusses the structural connectivity of sediment of single check dams and dam systems, and the evaluation of functional connectivity based on structural connectivity needs to be further analyzed in future research.

Author Contributions: Conceptualization, L.B. and J.J.; investigation, all authors; data curation, L.B.; writing—original draft preparation, L.B.; writing—review and editing, all authors. All authors have read and agreed to the published version of the manuscript.

Funding: This study was supported by Open Project Fund of Key Laboratory of the Loess Plateau Soil Erosion and Water Loss Process and Control, Ministry of Water Resources (Grant Number 2019001) and National Natural Science Foundation of China (Grant Number 41771319, 42077078).

Institutional Review Board Statement: Not applicable.

Informed Consent Statement: Not applicable.

Data Availability Statement: The data presented in this study are available on request from the corresponding author.

Acknowledgments: The authors are grateful to the Editor and reviewers for their constructive comments and suggestions.

References

1. Ministry of Water Resources of the People's Republic of China. *Zhongguo Heliu Nisha Gongbao*; China Water&Power Press: Beijing, China, 2020.
2. García-Ruiz, J.M.; Beguería, S.; Nadal-Romero, E.; González-Hidalgo, J.C.; Lana-Renault, N.; Sanjuán, Y. A meta-analysis of soil erosion rates across the world. *Geomorphology* **2015**, *239*, 160–173. [CrossRef]
3. Shi, Z.H.; Song, C.Q. Water erosion processes: A historical review. *J. Soil Water Conserv.* **2016**, *30*, 1–10. [CrossRef]
4. García-Ruiz, J.M.; Beguería, S.; Lana-Renault, N.; Nadal-Romero, E.; Cerdà, A. Ongoing and emerging questions in water erosion studies. *Land Degrad. Develop.* **2017**, *28*, 5–21. [CrossRef]
5. Marchamalo, M.; Hooke, J.M.; Sandercock, P.J. Flow and sediment connectivity in semi-arid landscapes in SE Spain: Patterns and controls. *Land Degrad. Develop.* **2016**, *27*, 1032–1044. [CrossRef]
6. Xu, X.Z.; Zhang, H.W.; Zhang, O.Y. Development of check-dam systems in gullies on the Loess Plateau, China. Environ. *Sci. Policy.* **2004**, *7*, 79–86. [CrossRef]
7. Liu, X.Y.; Gao, Y.F.; Ma, S.B.; Dong, G.T. Sediment reduction of warping dams and its timeliness in the Loess Plateau. *SHUILI XUEBAO* **2018**, *49*, 145–155. [CrossRef]
8. Chen, Z.Y.; Li, Z.B.; Wang, Z.Y. Some thoughts on the strategic positioning of check dam construction in the Loess Plateau. *Soil Water Conserv. China.* **2020**, *9*, 32–38. [CrossRef]
9. Li, J.Z.; Liu, L.B. Analysis on the sediment retaining amount by warping dams above Tongguan section of the Yellow River in recent years. *Yellow River* **2018**, *40*, 1–6. [CrossRef]
10. Conesa-García, C.; García-Lorenzo, R. Bed texture changes caused by check dams on ephemeral channels in Mediterranean semiarid environments. *Z. Geomorpho.* **2008**, *52*, 437–461. [CrossRef]
11. Zema, D.A.; Bombino, G.; Denisi, P.; Lucas-Borja, M.E.; Zimbone, S.M. Evaluating the effects of check dams on channel geometry, bed sediment size and riparian vegetation in Mediterranean mountain torrents. *Sci. Total Environ.* **2018**, *642*, 327–340. [CrossRef] [PubMed]
12. Galia, T.; Škarpich, V.; Ruman, S. Impact of check dam series on coarse sediment connectivity. *Geomorphology* **2021**, *377*, 107595. [CrossRef]
13. Surian, N.; Ziliani, L.; Comiti, F.; Comiti, F.; Lenzi, M.A.; Mao, L. Channel adjustments and alteration of sediment fluxes in gravel-bed rivers of North-Eastern Italy: Potentials and limitations for channel recovery. *River. Res. Applic.* **2010**, *25*, 551–567. [CrossRef]
14. Wang, H.W.; Kondolf, G.M. Upstream sediment-control dams: Five decades of experience in the rapidly eroding Dahan River basin, Taiwan. *J. Am. Water Resour. Assoc.* **2014**, *50*, 735–747. [CrossRef]
15. Marchi, L.; Comiti, F.; Crema, S.; Cavalli, M. Channel control works and sediment connectivity in the European Alps. *Sci. Total Environ.* **2019**, *668*, 389–399. [CrossRef] [PubMed]
16. Gao, H.D.; Jian, L.L.; Li, Z.B.; Xu, G.C.; Zhao, B.H. Mechanism underlying impact of check damson runoff based graph theory in the hilly-gully loess region. *Sci. Soil Water Conserv.* **2015**, *13*, 1–8. [CrossRef]
17. Wei, X.; Li, Z.B.; Shen, B.; Li, X.G.; Li, P. The water damage and the prevention measures in construction of check dam. *J. Water Resour. Water Eng.* **2004**, *15*, 55–59.
18. Wang, Z.L.; Zhang, B.S.; Liu, H.Z.; Li, C.J.; Lin, X.Z.; Yang, J.S. Damage causes and sand-blocking effects of warping dams in Dalat Banner in 2016. *Adv. Sci. Technol. Water Resour.* **2019**, *39*, 1–6. [CrossRef]
19. Wang, N.; Chen, Y.X.; Bai, L.C.; Wang, H.L.; Jiao, J.Y. Investigation on soil erosion in small watersheds under "7·26" extreme rainstorm in Zizhou county, northern Shaanxi province. *Bull. Soil Water Conserv.* **2017**, *37*, 338–344. [CrossRef]
20. Yu, G.Q.; Zhang, M.S.; Li, Z.B.; Li, P.; Zhang, X.; Cheng, S.D. Piecewise prediction model for watershed-scale erosion and sediment yield of individual rainfall events on the Loess Plateau, China. *Hydrol. Process.* **2015**, *28*, 5322–5336. [CrossRef]
21. Yang, J.S.; Yao, W.Y.; Zheng, M.G.; Li, L. Analysis on gravitational sediment yield in the check-dam controlled basins of Chabagou watershed. *SHUILI XUEBAO* **2017**, *48*, 241–245. [CrossRef]
22. Li, C.Z.; Wang, H.; Yu, F.L.; Yang, A.M.; Yan, D.H. Impact of soil and water conservation on runoff and sediment in Yanhe River basin. *Sci. Soil Water Conserv.* **2011**, *9*, 1–8. [CrossRef]
23. Wang, H.L.; Jiao, J.Y.; Tang, B.Z.; Chen, Y.X.; Bai, L.C.; Wang, N.; Zhang, Y.F. Characteristics of rill erosion and its influencing factors in slope farmland after "7·26" rainstorm in Zizhou County, Shaanxi Province. *Trans. Chin. Soc. Agric. Eng.* **2019**, *35*, 122–130. [CrossRef]

24. Bai, L.C.; Wang, N.; Jiao, J.Y.; Chen, Y.X.; Tang, B.Z.; Wang, H.L.; Chen, Y.X.; Yan, X.Q.; Wang, Z.J. Soil erosion and sediment interception by check dams in a watershed for an extreme rainstorm on the Loess Plateau, China. *Int. J. Sediment Res.* **2020**, *35*, 408–416. [CrossRef]

25. Yu, H. The period and trend analysis of streamfall and sediment load in yan river based on time series. Master's Thesis, Northwest Agricultural and Forestry University, Yangling, China, 2008.

26. Yu, T.; Li, Z.B.; Chen, Y.T.; Yuan, S.L.; Wang, W. Analysis of structural characteristics of typical check dam system in the third subregion of loess hilly region. *Res. Soil Water Conserv.* **2019**, *26*, 26–30. [CrossRef]

27. Hassanli, A.M.; Nameghi, A.E.; Beecham, S. Evaluation of the effect of porous check dam location on fine sediment retention (a case study). *Environ. Monit. Assess.* **2009**, *152*, 319–326. [CrossRef] [PubMed]

28. Wei, X.; Li, Z.B.; Wu, J.H.; Li, B.B.; Du, Z. A Discussion on some deological problems in research of water damage hazards of check dam. *Res. Soil Water Conserv.* **2007**, *14*, 235–237.

29. Luo, Z.B.; Yong, C.X.; Fan, J.; Shao, M.A.; Wang, S.; Jin, M. Precipitation recharges the shallow groundwater of check dams in the loessial hilly and gully region of China. *Sci. Total Environ.* **2020**, *742*, 140625. [CrossRef]

30. Jiao, J.Y.; Wang, Z.J.; Wei, Y.H. Characteristics of erosion sediment yield with extreme rainstorms in Yanhe Watershed based on field measurement. *Trans. Chin. Soc. Agric. Eng.* **2017**, *33*, 159–167. [CrossRef]

31. Li, L.; Wang, F.; Sun, W.Y.; Shi, X.J. Analysis of water damage of check dam in Loess Plateau. *Soil Water Conserv. China* **2014**, *10*, 20–22. [CrossRef]

32. Yuan, S.L.; Li, Z.B.; Li, P.; Gao, H.D.; Wang, D.; Zhang, Z.Y. MIKE coupling model simulating effect of check dam construction on storm flood process in small watershed. *Trans. Chin. Soc. Agric. Eng.* **2018**, *34*, 152–159. [CrossRef]

33. Zhang, Y.F.; Jiao, J.Y.; Tang, B.Z.; Chen, Y.X.; Wang, N.; Bai, L.C.; Wang, H.L. Channel sediment connectivity and influence factors in small watersheds under extremely rainstorm–a case study at Zizhou county, Shaanxi province. *Bull. Soil Water Conserv.* **2019**, *39*, 302–309. [CrossRef]

34. Zhou, Z.Y.; Wang, X.L.; Chen, W.L.; Deng, S.H.; Liu, M.H. Numerical simulation of dam-break flooding of cascade reservoirs. *Trans. Tianjin Univ.* **2017**, *23*, 570–581. [CrossRef]

Article

Comparing Rainfall Erosivity Estimation Methods Using Weather Radar Data for the State of Hesse (Germany)

Jennifer Kreklow [1,*], Bastian Steinhoff-Knopp [1], Klaus Friedrich [2] and Björn Tetzlaff [3]

[1] Institute of Physical Geography and Landscape Ecology, Leibniz Universität Hannover, Schneiderberg 50, 30167 Hannover, Germany; steinhoff-knopp@phygeo.uni-hannover.de
[2] Hessian Agency for Nature Conservation, Environment and Geology, 65203 Wiesbaden, Germany; klaus.friedrich@hlnug.hessen.de
[3] Institute of Bio- and Geosciences IBG-3, Forschungszentrum Jülich GmbH, 52425 Jülich, Germany; b.tetzlaff@fz-juelich.de
[*] Correspondence: kreklow@phygeo.uni-hannover.de; Tel.: +49-511-762-19798

Received: 9 April 2020; Accepted: 14 May 2020; Published: 16 May 2020

Abstract: Rainfall erosivity exhibits a high spatiotemporal variability. Rain gauges are not capable of detecting small-scale erosive rainfall events comprehensively. Nonetheless, many operational instruments for assessing soil erosion risk, such as the erosion atlas used in the state of Hesse in Germany, are still based on spatially interpolated rain gauge data and regression equations derived in the 1980s to estimate rainfall erosivity. Radar-based quantitative precipitation estimates with high spatiotemporal resolution are capable of mapping erosive rainfall comprehensively. In this study, radar climatology data with a spatiotemporal resolution of 1 km^2 and 5 min are used alongside rain gauge data to compare erosivity estimation methods used in erosion control practice. The aim is to assess the impacts of methodology, climate change and input data resolution, quality and spatial extent on the R-factor of the Universal Soil Loss Equation (USLE). Our results clearly show that R-factors have increased significantly due to climate change and that current R-factor maps need to be updated by using more recent and spatially distributed rainfall data. Radar climatology data show a high potential to improve rainfall erosivity estimations, but uncertainties regarding data quality and a need for further research on data correction approaches are becoming evident.

Keywords: R-factor; soil erosion; USLE; rainfall intensity; modeling; radar climatology; RADKLIM; rain gauge

1. Introduction

The R-factor is a measure of rainfall erosivity and an important input variable for estimating soil losses by water using the Universal Soil Loss Equation (USLE) and its many variations [1]. Based on the documented relationship between the amount of soil erosion and the kinetic energy of precipitation, the rainfall erosivity can be derived directly from temporally highly resolved precipitation time series [1–3]. The R-factor of one event is defined as the product of the kinetic energy and the maximum 30-min intensity of an erosive rainfall event. The R-factors of all events throughout a year are added to obtain the annual R-factor, which is usually averaged over a period of at least ten years as an input to the USLE.

In the past, measurement data from rain gauges or, more recently, from automated rain gauges were used for estimating rainfall erosivity. Still today, the R-factors calculated from these point-scale data for every station are spatially interpolated to derive maps of rainfall erosivity. This approach has also been recently applied to generate a European erosivity map [4]. However, due to the small spatial extent of convective precipitation cells and a high variability of precipitation intensity within these cells, which contributes significantly to rainfall erosivity, the spatial recording of the rainfall erosivity is incomplete and patchy [5]. Rain gauges are not capable of detecting the spatial distribution of local heavy rainfall hot spots or individual heavy rainfall events, which are highly relevant for erosion modelling. Interpolating R-factors calculated from point measurements therefore results in a smoothing and an underestimation of erosivity [6]. In order to capture the highly variable spatiotemporal distribution of rainfall intensity during erosive rainfall events, highly resolved precipitation data, both spatial and temporal, are needed. Weather radars are capable of providing such data, but the number of studies deriving erosivity directly from such highly resolved datasets is still rather low [4].

In practice, R-factor maps are frequently derived by regression equations from spatially interpolated summer precipitation sums or annual precipitation sums in order to obtain comprehensive erosivity information. This methodology is much easier to apply than the direct event-based derivation of the R-factor from gauge data, but it suffers from representativity issues. Again, data smoothing by spatial interpolation and regression equations lead to smoothed R-factors. High R-factors often remain limited to mountain tops, while the actual occurrence of heavy rainfall as a consequence of convective events in the lowlands is not taken into account [7].

In Germany, for instance, the R-factor is derived by regional authorities for each federal state according to the technical standard DIN 19708 [8], whereby most federal states use regional adjusted regression equations. The derived erosivity maps serve inter alia as an input for soil erosion modelling in order to evaluate the fulfilment of EU Cross-Compliance soil protection regulations. Based on these evaluation outcomes, income support for farmers is calculated and requirements for erosion control are imposed. However, the applied regression equations were usually derived based on data from a few rain gauge recorders (usually < 20) integrating rainfall data from the 1960s to the 1980s [9]. The regression equations are only rarely updated (e.g., in North-Rhine Westfalia [10]) or, in many federal states, not at all. However, several studies indicate spatial and temporal changes in precipitation distribution and quantities as well as an increase and intensification of heavy rainfall and thus an increase in precipitation erosivity due to climate change [6,11,12]. Consequently, the validity of the currently applied regression equations, which were determined based on precipitation data of the last climate period or even older data, must also be questioned, especially in regard to the current atmospheric conditions.

In the German federal state of Hesse, a lot of information on soil quality and degradation, including the R-factor, is collected in the technical information system "Erosion Atlas Hesse" [13,14]. The erosion atlas is an important instrument for precautionary soil protection in Hesse since it shows areas with a high risk of erosion and helps farmers to plan erosion control measures. Furthermore, it supports urban land-use planning through the identification of sites that require additional protection measures. The estimation of the R-factor for the erosion atlas is currently based on a regression equation derived in 1981 from data of 18 rain gauges in Bavaria, which comprise time series of up to 14 years throughout the period of 1958–1977 [9,15]. The precipitation data used for calculating the R-factor are spatially interpolated mean summer precipitation sums (May to October) for the period of 1971–2000 on a 1 km^2 grid [16]. There is evidence that rainfall distribution and intensity has changed since this time period [12,17], emphasising the need for updated precipitation datasets and methods that estimate rainfall erosivity.

The radar climatology dataset RADKLIM ("RADarKLIMatologie") [18] addresses the need for updating precipitation data. RADKLIM is a radar-based quantitative precipitation estimation dataset provided by the German Weather Service (Deutscher Wetterdienst, DWD). It is available for the whole of Germany starting from 2001 with a high spatial (1 km^2) and temporal (up to 5 min) resolution [19]. The largely comprehensive nationwide detection of all precipitation events indicates a high potential for the derivation of spatial information to calculate the R-factor. The high temporal resolution of the data as well as recent advances in computer hardware enable the direct event-based calculation of the R-factor. However, the differences in measurement method and scale between radar and rain gauges, especially in detecting heavy rainfall, must be taken into account when interpreting the results. The precipitation totals in radar climatology tend to be slightly lower than the precipitation amounts measured by rain gauges and this underestimation by radar climatology is particularly pronounced for high precipitation intensities [20]. This is due to the averaging of precipitation over the area of the radar pixels and path-integrated rainfall-induced attenuation of the radar beam [21].

For the direct event-based calculation of the R-factor based on radar data, Fischer et al. [22] found similar effects and derived correction factors to compensate for the underestimation of the R-factor calculated with radar climatology data. The proposed factors include a spatial scaling factor, which reflects the attenuation of intensity peaks by averaging the precipitation over the radar pixel area, and a method factor, which should compensate for the systematic underestimation of erosion by the radar data compared to rain gauge measurements.

In addition, several studies have recently investigated the influence of the temporal resolution of precipitation data on the calculation of the R-factor [22,23]. In principle, the authors agree that the level of the R-factor decreases with decreasing temporal resolution. The intensity peaks, which are decisive for determining the kinetic energy of the precipitation, are detected less accurately with decreasing temporal resolution and are thus attenuated. However, authors disagree about the correction of this effect, since the level of any correction factor depends on the temporal resolution of the rainfall data that is used as a reference. Based on rain gauge and RADKLIM data for Germany, Fischer et al. [22] use one minute as the highest possible resolution for a factor value of 1. Panagos et al. [23], on the other hand, use a reference of 30 min as factor value of 1 in their European-wide study based on rain gauge data. For the RADKLIM product with a 5-min resolution, this results in a temporal correction factor of 1.05 [22] or 0.7984 [23], and for the RADKLIM product with hourly resolution, the temporal correction factors are 1.9 and 1.5597, respectively.

The goal of this study was to compare the performance of different calculation methods for the R-factor using rain gauge and radar rainfall data. The impacts, advantages, disadvantages and correction approaches for several input datasets were analysed; additionally, updated regression equations were derived. Taking the improvement in monitoring systems through a higher coverage by measurements and discrepancies concerning methodology, input data quality and resolution, observation period and correction approaches into account, the paper proposed these hypotheses for the derivation of R-factors from radar climatology and rain gauge data for the period 2001–2016:

1. The newly calculated R-factors from both datasets are higher than the R-factors from earlier calculations due to changes in climate, interannual rainfall distribution and rainfall intensity.
2. Since radar data include small-scale convective cells without gaps, the R-factors derived from the radar climatology should be higher on average than those calculated from rain gauge measurements. At the same time, the radar measurements underestimate the maximum precipitation intensities. The latter can be compensated by the correction factors according to Fischer et al. [22].
3. The spatial distribution of the R-factors derived from the radar climatology deviates from the patterns of the R-factors calculated and interpolated by means of the regression equation due to the comprehensive coverage of all heavy rainfall events.

2. Materials and Methods

2.1. Study Area

For this study, the federal state of Hesse was selected as the investigation area due to its central location within Germany and its complex terrain, which allow for a good transferability of the outcomes. The federal state of Hesse has a total area of approximately 21,115 km^2. The area is characterised by a diverse topography with several low mountain ranges and highlands crossed by depressions and river valleys (see Figure 1). The highest elevation is 950 m.a.s.l., whereas the lowest elevation is about 73 m.a.s.l. A large portion of the intensively used agricultural areas in the lowlands are oriented in Rhenish direction (SSW-NNO) [24]. The study area is located in the humid midlatitudes in a transition zone between a maritime climate in north-western Germany and a more continental climate in the south and east of Germany. Westerly winds influence the distribution of precipitation and, thus, many of the intensively used agricultural areas are located in the rain shadow on the lee side east of the mountain ranges.

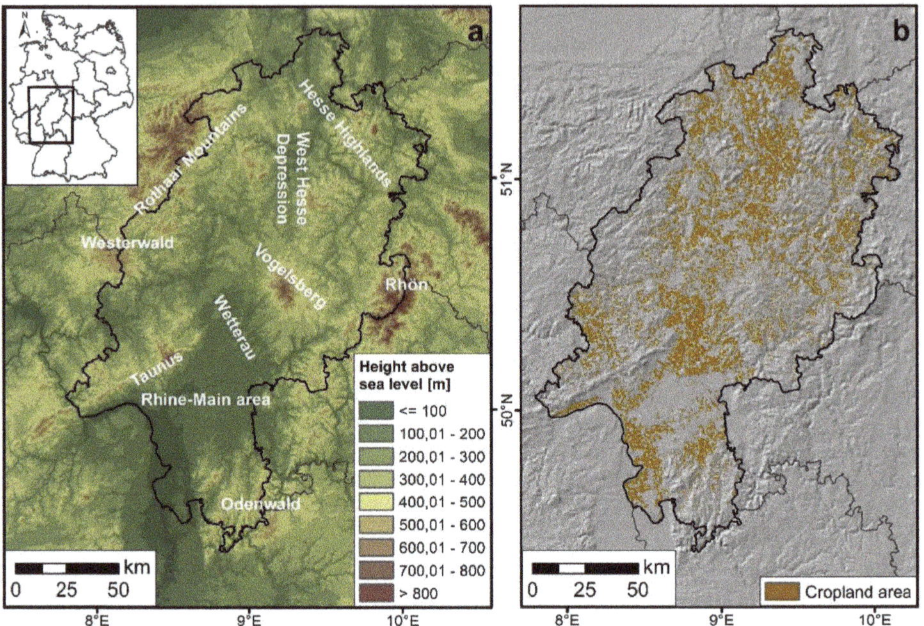

Figure 1. (**a**) Location, height above sea level [m] and selected landscape units of the federal state of Hesse, (**b**) spatial distribution of cropland areas in the study area.

2.2. Data Basis

2.2.1. Radar Climatology Data

The DWD currently operates 17 ground-based C-band weather radars. The nationwide coverage was established in 2001. In 2018, the DWD published the radar climatology dataset RADKLIM, which consists of gridded nationwide quantitative precipitation estimate composites with a spatial resolution of 1 km^2 and a temporal resolution of up to 5 min starting from 2001. For this study, we used the YW product in 5-min resolution [18] and the RW product [25] in hourly resolution for the period 2001–2016. Their derivation procedure comprises various correction algorithms to compensate for typical radar-related errors and artefacts such as clutter, spokes, signal attenuation and bright band effects. Ground clutter can be caused by non-meteorological objects such as mountains, buildings, wind energy plants or trees that disturb the radar signal and cause non-precipitation echoes. If the radar beam is blocked in whole or in part by such objects, the sector behind these obstacles is shielded, which causes a linear artefact, the so-called negative spoke. Signal attenuation may cause significant underestimation of rainfall rates. It can be caused by a wet radome, by heavy precipitation events that shield the sector behind or by range degradation at far range from the radar. Bright Band effects occur in the melting layer where the comparatively large surface of melting snowflakes is covered by a film of water, which may cause very strong radar signals.

For the derivation of the radar climatology, the reflectivity is converted to rain rates, and the local radar station data are merged and transformed to a cartesian grid. Aggregated hourly rain rates are adjusted to ground-truth automated rain gauge measurements, which yields the RW product. Finally, the hourly rain rates are disaggregated to the original 5-min intervals in order to obtain the quasi-adjusted YW product [19]. For disaggregation, the hourly precipitation sum of the adjusted RW product is distributed to the twelve 5-min intervals based on the temporal rainfall distribution throughout the respective hour. The data processing was conducted by DWD. In the state of Hesse, only the stations operated by DWD are used for radar data adjustments.

2.2.2. Rain Gauge Data

For this study, we combined two different rain gauge datasets in 1-min resolution. We used data from 76 automated rain gauges throughout Hesse operated by DWD, which are freely available in the DWD Open Data Portal [26], as well as from 52 rain gauges of the Hessian monitoring network operated by HLNUG, which are not publicly available. Both datasets were carefully checked for plausibility and a cleaning procedure was implemented to remove erroneous values. For a detailed description of the data processing and cleaning procedure please refer to [27].

In general, the DWD rain gauge data are available for the period 2001–2016, whereas those of the HLNUG stations only cover the period 2001–2015. However, the time series of the combined rain gauge dataset varies strongly between stations. In this study, 21 stations with time series shorter than nine years were excluded. The final dataset used for analysis consisted of 110 rain gauge stations. Finally, the 1-min rain gauge data were aggregated to a temporal resolution of 5 min in order to match the temporal resolution of the radar climatology data.

2.3. Methodology

2.3.1. R-factor Calculation According to DIN 19708

The R-factors were calculated according to the specifications of DIN 19708 [8] for the RADKLIM YW product and the rain gauge data, both in 5-min resolution. According to DIN 19708 [8], which is based on the results of Schwertmann et al. [2], erosive precipitation events have a precipitation sum of at least 10 mm or a precipitation intensity exceeding 10 mm/h within a time window of 30 min (i.e., an actual precipitation quantity of 5 mm in 30 min). The maximum precipitation sum occurring within a 30-min window of a rainfall event is identified by applying a moving window of six 5-min intervals and is related to one hour by doubling it. This value is referred to as maximum 30-min intensity I_{30}. As defined by DIN 19708 [8], the total amount of precipitation is doubled and assigned to I_{30} if an event lasts less than 30 min. Rainfall events are separated by a precipitation pause of at least 6 h.

The R-factor of a specific precipitation event results from the product of the maximum 30-min intensity I_{30} [mm/h] and the kinetic energy E [kJ/m^2] of the total rainfall during the event.

$$R_{event} = E \cdot I_{30} \tag{1}$$

The kinetic energy E of an erosive rainfall event was calculated with the following equation from DIN 19708:

$$E = \sum_{i=1,n}^{i=n} E_i \tag{2}$$

with

$$E_i = (11.89 + 8.73 \cdot log_{10}(I_i)) \cdot N_i \cdot 10^{-3} \, for \, 0.05 \, \leq \, I_i \, \leq 76.2$$

$$E_i = 0 \, for \, I_i < 0.05$$

$$E_i = 28.33 \cdot N_i \cdot 10^{-3} \, for \, I_i > 76.2$$

Thereby is

i 5-min interval of the rainfall event

E_i kinetic energy of the rainfall in period i [kJ/m^2]

N_i rainfall depth in period i, [mm]

I_i rainfall intensity in period i, [mm/h], that is $I_i = N_i \cdot \frac{60\ \text{Min}}{5\ \text{Min}}$

Finally, the R-factor per year for a given location is the sum of the R_{event} products [kJ/m^2 mm/h = N/(ha a)] of all erosive rainfall events in a year. Due to the great interannual variability of erosivity, it is recommended to average the annual R-factors over a period of at least ten years [8]. For the calculations based on the radar climatology this criterion was fulfilled everywhere, whereas the time series of five rain gauges was limited to nine years.

For the calculation of the R-factor from both data sets, the development of new routines was necessary. One difficulty is the large data volume of the YW product for the whole of Hesse, which required a balancing of memory requirements and computing efficiency. The developed Python routines are based on the HDF5 file format [28] with monthly pandas [29] DataFrames introduced by Kreklow [30]. This enables a continuous calculation of the R-factor over all days of a month. However, for reasons of efficiency, no smooth transitions between months were implemented. The routine assumes an end of the precipitation event at the end of each month and carries out the calculation for the amount of precipitation that has fallen up to that point. Thus, long lasting nightly precipitation events may be divided into two events or one event can be classified as non-erosive due to the interruption. However, since erosivity shows a clearly pronounced maximum in the late afternoon [5], when convection is usually strongest, the inaccuracy in the calculation due to the interruption at the turn of the month was regarded as negligible.

2.3.2. R-factor Calculation Using Regression

For the erosion atlas Hesse [13], the R-factor was derived using the following regression equation from the mean long-term precipitation of the summer months May–October N_{su}:

$$R_{EA} = 0.141 \cdot N_{Su} - 1.48 \tag{3}$$

For comparison of methodologies and effects of precipitation changes, additional R-factors were calculated using this regression equation based on the hourly RW product of the radar climatology and the condensed rain gauge dataset. In conjunction with the R-factors calculated according to DIN 19708 (see Section 2.3.1) and the erosion atlas Hesse, these additional R-factor estimates based on regression allow to compare different combinations of input data and derivation methods.

All calculated R-factor derivatives are summarised in Table 1.

Since the R-factor is only important for estimating soil loss from agricultural land and not in forests or urban areas, we conducted an additional analysis of all of the abovementioned R-factor derivatives that only considered cropland areas. For this, all data pairs for which the respective RADKLIM pixel contained less than ten hectares of cropland were removed. The resulting datasets are marked by the appendix "*Agri*" in the R-factor index, e.g., $R_{YW,DIN,Agri}$.

Consequently, the analyses of this study cover three different spatial extents for which data pairs of all available datasets were created in order to enable meaningful comparisons for similar spatial scales:

(a)   all 1 km^2 pixels within Hesse ($n = 23{,}320$)

(b)   all pixels containing at least ten hectares of cropland ($n = 11{,}555$)

(c)   all rain gauge stations ($n = 110$)

In addition, the summer precipitation sums of RADKLIM and the rain gauges and their respective R-factor derivatives $R_{YW,DIN}$ and $R_{G,DIN}$ were used to determine two new regression equations. These serve to assess the following: the changes in the correlation between rainfall erosivity and precipitation sums, changes in comparison to the existing regression equation used for the erosion atlas, and the impact of sample size.

Table 1. Overview of all R-factor derivatives analysed in this study.

Name	Derivation Method	Input Dataset	Spatial Extent	n
$R_{YW,DIN}$	DIN 19708	RADKLIM YW (5 min)	All radar pixels in Hesse (1×1 km)	23,320
$R_{YW,DIN,Agri}$	DIN 19708	RADKLIM YW (5 min)	Radar pixels containing $\geq$ 10 ha of cropland	11,555
$R_{G,DIN}$	DIN 19708	Rain gauge data (5 min)	All rain gauges	110
$R_{YWG,DIN}$	DIN 19708	RADKLIM YW (5 min)	Pixels containing a rain gauge	110
R_{EA}	$0.141 \cdot N_{Su} - 1.48$	Interpolated rain gauge data (1971–2000)	1×1 km grid for Hesse	23,320
$R_{EA,Agri}$	$0.141 \cdot N_{Su} - 1.48$	Interpolated rain gauge data (1971–2000)	Grid cells containing $\geq$ 10 ha of cropland	11,555
$R_{RW,Reg}$	$0.141 \cdot N_{Su} - 1.48$	RADKLIM RW (1 h)	All radar pixels in Hesse	23,320
$R_{RW,Reg,Agri}$	$0.141 \cdot N_{Su} - 1.48$	RADKLIM RW (1 h)	Radar pixels containing $\geq$ 10 ha of cropland	11,555
$R_{G,Reg}$	$0.141 \cdot N_{Su} - 1.48$	Rain gauge data	All rain gauges in Hesse	110
$R_{RWG,Reg}$	$0.141 \cdot N_{Su} - 1.48$	RADKLIM RW (1 h)	Pixels containing a rain gauge	110
R_{EAG}	$0.141 \cdot N_{Su} - 1.48$	Interpolated rain gauge data (1971–2000)	Grid cells containing a rain gauge	110
$R_{YW,F}$	$R_{DIN} \cdot ((1.13 + 0.35) \cdot 1.05)$	RADKLIM YW (5 min)	All radar pixels in Hesse	23,320
$R_{YW,F,Agri}$	$R_{DIN} \cdot ((1.13 + 0.35) \cdot 1.05)$	RADKLIM YW (5 min)	Radar pixels containing $\geq$ 10 ha of cropland	11,555
$R_{G,F}$	$R_{DIN} \cdot 1.05$	Rain gauge data	All rain gauges	110
$R_{YWG,F}$	$R_{DIN} \cdot ((1.13 + 0.35) \cdot 1.05)$	RADKLIM YW (5 min)	Pixels containing a rain gauge	110
$R_{G,P}$	$R_{DIN} \cdot 0.7984$	Rain gauge data	All rain gauges	110

2.3.3. Application of Scaling Factors

Recent studies propose various scaling/correction factors to compensate for the temporal resolution of the input data and the differences between rain gauge and radar data. In order to be able to estimate the influence of the correction factors and to compensate for the presumed underestimation of the R-factor by the radar climatology, these factors were applied to the R-factors that were calculated according to DIN 19708.

The scaling according to [22] for the R-factors calculated from radar climatology results from

$$R_{YW,\,F} = R_{YW,DIN} \cdot ((spatial\ scaling + method\ factor) \cdot temporal\ scaling) \tag{4}$$

with

$$spatial\ scaling\ factor = 1.13;\ for\ a\ spatial\ resolution\ of\ 1\ km^2$$

$$method\ factor = 0.35$$

$$temporal\ scaling\ factor = 1.05;\ for\ a\ temporal\ resolution\ of\ 5\ min$$

For the rain gauge data, the scaling reduces to

$$R_{G,F} = R_{G,DIN} \cdot temporal\ scaling\ factor \tag{5}$$

with

$$temporal\ scaling\ factor = 1.05;\ for\ a\ temporal\ resolution\ of\ 5\ min$$

In order to include the strongly deviating temporal correction factor proposed by Panagos et al. [23], a further calculation was performed for $R_{G,DIN}$:

$$R_{G,P} = R_{G,DIN} \cdot temporal\ scaling\ factor \tag{6}$$

with

$$temporal\ scaling\ factor = 0.7984;\ for\ a\ temporal\ resolution\ of\ 5\ min$$

3. Results

3.1. Statistical Comparison of the Calculated R-Factors

The R-factor $R_{YW,DIN}$ calculated from the original unscaled RADKLIM YW product according to DIN 19708 ranges between 28.8 and 173.2 kJ/m^2 mm/h with an average value of 58.0 kJ/m^2 mm/h (see Table 2 and Figure 2). It is thus 6.4% higher on average than the values of the erosion atlas R_{EA}, whereas its range is 263.7% higher and its standard deviation is 122.7% higher. $R_{YW,DIN}$ shows thus a much higher variability than the strongly smoothed R_{EA} which was derived from spatially interpolated rainfall data using a regression equation (Equation (3)).

The R-factor calculated from the gauge dataset $R_{G,DIN}$ has an average of 80.6 kJ/m^2 mm/h, which is 47.8% higher than the average value of R_{EA} and 39% higher than the average of $R_{YW,DIN}$. At 107 of 110 stations the rain gauges show higher R-factors than the corresponding pixels of the radar climatology. The average R-factor difference for all point-pixel pairs amounts to 20.5 kJ/m^2 mm/h between $R_{YWG,DIN}$ and $R_{G,DIN}$. For the 72 stations operated by DWD, which were used for radar data adjustments, the average difference between $R_{YWG,DIN}$ and $R_{G,DIN}$ amounts to 19.1 kJ/m^2 mm/h, whereas the average difference at the 38 stations operated by HLNUG is slightly higher with 23.1 kJ/m^2 mm/h. Compared to the erosion atlas, all 110 rain gauge stations show higher R values with an average difference of 24.7 kJ/m^2 mm/h.

Table 2. Statistical summary of all R-factor derivatives.

R-factor	n	Method	Data Source	Mean	Standard Deviation	Min	Median	Max
$R_{YW,DIN}$	23,320	DIN 19708	RADKLIM	58.0	14.7	28.8	54.6	173.2
$R_{YW,DIN,Agri}$	11,555	DIN 19708	RADKLIM	54.2	12.0	28.8	52.3	146.1
$R_{G,DIN}$	110	DIN 19708	Gauges	80.6	20.6	53.4	75.3	157.2
$R_{YWG,DIN}$	110	DIN 19708	RADKLIM	60.1	15.8	31.0	57.8	104.7
R_{EA}	23,320	Regression	Erosion atlas	54.5	6.6	42.1	52.8	81.8
$R_{EA,Agri}$	11,555	Regression	Erosion atlas	52.8	5.3	42.1	51.7	81.0
$R_{RW,Reg}$	23,320	Regression	RADKLIM	53.2	6.8	32.8	53.0	77.0
$R_{RW,Reg,Agri}$	11,555	Regression	RADKLIM	51.9	6.4	32.8	52.1	71.4
$R_{G,Reg}$	110	Regression	Gauges	57.0	8.8	44.7	55.0	84.7
$R_{RWG,Reg}$	110	Regression	RADKLIM	53.1	7.8	35.9	52.4	73.0
R_{EAG}	110	Regression	Erosion atlas	55.9	8.1	45.2	53.7	81.8
$R_{YW,F}$	23,320	DIN scaled	RADKLIM	90.1	22.8	44.5	84.8	269.1
$R_{YW,F,Agri}$	11,555	DIN scaled	RADKLIM	84.2	18.6	44.5	81.3	227.0
$R_{G,F}$	110	DIN scaled	Gauges	84.6	21.6	56.1	79.1	165.1
$R_{YWG,F}$	110	DIN scaled	RADKLIM	93.4	24.6	48.0	89.8	162.7
$R_{G,P}$	110	DIN scaled	Gauges	64.4	16.4	42.6	60.1	125.5

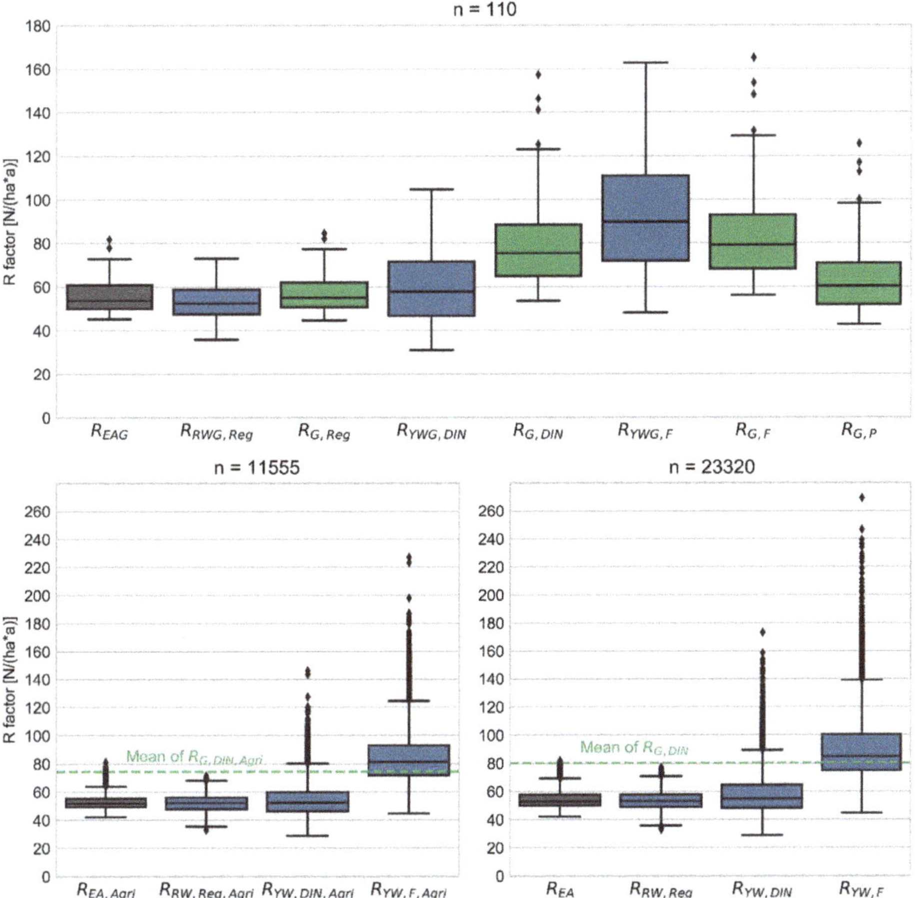

Figure 2. Boxplots of all R-factor derivatives grouped by spatial extent. In the lower subplots, the average of the rain gauges ($R_{G,DIN}$) and the rain gauges in pixels with cropland ($R_{G,DIN,Agri}$) have been added as a ground-truth reference. See Table 1 for explanation of the used abbreviations.

Using the regression equation from the erosion atlas (Equation (3)) and the RADKLIM RW product to derive $R_{RW,Reg}$ yielded comparable values as R_{EA} with a slightly lower mean and maximum, significantly lower minimum, but a slightly higher median and standard deviation. For $R_{G,Reg}$, all statistical values were slightly higher than for R_{EA} and $R_{RW,Reg}$. Consequently, before scaling, the rain gauge dataset consistently produces the highest R-factors, but the magnitude of the differences is governed by the derivation method. The input dataset has little influence on the statistical characteristics of the outcome when using a regression equation and the major differences between these regression-based derivatives are the spatial resolutions and spatial distributions (see Section 3.2). When grouping all R-factor derivatives by the calculation method—irrespective of input data and spatial extent—the mean of those R-factors derived according to DIN 19708 (without scaling) is 9.1 kJ/m^2 mm/h higher than the mean of all R-factors derived using the regression equation. Furthermore, with 15.8 kJ/m^2 mm/h, the DIN method group showed on average a 122.2% higher standard deviation than the regression method group (7.1 kJ/m^2 mm/h), which underlines the smoothing effect that can be obtained by using a regression equation instead of the event-based method according to DIN 19708. The difference between both methods is particularly well illustrated by the very steep empirical cumulative distribution functions (ECDF) of all regression-based derivatives (see Figure 3).

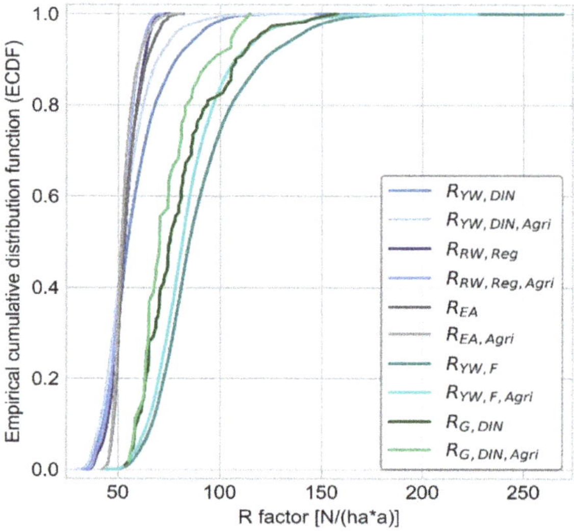

Figure 3. Empirical cumulative distribution functions (ECDF) for all spatially highly resolved R-factor derivatives. The ECDFs for the rain gauges ($R_{G,DIN}$) and the rain gauges in pixels with cropland ($R_{G,DIN,Agri}$) have been added as a ground-truth reference.

Selecting pixels with cropland leads to an average decrease of $R_{YW,DIN}$ by 3.8 kJ/m^2 mm/h (−6.6%). The minimum did not change, while the maximum decreased by 27.1 to 146.1 kJ/m^2 mm/h (see Figures 2 and 3). Taking into account only the pixels with cropland and rain gauges, the count was reduced to 54 (a total of 54 rain gauges are located in radar pixels with cropland), the average R-factor ($R_{YWG,DIN,Agri}$) decreased also by 3.8 to 56.3 kJ/m^2 mm/h and the maximum decreased by 12.5 to 92.2 kJ/m^2 mm/h. For $R_{G,DIN}$, the impact of the data selection on the statistical distribution is considerably higher due to the smaller sample size. Its average decreased by 6.1 to 74.5 kJ/m^2 mm/h, whereby the maximum decreased by 42.5 to 114 kJ/m^2 mm/h when selecting only pixels with cropland. Consequently, the removal of many high erosivity values in the mountainous regions (see Figure A1), for which the uncertainty and underestimation of the radar data is particularly high, leads to a slightly better agreement of the R-factors calculated according to DIN 19708 from RADKLIM and rain gauge data. Grouping the nine R-factor derivatives based on $R_{YW,DIN}$, $R_{RW,Reg}$ and R_{EA} by spatial extent resulted in a mean of 55.2 kJ/m^2 mm/h for all pixels of the study area, 52.9 kJ/m^2 mm/h for pixels with cropland and 56.3 kJ/m^2 mm/h for pixels with a rain gauge.

In regard to the data source, the results showing an underestimation of rainfall erosivity by the radar climatology compared to rain gauge data are in line with the outcomes of Fischer et al. (2018), thus the application of the proposed correction factors was considered to be useful and necessary. After scaling, the R-factors of the radar climatology and rain gauges correspond much better (see Figures 2 and 3). The difference between the two datasets shifts in favour of the radar climatology, since on average $R_{YW,F}$ is 8.8 kJ/m^2 mm/h higher than $R_{G,F}$ (see Table 2). In comparison to R_{EA}, both R-factors were significantly higher after scaling. On average, $R_{YW,F}$ was 65.3% higher and $R_{G,F}$ was 58.5% higher than the R-factor R_{EA} of the erosion atlas. Although the correction factor proposed by Panagos et al. [23] reduces the R-factor to a level close to $R_{YW,DIN}$, $R_{G,P}$ still showed an 18.2% higher mean than R_{EA}. Irrespective of the dataset used for derivation and the application of correction procedures, an increase of the R-factor compared to R_{EA} can thus be determined without doubt.

3.2. Spatial Distribution

For erosion control applications at a federal state scale which aim to identify regions with a particularly high risk of erosion, the spatial distribution of rainfall erosivity is actually more relevant than the absolute erosivity values. The lowest values of $R_{YW,DIN}$ occur in the north of Hesse, around the West Hesse Depression, in an area for which no radar measurements were available during some months of the years 2007 and 2014 due to radar hardware upgrades. The average value of the annual R-factor without these two years shows that the minimum is nevertheless located in this area. This is therefore in accordance with the R-factor R_{EA} (calculation based on regression), which also shows a minimum in this area (see Figure 4). The areas of relatively low R-factors northwest of Fulda and in the Upper Rhine Plain correspond well in both datasets, too. In the north-east of Hesse, however, the newly calculated R-factor $R_{YW,DIN}$ showed slightly lower erosivity over a large area with a similar spatial distribution. Both datasets showed an increase of the R-factor with increasing terrain height, whereby $R_{YW,DIN}$ showed significantly higher values over a large area, especially in the Odenwald, Taunus, Westerwald and at Vogelsberg. However, at Vogelsberg, a weakness of the radar climatology to correctly quantify precipitation at higher altitudes was evident as the increase of the R-factor in the lower slope areas was considerably higher than in the summit area. In the area of Wetterau, a negative spoke of the Frankfurt Radar was clearly visible in $R_{YW,DIN}$ and all other R-factors derived from RADKLIM. Still, in this area an increase of the R-factor compared to the erosion atlas can be seen in most of the grid cells, in some places even up to 45% (see Figure 5a).

The scaling is able to compensate for the underestimation of the R-factor by the radar climatology, which becomes particularly obvious in the northern parts of Hesse where the difference to the erosion atlas shows mostly positive values except for a few single pixels (see Figure 5c). Moreover, Figure 5b shows much better conformity with $R_{G,DIN}$ in the entire study area, which has already been indicated in the previous section.

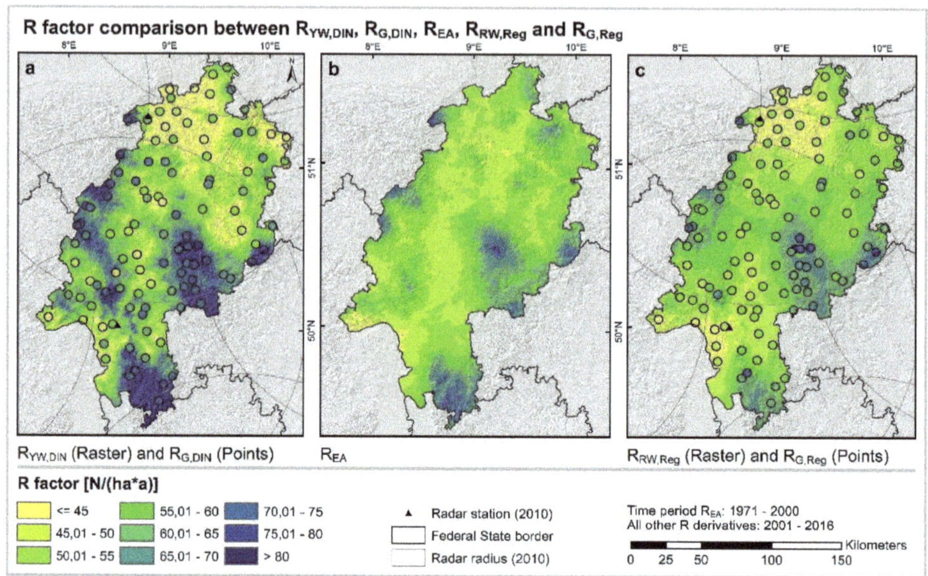

Figure 4. R-factor comparison between $R_{YW,DIN}$, $R_{G,DIN}$ (**a**), R_{EA} (**b**), $R_{RW,Reg}$ and $R_{G,Reg}$ (**c**).

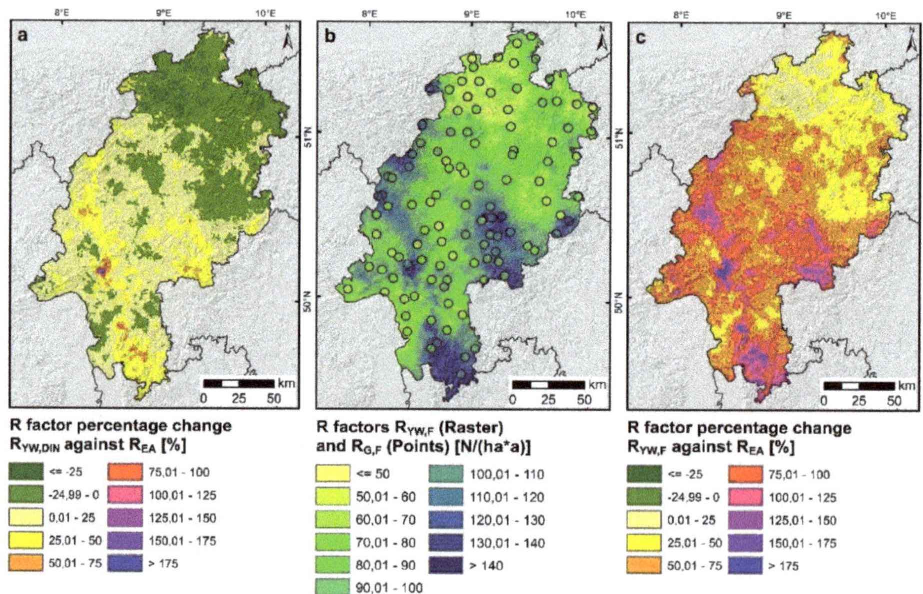

Figure 5. R-factor percentage change of $R_{YW,DIN}$ against R_{EA} (**a**), scaled R-factors $R_{YW,F}$ and $R_{G,F}$ (**b**) and percentage change of $R_{YW,F}$ against R_{EA} (**c**).

3.3. Derivation of Updated Regression Equations

The statistical comparisons in Section 3.1 show consistently lower values for all R-factors derived by means of regression. Besides the method itself and the input data source, the observation time period of the data used for the derivation of the regression equation might play a role due to climate change, which is why we derived an updated regression equation for comparison.

The new regression equations derived from the rain gauge data and the radar climatology both show a strong correlation between summer precipitation and R-factor. The fitted regression line has a considerably higher slope than the original one used for the erosion atlas (Equation (3)) (see Figure 6). Some data points of $R_{YW,DIN}$, which are mainly located in the area of the radar gap in northern Hesse, are still below the regression line from the erosion atlas. For $R_{G,DIN}$, however, all data points are above. Consequently, for the period 2001–2016, the regression equation used in the erosion atlas provides a value deviating from the R-factor according to DIN 19708 for all of the rain gauges.

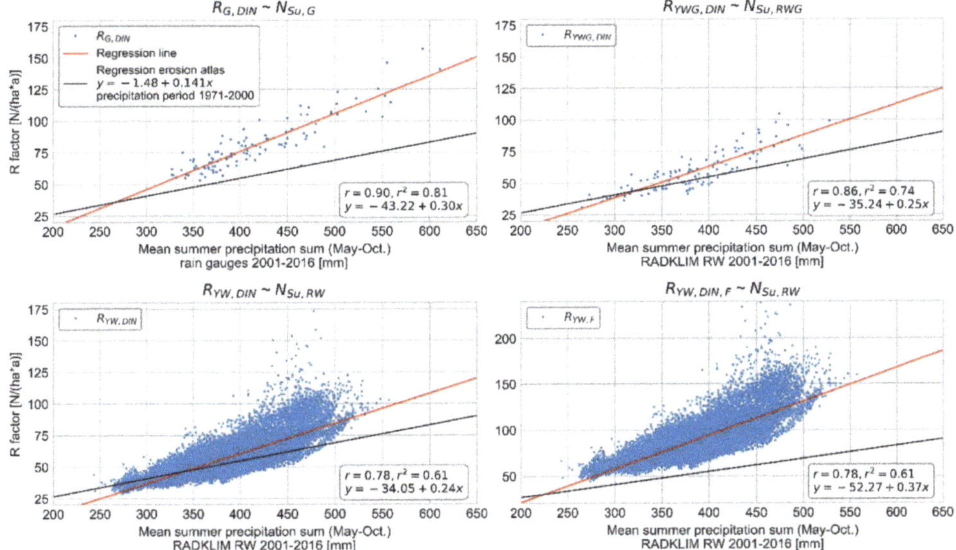

Figure 6. Comparison of regression models between different R-factors and the respective mean summer precipitation sums.

When considering the newly derived regression equations from the radar climatology, it is striking that the equations for the entire data set and the pixels at the rain gauge locations are almost identical (see Figure 6). Consequently, the spatial distribution of the rain gauge locations can be regarded as very representative for mapping the overall distribution of rainfall erosivity in Hesse.

Another striking difference with regard to the sample size, however, is a series of several very high values of $R_{YW,DIN}$ in the range between 400 and 500 mm summer precipitation. These are only included in the R-factor of the entire radar climatology dataset, but are not significantly reflected in the regression due to their relatively small number. Therefore, it can be assumed that extraordinarily intensive individual events have a strong impact due to the comparatively short time series. These events could only be detected by the high spatial resolution of the radar climatology and are not included in the rain gauge dataset.

Using the new regression equation derived from the rain gauges ($R = -43.22 + 0.3\ N_{Su}$) with the summer precipitation sums of the RADKLIM RW product for the federal state of Hesse leads to a R-factor value range between 29.7 and 123.9 kJ/m^2 mm/h with an average of 73.2 kJ/m^2 mm/h. It has thus a significantly lower maximum than all event-based R-factor derivatives. Its mean value is slightly lower than that of $R_{G,DIN}$ (80.6 kJ/m^2 mm/h) due to the slight overall underestimation of precipitation by the radar climatology, and lies approximately in the centre between the averages of $R_{YW,DIN}$ (58 kJ/m^2 mm/h) and the corrected $R_{YW,F}$ (90.1 kJ/m^2 mm/h).

4. Discussion

An evaluation of the radar climatology dataset revealed that it slightly underestimates precipitation quantities. This underestimation is particularly pronounced at higher altitudes and at high rainfall intensities [31]. In particular, the latter plays a decisive role for rainfall erosivity since rainfall intensity is directly linked to the kinetic energy of rainfall and, thus, its ability to detach soil particles. The assumption that the R-factor calculated from the radar climatology according to DIN 19708 without input data correction is too low could be confirmed by comparing it with the R-factor derived from the rain gauge dataset. However, irrespective of the dataset used for derivation, the spatial scale and the application of correction procedures, an increase of the R-factor compared to R_{EA} which is currently used in the technical information system erosion atlas Hesse [13] can be determined without doubt. This result highlights the need of updated R-factor methods for consultation and planning in Hesse.

The R-factors calculated by the regression equation from the erosion atlas, the summer precipitation sums of radar climatology, and rain gauges showed only slightly higher average values than the erosion atlas. Considering the significant differences to the R-factor derivations according to DIN 19708, this indicates that the regression equation used for the erosion atlas, which was derived from precipitation data of the 1960s, 70s and 80s, is no longer representative of the current climate conditions. Apparently, although there has only been a small increase in summer precipitation, there is a change in the heavy rainfall characteristics and/or in the relationship between erosive rainfall and total precipitation amount. This observation is in line with the projected changes in precipitation characteristics with regard to climate change. For most of Europe, it is expected that precipitation will increase during winter and decrease during summer [32,33]. Furthermore, the number of wet days is expected to decrease, whereas the intensity and the return levels of daily precipitation events will increase [12,32,34,35]. The combination of increasingly intense heavy rainfall and the reduced water infiltration capacity of dry soils is expected to amplify the risk of floods [36] and is also very likely to increase soil erosion. These observations indicate that the validity of regression equations for R-factor calculation might decrease, particularly if mean summer precipitation sums are used instead of mean annual sums. An additional influencing factor for higher R-factors calculated from rain gauge data could be the better recording of intensity peaks by more accurate modern rain gauges as opposed to the less accurate rain gauges used to collect the data for the 1971–2000 dataset [37].

Despite the discussed limitations, the regression-based approach has the advantage that it is much easier to apply in practice than the method according to DIN 19708, which is computationally much more expensive, especially when using it on spatially highly resolved data such as the radar climatology. Moreover, the use of a regression equation with precipitation sums always leads to a certain smoothing and is thus more robust against outliers than the event-based method when only comparatively short precipitation time series are available. However, as our results have clearly shown, the regression approach also requires frequent updates of the equations and hence a certain maintenance of the methodology. Obviously, updates to the equations rely on the availability of rain gauge data. For Germany, this is not a major issue anymore since temporally highly resolved rain gauge data are freely available at the DWD Open Data Portal. In other countries, however, this may be a greater obstacle.

With regard to the scaling of the R-factors which was proposed in recent studies [22,23], it should be noted that a correction that increases the RADKLIM R-factor is undoubtedly necessary to compensate for the systematic underestimation of precipitation data obtained from radar climatology. However, the degree of correction is difficult to estimate due to a lack of reference. If the scaled R-factor of the rain gauge dataset $R_{G,F}$ is regarded as a correct reference for validation, the correction applied for $R_{YW,F}$ and $R_{YWG,F}$ appears somewhat too high, especially when looking at Figure 2. When considering the identical sample size and the largely consistent location of the point-pixel data pairs of $R_{YWG,F}$, the advantage of the radar and the fact that more events tend to be recorded hardly matters. However, the median of $R_{YWG,F}$ almost corresponds to the third quartile of $R_{G,F}$. Here, a direct transferability of the correction factors, which were derived from a four-year series of measurements of 12 rain gauges within one square kilometre in Bavaria [22], may be limited. Further research efforts and measurements to extend these time series and derive correction factors of higher spatial representativity from more than one single raster cell would have the potential to significantly reduce the uncertainty when using radar climatology data—not only for rainfall erosivity estimation but for applications related to heavy rainfall in general.

In contrast, the scaling according to Panagos et al. [23] to compensate for the temporal resolution of the input data provides very questionable results. Taking into account the conducted plausibility check of the radar climatology and the comparisons with the rain gauge data by Kreklow et al. [31], an underestimation of the R-factor by the radar data is clearly demonstrated. Since the correction factor proposed by Panagos et al. [23] reduces the R-factor of the rain gauges to a level almost identical to that of the radar climatology, a correction factor that is too low must be assumed. The correction factor does not appear to be representative for Hesse, due to the fact that its derivation is based on a rain gauge dataset for the whole of Europe and equally includes data from maritime, continental, temperate, subpolar and Mediterranean climates. Already for the two neighbouring countries Austria and Italy, Fiener et al. [38] found significant differences in the magnitude and monthly distribution of the R-factor, which indicates a lack of spatial representativity of the temporal scaling factor proposed by Panagos et al. [23]. Such representativity issues have been subject to discussion between the authors [4,38,39]. In addition, the original methodology for the calculation of the R-factor is based on continuous precipitation recordings, which were aggregated to intervals of constant intensity [1]. Consequently, a temporal resolution of 1 min as a lowest reference chosen by Fischer et al. [22] is much closer to the original method than the reference resolution of 30 min used by Panagos et al. [23]. The much lower reference resolution used by Panagos et al. [23] thus explains the significantly lower temporal correction factor compared to the factor proposed by Fischer et al. [22].

With regard to practical application, it is recommended that the R-factor map currently used in the erosion atlas should be updated. Our results show that the first and most important step is to use more recent precipitation data for derivation, which are more representative under current climate conditions. Obviously, using the event-based method according to DIN 19708 with radar climatology, which was proposed by Auerswald et al. [6], provides the R-factor with the highest spatial detail, but it may be locally biased by some extreme rainfall events or radar artefacts which are not balanced out in the comparatively short radar time series. Moreover, a correction of R-factors derived from radar climatology according to DIN 19708 is necessary to compensate for underestimation, but the level of correction required is still subject to discussion. However, the radar climatology time series is still considerably longer than the time series used for deriving the original regression equations by Sauerborn [9], of which one was also used in the erosion atlas Hesse. Consequently, during a transition period, the most robust and easy-to-use approach to obtain updated R-factors is by using an updated regression equation derived from recent rain gauge data with summer precipitation sums calculated from radar climatology data. On the one hand, this approach accounts for climate change by increasing the R-factors according to reliable rain gauge observations. On the other hand, it makes use of the high spatial resolution of radar data and comprises a certain smoothing, since precipitation sums are less biased by local extreme events and by the underestimation of high rainfall intensities by weather radar in comparison to the event-based R-factors derived according to DIN 19708. Moreover, due to less snowfall and thus fewer uncertainties in the radar climatology data during the summer half-year, the use of radar-based summer precipitation sums increases the robustness of the recommended method compared to the use of radar-based annual precipitation sums.

Due to the central location of Hesse within Germany, the recommended updated regression equation based on rain gauge data for Hesse ($R = -43.22 + 0.3\ N_{Su}$) has a high transferability for most of Germany. However, for federal states in northern and eastern Germany which have a more maritime or continental climate, regional regression equations should be calculated from recent local rain gauge data.

5. Conclusions

In this study, we compared several derivation approaches for the R-factor of the USLE and evaluated the performance of radar climatology and rain gauge data for different methods and three spatial extents. Moreover, two correction factors proposed in other studies were tested and updated regression equations were derived for the German federal state of Hesse.

Regarding the three hypotheses put forward at the beginning of this study, our results can be summarised as follows:

1. The newly derived R-factors from rain gauge and radar climatology data are indeed higher than the R-factors from existing calculations due to climate and weather changes. For the study period of 2001–2016, the regression equation used in the erosion atlas provides a lower R-factor than DIN 19708 for all of the rain gauges.

2. The contradiction between the theoretically higher R-factor of the radar climatology due to the more complete recording of all erosive rainfall events on the one hand and the underestimation of the R-factor due to the attenuation of intensity peaks, on the other hand, could be established. In the spatial average as well as when looking at the point-pixel data pairs, which largely eliminates the influence of the higher spatial resolution of the radar climatology data, the R-factors of the rain gauges are significantly higher. However, when looking at the entire radar data set, some strikingly high R-factor values, which were not captured by the rain gauges, become apparent. Due to their comparatively small number, however, they have no significant influence on the spatial mean value. In addition, these extraordinary high R-factors can also be a result of very intensive rainfall events in the comparatively short observation period that might be smoothed by prolonging the radar climatology dataset. The correction of the R-factors according to Fischer et al. [22] provides an improvement of the results for the radar climatology, although a possible overcorrection cannot be excluded.

3. The spatial distribution of the newly calculated R-factor according to DIN 19708 and that from the erosion atlas show a relatively good conformity with minima and maxima in similar regions as well as a consistent mapping of a relief dependency. In the northeast of Hesse, the R-factor calculated from the uncorrected radar climatology according to DIN 19708 shows comparatively lower values than the erosion atlas. In contrast, it also shows large areas of higher R-factor values than the erosion atlas, especially in the ridges of the low mountain ranges and in the central lowland areas of Hesse, for example, the Wetterau. The updated regression equations, which are almost identical for all radar pixels and the point-pixel data pairs, indicate that the rain gauge locations are very representative for mapping the overall spatial distribution of rainfall erosivity in the study area.

The results of this study clearly indicate that the R-factor map currently used in the erosion atlas should be updated. For a transition period until the radar climatology time series is long enough to compensate for bias from extraordinarily intensive rainfall events, it is recommended to apply a new regression equation derived from recent rain gauge measurements with summer precipitation sums calculated from radar climatology data.

With the progressive improvement of the data basis (time series, quality and correction), however, radar climatology data will be further incorporated into operational applications such as risk management and erosion consulting.

Author Contributions: Conceptualisation, J.K. and B.T.; methodology, J.K.; software, J.K.; validation, J.K., B.T. and B.S.-K.; formal analysis, J.K.; data curation, J.K.; writing—original draft preparation, J.K.; writing—review and editing, B.S.-K., B.T., J.K. and K.F.; visualisation, J.K.; funding acquisition, B.T. and K.F. All authors have read and agreed to the published version of the manuscript.

Funding: This research was funded by the Hessian Agency for Nature Conservation, Environment and Geology (HLNUG) within the project "KLIMPRAX–Starkregen," working package 1.4.

Acknowledgments: The authors are grateful to DWD for providing open access radar and rain gauge data. Thank you to Erik Jähnke and Angie Faust for proofreading. Finally, the authors would like to thank the four anonymous reviewers for their proficient and constructive comments which helped improve the manuscript.

Conflicts of Interest: The authors declare no conflict of interest. The funders had no role in the design of the study; in the collection, analyses, or interpretation of data; in the writing of the manuscript, or in the decision to publish the results.

Appendix A

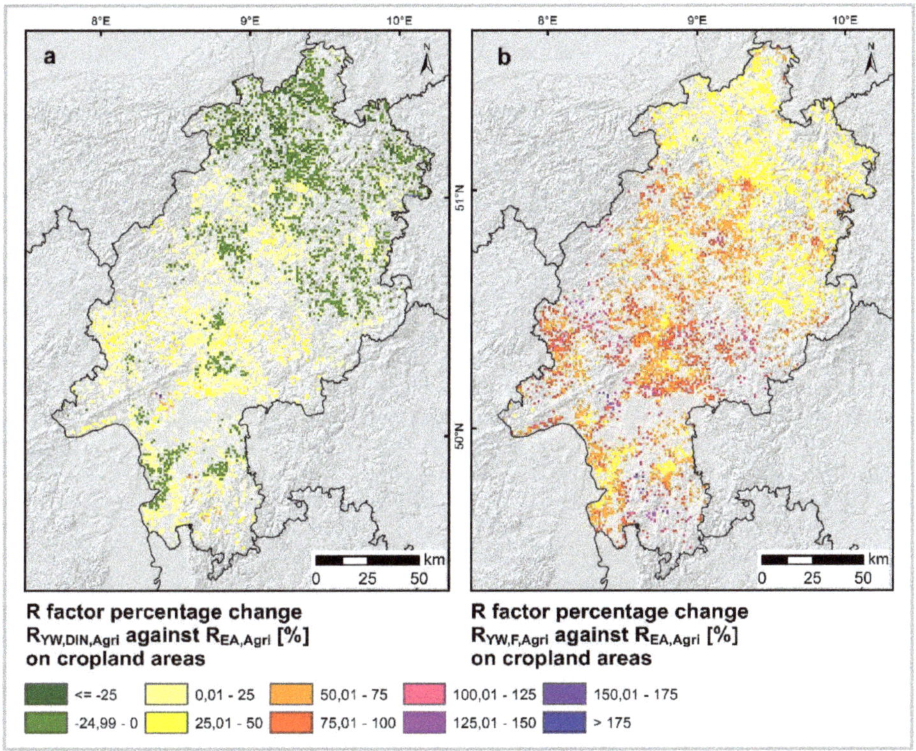

Figure A1. R-factor percentage change of $R_{YW,DIN,Agri}$ against $R_{EA,Agri}$ (**a**), and percentage change of $R_{YW,F,Agri}$ against $R_{EA,Agri}$ (**b**).

References

1. Wischmeier, W.H.; Smith, D.D. *Predicting Rainfall Erosion Losses. A Guide to Conservation Planning*; Agriculture handbook number 537; U.S. Department of Agriculture: Washington, DC, USA, 1978.
2. Schwertmann, U.; Vogl, W.; Kainz, M. *Bodenerosion durch Wasser. Vorhersage des Abtrags und Bewertung von Gegenmaßnahmen, 2. Aufl.*; Ulmer: Stuttgart, Germany, 1990; ISBN 3800130882.
3. Deutsches Institut für Normung. *DIN 19708. Bodenbeschaffenheit—Ermittlung der Erosionsgefährdung von Böden durch Wasser mithilfe der ABAG*; Normenausschuss Wasserwesen im DIN: Berlin, Germany, 2005; (19708).
4. Panagos, P.; Ballabio, C.; Borrelli, P.; Meusburger, K.; Klik, A.; Rousseva, S.; Tadic, M.P.; Michaelides, S.; Hrabalikova, M.; Olsen, P.; et al. Rainfall erosivity in Europe. *Sci. Total Environ.* **2015**, *511*, 801–814. [CrossRef] [PubMed]
5. Fischer, F.; Hauck, J.; Brandhuber, R.; Weigl, E.; Maier, H.; Auerswald, K. Spatio-temporal variability of erosivity estimated from highly resolved and adjusted radar rain data (RADOLAN). *Agric. For. Meteorol.* **2016**, *223*, 72–80. [CrossRef]
6. Auerswald, K.; Fischer, F.K.; Winterrath, T.; Brandhuber, R. Rain erosivity map for Germany derived from contiguous radar rain data. *Hydrol. Earth Syst. Sci.* **2019**, *23*, 1819–1832. [CrossRef]

7. Tetzlaff, B.; Friedrich, K.; Vorderbrügge, T.; Vereecken, H.; Wendland, F. Distributed modelling of mean annual soil erosion and sediment delivery rates to surface waters. *CATENA* **2013**, 13–20. [CrossRef]

8. Deutsches Institut für Normung. *DIN 19708. Bodenbeschaffenheit—Ermittlung der Erosionsgefährdung von Böden durch Wasser mit Hilfe der ABAG, 2017–08*; Beuth Verlag GmbH: Berlin, Germany, 2017.

9. Sauerborn, P. Die Erosivität der Niederschläge in Deutschland. Ein Beitrag zur quantitativen Prognose der Bodenerosion durch Wasser in Mitteleuropa. Ph.D. Thesis, Inst. für Bodenkunde, Bonn, Germany, 1994.

10. Elhaus, D. Erosionsgefährdung. Informationen zu den Auswertungen der Erosionsgefährdung durch Wasser, Germany. 2015. Available online: https://www.gd.nrw.de/zip/erosionsgefaehrdung.pdf (accessed on 11 March 2020).

11. Burt, T.; Boardman, J.; Foster, I.; Howden, N. More rain, less soil: Long-term changes in rainfall intensity with climate change. *Earth Surface Process. Landf.* **2016**, *41*, 563–566. [CrossRef]

12. Fiener, P.; Neuhaus, P.; Botschek, J. Long-term trends in rainfall erosivity–analysis of high resolution precipitation time series (1937–2007) from Western Germany. *Agric. For. Meteorol.* **2013**, *171-172*, 115–123. [CrossRef]

13. Friedrich, K.; Schmanke, M.; Tetzlaff, B.; Vorderbrügge, T. Erosionsatlas Hessen. In Proceedings of the Tagungsband d. Jahrestagung der DBG/BGS, Kommission VI, "Erd-Reich und Boden-Landschaft", Bern, Switzerland, 24–27 August 2019. DBG/BGS, Ed..

14. Hessisches Landesamt für Naturschutz, Umwelt und Geologie (HLNUG). BodenViewer Hessen. Available online: http://bodenviewer.hessen.de/mapapps/resources/apps/bodenviewer/index.html?lang=de (accessed on 21 January 2020).

15. Rogler, H.; Schwertmann, U. Erosivität der Niederschläge und Isoerodentkarte Bayerns. *Z. für Kult. und Flurberein.* **1981**, *22*, 99–112.

16. Hessisches Landesamt für Naturschutz, Umwelt und Geologie (HLNUG). R-Faktor. Available online: https://www.hlnug.de/themen/boden/auswertung/bodenerosionsbewertung/bodenerosionsatlas/r-faktor (accessed on 16 January 2020).

17. Donat, M.G.; Alexander, L.V.; Yang, H.; Durre, I.; Vose, R.; Dunn, R.J.H.; Willett, K.M.; Aguilar, E.; Brunet, M.; Caesar, J.; et al. Updated analyses of temperature and precipitation extreme indices since the beginning of the twentieth century: The HadEX2 dataset. *J. Geophys. Res. Atmos.* **2013**, *118*, 2098–2118. [CrossRef]

18. Winterrath, T.; Brendel, C.; Hafer, M.; Junghänel, T.; Klameth, A.; Lengfeld, K.; Walawender, E.; Weigl, E.; Becker, A. Radar climatology (RADKLIM) version 2017.002 (YW); gridded precipitation data for Germany. Available online: https://search.datacite.org/works/10.5676/dwd/radklim_yw_v2017.002 (accessed on 25 June 2019).

19. Winterrath, T.; Brendel, C.; Hafer, M.; Junghänel, T.; Klameth, A.; Walawender, E.; Weigl, E.; Becker, A. *Erstellung einer radargestützten Niederschlagsklimatologie*; Berichte des Deutschen Wetterdienstes No. 251; 2017; Available online: https://www.dwd.de/DE/leistungen/pbfb_verlag_berichte/pdf_einzelbaende/251_pdf.pdf?__blob=publicationFile&v=2 (accessed on 29 March 2019).

20. Kreklow, J.; Tetzlaff, B.; Burkhard, B.; Kuhnt, G. Radar-Based Precipitation Climatology in Germany—Developments, Uncertainties and Potentials. *Atmosphere* **2020**, *11*. [CrossRef]

21. Bronstert, A.; Agarwal, A.; Boessenkool, B.; Crisologo, I.; Fischer, M.; Heistermann, M.; Köhn-Reich, L.; López-Tarazón, J.A.; Moran, T.; Ozturk, U.; et al. Forensic hydro-meteorological analysis of an extreme flash flood: The 2016-05-29 event in Braunsbach, SW Germany. *Sci. Total Environ.* **2018**, 977–991. [CrossRef]

22. Fischer, F.K.; Winterrath, T.; Auerswald, K. Temporal- and spatial-scale and positional effects on rain erosivity derived from point-scale and contiguous rain data. *Hydrol. Earth Syst. Sci.* **2018**, *22*, 6505–6518. [CrossRef]

23. Panagos, P.; Borrelli, P.; Spinoni, J.; Ballabio, C.; Meusburger, K.; Beguería, S.; Klik, A.; Michaelides, S.; Petan, S.; Hrabalíková, M.; et al. Monthly Rainfall Erosivity: Conversion Factors for Different Time Resolutions and Regional Assessments. *Water* **2016**, *8*, 119. [CrossRef]

24. Semmel, A. Hessisches Bergland. In *Physische Geographie Deutschlands*, 2nd ed.; Liedtke, H., Marcinek, J., Eds.; Justus Perthes Verlag: Gotha, Germany, 1995; pp. 340–352. ISBN 3-623-00840-0.

25. Winterrath, T.; Brendel, C.; Hafer, M.; Junghänel, T.; Klameth, A.; Lengfeld, K.; Walawender, E.; Weigl, E.; Becker, A. Radar climatology (RADKLIM) version 2017.002 (RW); gridded precipitation data for Germany. Available online: https://search.datacite.org/works/10.5676/dwd/radklim_rw_v2017.002 (accessed on 25 June 2019).

26. Deutscher Wetterdienst Open Data Portal. Rain Gauge Precipitation Observations in 1-Minute Resolution. Available online: https://opendata.dwd.de/climate_environment/CDC/observations_germany/climate/1_minute/precipitation/ (accessed on 20 February 2020).

27. Kreklow, J.; Tetzlaff, B.; Kuhnt, G.; Burkhard, B. A Rainfall Data Intercomparison Dataset of RADKLIM, RADOLAN, and Rain Gauge Data for Germany. *Data* **2019**, *4*. [CrossRef]

28. The HDF Group. Hierarchical Data Format. Available online: https://portal.hdfgroup.org (accessed on 18 December 2018).

29. Mckinney, W. pandas: A Foundational Python Library for Data Analysis and Statistics. *Python High-Perform. Sci. Comput.* **2011**, *14*.

30. Kreklow, J. Facilitating radar precipitation data processing, assessment and analysis: A GIS-compatible python approach. *J. Hydroinformatics* **2019**, *21*, 652–670. [CrossRef]

31. Kreklow, J.; Tetzlaff, B.; Burkhard, B.; Kuhnt, G. Radar-Based Precipitation Climatology in Germany - Developments, Uncertainties and Potentials. Available online: https://www.preprints.org/manuscript/202002.0044/v1 (accessed on 10 February 2020).

32. Field, C.B.; Barros, V.R.; Dokken, D.J.; Mach, K.J.; Mastrandrea, M.D. *Climate Change 2014 – Impacts, Adaptation and Vulnerability: Part A: Global and Sectoral Aspects*; Working Group II Contribution to the IPCC Fifth Assessment Report; IPCC: Geneva, Switzerland, 2015. Available online: https://www.cambridge.org/core/books/climate-change-2014-impacts-adaptation-and-vulnerability-part-a-global-and-sectoral-aspects/1BE4ED76F97CF3A75C64487E6274783A (accessed on 7 February 2018).

33. Giorgi, F.; Bi, X.; Pal, J. Mean, interannual variability and trends in a regional climate change experiment over Europe. II: climate change scenarios (2071–2100). *Clim. Dyn.* **2004**, *23*, 839–858. [CrossRef]

34. Semmler, T.; Jacob, D. Modeling extreme precipitation events—A climate change simulation for Europe. *Glob. Planet. Chang.* **2004**, *44*, 119–127. [CrossRef]

35. Frei, C.; Schöll, R.; Fukutome, S.; Schmidli, J.; Vidale, P.L. Future change of precipitation extremes in Europe: Intercomparison of scenarios from regional climate models. *J. Geophys. Res.* **2006**, *111*, 224. [CrossRef]

36. Kyselý, J.; Beranová, R. Climate-change effects on extreme precipitation in central Europe: Uncertainties of scenarios based on regional climate models. *Theor. Appl. Clim.* **2009**, *95*, 361–374. [CrossRef]

37. Quirmbach, M.; Einfalt, T.; Langstädtler, G.; Janßen, C.; Reinhardt, C.; Mehlig, B. Extremwertstatistische Untersuchung von Starkniederschlägen in NRW (ExUS). *Korresp. Abwasser und Abfall* **2013**, *60*, 591–599.

38. Fiener, P.; Auerswald, K. Comment on "The new assessment of soil loss by water erosion in Europe" by Panagos et al. (Environmental Science & Policy 54 (2015) 438–447). *Environ. Sci. Policy* **2016**, *57*, 140–142. [CrossRef]

39. Panagos, P.; Meusburger, K.; Ballabio, C.; Borrelli, P.; Beguería, S.; Klik, A.; Rymszewicz, A.; Michaelides, S.; Olsen, P.; Tadić, M.P.; et al. Reply to the comment on "Rainfall erosivity in Europe" by Auerswald et al. *Sci. Total Environ.* **2015**, *532*, 853–857. [CrossRef] [PubMed]

Article

Impact of Land Cover Change Due to Armed Conflicts on Soil Erosion in the Basin of the Northern Al-Kabeer River in Syria Using the RUSLE Model

Hussein Almohamad

Department of Geography, College of Arabic Language and Social Studies, Qassim University, Buraydah 51452, Saudi Arabia; H.Almohamad@qu.edu.sa

Received: 17 October 2020; Accepted: 22 November 2020; Published: 26 November 2020

Abstract: Due to armed conflicts, the sudden changes in land cover are among the most drastic and recurring shocks on an international scale, and thus, have become a major source of threat to soil and water conservation. Throughout this analysis, the impact of land cover change on spatio-temporal variations of soil erosion from 2009/2010 to 2018/2019 was investigated using the Revised Universal Soil Loss Equation (RUSLE) model. The goal was to identify the characteristics and variations of soil erosion under armed conflicts in the basin of the Northern Al-Kabeer river in Syria. The soil erosion rate is 4 t ha^{-1} year^{-1} with a standard deviation of 6.4 t ha^{-1} year^{-1}. In addition, the spatial distribution of erosion classes was estimated. Only about 10.1% of the basin is subject to a tolerable soil erosion rate and 79.9% of the study area experienced erosion at different levels. The soil erosion area of regions with no changes was 10%. The results revealed an increase in soil erosion until 2013/2014 and a decrease during the period from 20013/2014 to 2018/2019. This increase is a result of forest fires under armed conflict, particularly toward the steeper slopes. Coniferous forest as well as transitional woodland and scrub are the dominant land cover types in the upper part of the basin, for which the average post-fire soil loss rates (caused by factor C) were 200% to 800% higher than in the pre-fire situation. In the period from 2013/2014 to 2019/2020, soil erosion was mitigated due to a ceasefire that was agreed upon after 2016, resulting in decreased human pressures on soils in contested areas. By comparing 2009/2010 (before war) with 2018/2019 (at the end of the war stage), it can be concluded that the change in C factors slowed down the deterioration trend of soil erosion and reduced the average soil erosion rate in more than half of the basin by about 10–75%. The area concerned is located in the western part of the basin and is relatively far from the centers of armed conflicts. In contrast, the areas with increased soil erosion by about 60–400% are situated in the northeast and east, with shorter distances to armed conflict centers. These findings can be explained by forest fires, after which the burned forests were turned into agricultural land or refugee camps and road areas. Understanding the complex biophysical and socio-economic interactions of exposure to land loss is a key to guarantee regional environmental protection and to conserve the ecological quality of soil and forest systems.

Keywords: soil erosion; RUSLE; land cover change; armed conflict; Northern Al-Kabeer river Syria

---

1. Introduction

Soil erosion and its consequent land degradation in the marginal lands of the Mediterranean is a serious problem, as it directly affects the environment and sustainable development [1–5]. The eastern Mediterranean area is especially vulnerable to erosion, since it is subject to long dry spells accompanied by heavy rainfall falling on steep slopes with weak soil [2,6]. Erosion levels are predicted to rise in the 21st century due to climate change [7]. While soil erosion results from the interplay between soil erodibility and rainfall erosivity factors, maladjusted human activities such as slope agriculture, deforestation, expansion of urban areas and highways as well as overgrazing exacerbate the issue [8,9]. Erosion has been growing due to land cover change and unsustainable land cover transition management activities. The increase of wildfires are one of the most important sources of land-use change in the Middle East [10,11]. This problem has been intensified by the emergence of numerous armed conflicts. Syrian ecosystems are often vulnerable to wildfires. Accelerated deforestation cycles due to wildfires in these ecosystems are a significant limiting factor in their sustainability. In the mountainous regions of Syria, where dense forests cover much of the area, forest fires are the worst human intrusion [12]. These fires have increased dramatically over the past 10 years along with changes in land use due to the Civil War. Many regions were subject to extensive fires in conflict areas, which led to the destruction of crops and forests [13,14]. In 2010, Syria had 116 kha of natural forest, covering over 0.62% of its land area. From 2010 to 2019, Syria lost 16.18 kha of forest, which is equivalent to a 14% decrease in tree cover since 2010 [15]. Although fires are a natural occurrence that can have beneficial effects on the revitalization of vegetation, the high frequency and intensity of these fires have caused the destruction of forest habitats. Warfare and armed conflicts are among the most dramatic shocks, and can have tremendous impacts on communities and, hence, on land systems. Although widely believed to have significant consequences on land-use transition and soil erosion, research into how military wars influence land-use decisions and soil erosion as well as its trends is rare. Studies have indicated that the impact of armed conflicts on forests and land use can be broadly summarized as reaching in two directions: firstly, this interaction entails a shift in resources or land usages caused by the intensified or inappropriate utilization of natural resources throughout the war. Examples of this include intensified timber and fuelwood use close to refugee settlements [16,17], a lack of habitat, where protected areas are shelters for terrorists, or areas being left unguarded, leading to stolen natural resources throughout periods of fighting [18]. Secondly, the interaction between armed conflicts and forests and land use suggests that biodiversity and ecosystems may also benefit from military conflicts. This would, for example, be the case in the dispute areas where human activity is reduced or in landmine-contaminated areas (e.g., the demilitarized area between North and South Korea) [19,20]. Forest recovery has also been associated with more complex social and economic shifts tied to civil war and global trade in El Salvador [17,21].

The Syrian coastal mountain fires are one of the most important sources of land-use change in the Middle East. Erosion due to the wildfire impact is of great concern to land managers. Research on soil degradation due to water erosion in Syria is restricted to a few studies focused on experimental and modeling studies. For instance, Karydas et al. [22] found that the runoff coefficient was three times greater on burned watershed than the unburned part. In an experimental study, in Ein Al-Jaouz/Tartous the soil erosion rate was 0.1 t ha^{-1} year^{-1} for the area that was not burned and 7.2 t ha^{-1} year^{-1} when burned. Reference experimental sites in the Syrian coastal mountains region found that the soil loss rates ranged between 9 and 56.5 t ha^{-1} year^{-1} in the burned forest compared to 1.4 and 15 t ha^{-1} year^{-1} before burning and 165 t ha^{-1} year^{-1} when the slope was 45% in the agricultural system [23]. Mohammed et al. (2020) found that the average soil erosion rate ranged between 1.4 and 7.4 t ha^{-1} year^{-1} in the burned forest in five locations in the coastal region of Syria [24]. Abdo and Salloum (2017) found a high soil erosion rate of 4% in the Alqerdaha basin of north-west Syria [25]. Barakat et al. (2014) [2] produced a soil erosion risk map based on the COoRdination of Information on the Environment (CORINE) model for the middle and down the basin of the Northern Al-Kabeer River and showed that 2% of the study area had a high-risk soil erosion rate, 22% a moderate-risk soil erosion and 75% a low-risk soil erosion.

The impact of the war on erosion has not been addressed. Most studies were limited to the effect of the war on agricultural production [13], the impact of the war on green spaces in the city of Aleppo [26] and the impact of oil refining on the environment [27]. The conflict situation adversely affected the agricultural sector, leading to a change in the land use patterns and a reduction in both the cultivated land and forest areas.

There are several models for predicting the magnitude of the erosion caused by water. The models range from the empirical RUSLE (Revised Universal Soil Loss Equation [28]) and EPM (Erosion Potential Method [29]) to the physical PESERA (Pan European Soil Erosion Risk Assessment), Water Erosion Prediction Project [30,31] and WEPP (Water Erosion Prediction Project [32,33]).

These models vary considerably in factors and in the complexity involved in calculating each factor [2,34–38]. Among them, the RUSLE model is considered as one of the widely applied empirical model for estimating soil water erosion [39–41]. This model has not yet been verified in Syria [25], but it has been verified in a number of Mediterranean regions (Portugal, Italy and the Palestinian Autonomous Area) [42–45] that are similar to the study area. For example, Abu Hammad et al. [44] checked the RUSLE in the study region and used field plots soil erosion measurements in the southeast of the Ramallah District in the Palestinian Autonomous Area (65 km^2). The results showed the RUSLE soil loss was estimate to be three times the actual soil loss. By adjusting the RUSLE, according to the prevailing conditions of the Mediterranean area, they improved the performance of the model three times. Aiello et al. [45] also verified the rates of soil erosion at the sub-basin scale of the Bradano basin (1500 km^2) by comparison with the San Giuliano reservoir silting value. The total amount of gross soil loss ranges between ~1.04×10^6 t year^{-1}, as computed with the Revised Universal Soil Loss Equation for Complex Terrain (RUSLE3D), and ~1.33×10^6 t year^{-1}, as computed with the measured silting data. The RUSLE estimation showed a good match with the measured silting data. We also verified the rates of soil erosion in our study area at the sub-basin of the Northern Alkabeer basin by comparing them with an analysis of the sediments in the reservoir provided by Hasan et al. [46] on 16 November 2017. The RUSLE calculation showed good consistency with the measured sediment outputs. The total gross soil loss ranges from 465,785 t year^{-1}, as calculated with RUSLE, to 474,865 t year^{-1}, as sedimentation has been measured. The validation attempts of the RUSLE model showed its feasibility to estimate the spatial distribution of soil loss for a region in the Mediterranean areas, to provide estimates for soil erosion at the watershed scale and to ensure a good match with the measured silting data. Based on this, we used the RUSLE model, considering that its input parameters are easily available [6,47,48].

In view of the above, the primary objective of this article is to estimate the impact of land cover change on the spatio-temporal distribution and to identify characteristics of soil erosion for the basin of the Northern Al-Kabeer river from 2009/2010 (hydrological year from September to August) to 2018/2019 under armed conflict. Estimates are to be made using the RUSLE model Remote Sensing (RS) and Geographic Information Systems (GIS) technologies with free available data for conditions before and during armed conflict in Syria. The second goal is to provide decision-makers and planners with knowledge to take sufficient priority steps for forest and soil protection when the war is over.

2. Materials and Methods

2.1. Study Area

The research was performed in the basin of the Northern Al-Kabeer River of the Northern Province of Latakia (Syria), one of the main coastal rivers. The Northern Al-Kabeer watershed covers an 845 km^2 area (35°29′11.546″ to 35°53′48.59″ N, and 35°48′14.36″ to 36°15′7.47″ E) (Figure 1). The length of the river within the Syrian lands is about 60 km. It originates from the northern end of the western mountains of Latakia, specifically from the high mountains situated at the Turkish border and as the Ansari Mountains of the northern province of Latakia. The altitudinal range is from around 1 m above mean sea level in the western plain (the so-called Latakia Plains) to the northeastern coastal mountains (1551 m) (Figure 1). About 81% of the total basin area includes hilly and mountainous regions; the remaining 19% are flatlands. The region lies within the Mediterranean climate zone [49]. It is mainly characterized by seasonal atmospheric circulations, altitude and topography, has an average annual temperature of 12–28 °C and an annual precipitation of 550–1100 mm. The main land-use areas are forests and shrubland in the east basin, whereas components from agriculture field, forest groves of citrus and olive trees and other fruits can be found in the western basin [49]. Forests cover widespread areas throughout the upper basin and parts of the central basin. The most important of these host species are *Pinus bruti, Quercus calliprino* and *Pistasia palaestina* (Boiss) [50].

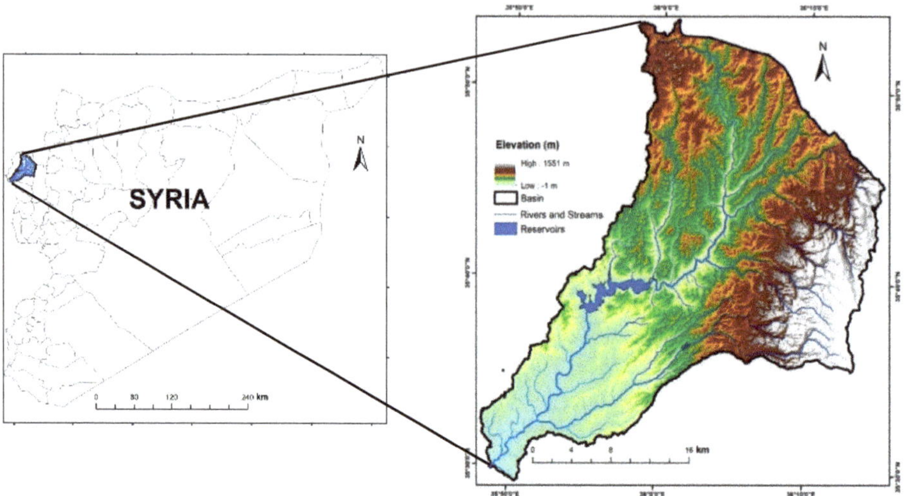

Figure 1. Location and topography of the study area in Syria.

2.2. Data

The key relevant data included in this study are based on five input parameters derivable from the RUSLE model's soil properties, precipitation, topography, cover and crop management as well as conservation practices as follows:

Due to the lack of climatic data from earth stations during the war, we selected daily Climate Hazards Group InfraRed Precipitation with Station data (CHIRP) V2.0 with a spatial resolution of 0.25° for this study. The rainfall is estimated from Rain Gauge and Satellite Observations. Monthly and yearly data were calculated according to the hydrological year (September–August) for the years 2009/2010, 2013/2014 and 2018/2019.

In this research, freely available data from the Panchromatic Remote-sensing Instrument for Stereo Mapping, Digital Elevation Model (PALSAR DEM) with a spatial resolution of 12.5 m were collected from the Alaska Satellite Facility Distributed Active Archive Data Center (ASF DAAC) in GIS-ready GeoTIFF format. Landsat TM remote sensing images in 2009/2010 and Landsat 8 Operational Land Imager (OLI) data from 2013/2014 and 2018/2019 were collected from the Data Sharing Infrastructure of Earth System Science (http://www.geodata.cn), with a spatial resolution of 30 m.

Soil data sets in SoilGrids system at 250 m, including soil type distribution map soil properties (sand, clay, silt and organic carbon fractions), were compiled by International Soil Reference and Information Centre (ISRIC)—World Soil Information.

2.3. Soil Erosion Using RUSLE Model

In the RUSLE model, the following five parameters were used to predict soil loss [51]:

$$A = R \times K \times LS \times C \times P \tag{1}$$

here, A: computed spatial yearly soil loss (t ha^{-1} year^{-1}); R: Rainfall erosivity factor (MJ mm ha h^{-1} year^{-1}); K = soil erodibility factor (t h^{-1}MJ^{-1} mm^{-1}); LS = slope length factor and slope steepness factor (unitless); C= land surface cover management factor (unitless); and P = conservation practice factor or erosion control (unitless).

Due to the ongoing war in the region and the high costs of field work measurements [52,53], the study relied on secondary data available in the Geographic Information System (GIS) and remote sensing. Field work was limited to field trips to verify fires, land use changes and soil erosion in 2012–2016.

The R factor determines the erosivity of rainfall at a given location, depending on the amount and intensity of rainfall and the rate of rainfall-related runoff [54]. The majority of sheet or rill erosion is caused by high runoff flow as a consequence of heavy storms. In this analysis, the approach of Arnoldus (1977) was employed, as it was derived under similar climate conditions as in the study region and is commonly used in Syria [49,55] (see Appendix A). The rainfall erosivity factor (R) was calculated using a method based on monthly average rainfall aggradations given by data from the Climate Hazards Group InfraRed Precipitation with Station (CHIRPS) including the interpolation tool Inverse Distance Weighted (IDW) in the program ArcGIS 10.7.

Soil erodibility factor (K) is a dynamic property that quantifies the susceptibility of soil particles in sheet flow and rills to detachment and transport by a splash during runoff, water flow or both [53,56–58]. Soil erodibility is related to the combined impact of rainfall, drainage and soil loss penetration, and is generally referred to as soil erosion factor (K). This study used the K factor (Figure 2a), estimated using soil properties (sand, silt, clay and organic carbon fractions) at 5 cm depth compiled by ISRIC—World Soil Information with a spatial resolution of 250 m [39]. The equation was used to estimate the erodibility of soil, as suggested by [59] (see Appendix A).

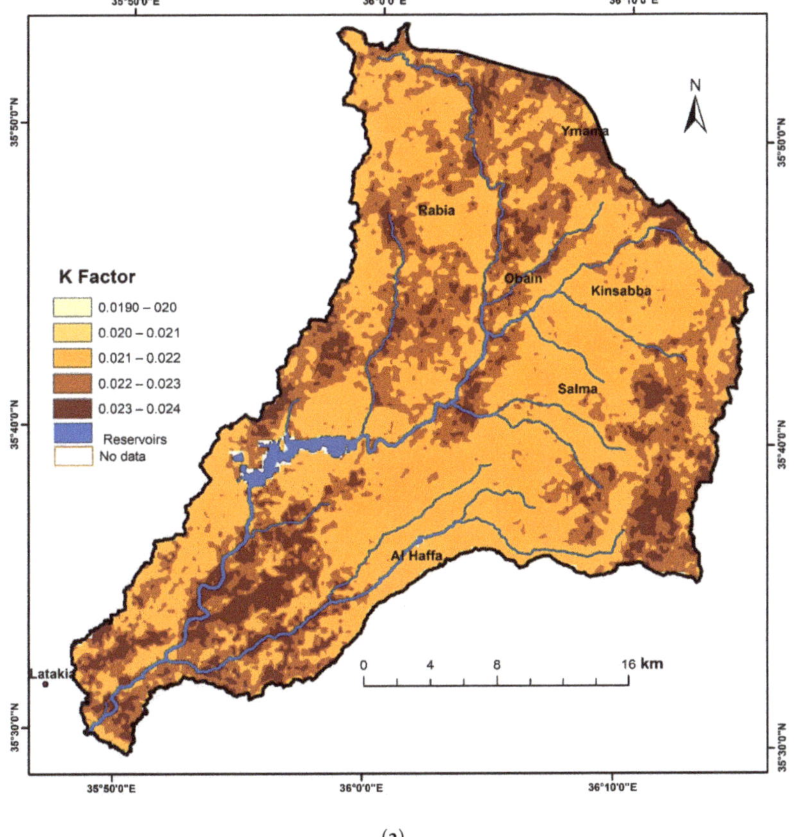

(a)

Figure 2. *Cont.*

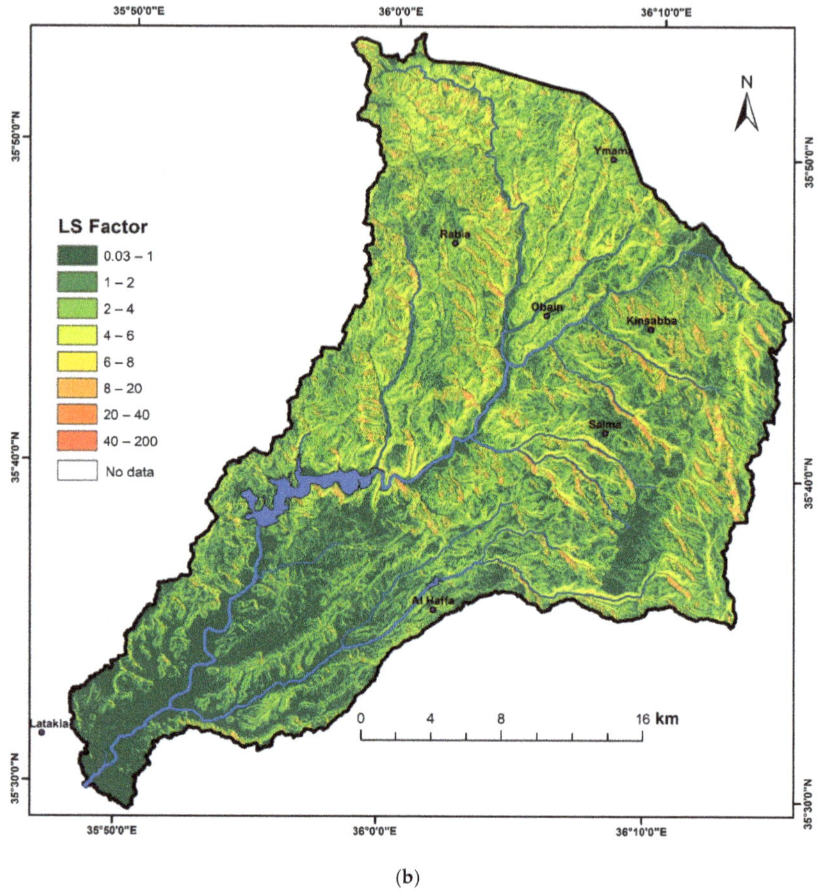

(**b**)

Figure 2. Factor maps of soil erosion modeling of the Northern Al-Kabeer river basin. (**a**) Soil erodibility factor map; (**b**) slope length and slope steepness (LS) factor map.

The LS factor reflects the influence of local topography on the soil erosion rate, integrating the effects of the slope length (L) and slope steepness (S). The LS factor has been generated using the PALSAR DEM with a spatial resolution of 12.5 m (2011), which has been collected from the Alaska Satellite Facility Distributed Active Archive Data Center (ASF DAAC).

The slope length factor (L) is provided by the Desmet and Govers (1996) [60], and is enhanced by the USLE estimation technique. It considers the upstream contribution area, where the impact of the slope length is a function of the ratio of rill erosion to inter-rill erosion (caused by raindrop impact) and is more appropriate for areas with complex slopes. Steep slopes (L) and rolling topography provide a vital medium for a lower position of the springing runoff water. The slope-steepness factor (S) shows how easily water can flow over a given surface that interacts with the ground angle and affects the soil erosion rate [40]. The estimation of the S-factor originally proposed by Wischmeier and Smith (1978) was proposed by McCool et al. (1987) in the RUSLE model to achieve an improved representation of the slope steepness factor, taking into account the ratio of the rill and inter-rill erosion. McCool et al. (1987) found that soil erosion occurred more rapidly on slopes with a steepness of more than 9%. Therefore, he used one algorithm for slopes < 9% and another for slopes > 9% (see Appendix A).

The Normalized Difference Vegetation Index (NDVI) is one method to estimate the C factor from remotely sensed data and is the most widely used vegetation index. In Europe, van der Knijff et al. [61]

developed the following relationship with the image satellite-based standardized vegetation difference index (NDVI) between field-calibrated C factor values to produce a continuous C factor surface (Equation (2)):

$$C = exp\left[-2\frac{NDVI}{(1-NDVI)}\right] \tag{2}$$

It was proposed to use the NDVI image acquired when soil erosion is strongly active during the rainy season. Thus, the C factor layer (Figure 2b) used in our study was created using Landsat TM remote sensing images in 2009/2010 and Landsat 8 Operational Land Imager (OLI) 2013/2014 and 2018/2019 for the rainy seasons (September–May) using Equation (2).

The Conservation Practice Factor in the RUSLE model expresses the effect of conservation practices mitigating erosion by minimizing the volume and rate of water runoff. It is the ratio of soil loss to the related loss of slope-parallel tillage with a particular support activity on cropland. As a result of the lack of support activities in place in the Northern Al-Kabeer river basin, a value of 1.0 for the entire region was assumed to support the conservation practice factor.

2.4. Role of Vegetation in the Soil Erosion Changes

In order to detect the role of vegetation in soil loss transition, the soil erosion modulus of various times has been determined in tow scenarios according to Wang and Su [62]. The natural condition and C factor fixation to detect the role of vegetation in soil loss transition and the contribution rate of vegetation to soil erosion were analyzed by comparing the average soil erosion modulus under C factor fixation scenarios with the real average soil erosion modulus in the natural condition:

- The first scenario showed the natural condition of the soil erosion modulus in the initial year in each period of 2009/2010–2013/2014, 2013/2014–2018/2019 and 2009/2010–2018/2019, i.e., the actual soil erosion modulus in each period.
- Scenario C factor fixation is the soil erosion modulus calculated by the C factor value using the end year of each period, i.e., 2013/2014 and 2018/2019, while other factors values remain as used at the initial year of each periods.

The wildfire inventory dataset was created using the GPS data from the field surveys in 2012–2016 and evaluated using the hotspots of MODIS and Google Earth images. These corrections were manually done in the geographic information system ArcGIS 10.7. However, they did not include all wildfire events due to difficulty in access as a result of fires, artillery and missile shelling.

3. Results

3.1. Distribution of Soil Erosion Factors in the Northern Al-Kabeer River Basin

The factors were mapped by ArcGIS environment in the Northern Al-Kabeer river basin. The erosion layers were generated at a cell size of 12.5 m, following the DEM resolution, which was the finest among the input datasets. The results indicate that the value of the K factor varied from 0.019 to 0.023 t h^{-1}MJ^{-1} mm^{-1} (Figure 2a). Much of the Northern Al-Kabeer basin areas are covered with a texture of sand, clay and loam. The value of the LS factor ranged from 0.03 to 200 for the entire region (Figure 2b). The cover management factor (C) value varied between 0 and 1. The R factor value ranges from 375 to 650 MJ mm ha^{-1} h^{-1} yr^{-1}, with the maximum values in the north-eastern part of the basin and the lowest values in the south-western part of the basin. (Figure 3a–c). The average R factor was 430 MJ mm ha h^{-1} year^{-1} in 2010 and then increased to 505 MJ mm ha h^{-1} year^{-1} in 2013/2014, but decreased again to 445 MJ mm ha h^{-1} year^{-1} in 2019/2020. The value of the C factor decreased in 2013/2014 in comparison to 2009/2010 due to the forest fire in 2012 and 2013. After that it increased again until 2019/2020. The soil that is covered by trees and forests, and hence protected from soil erosion, are classified as low (0.001–0.03). Built-up areas/settlements and barren land are associated with high soil loss and a value of 0.8–1. Similarly, there is less soil erosion associated with farming. These are ranked as 0.18 and 0.28, respectively (Figure 3d–f).

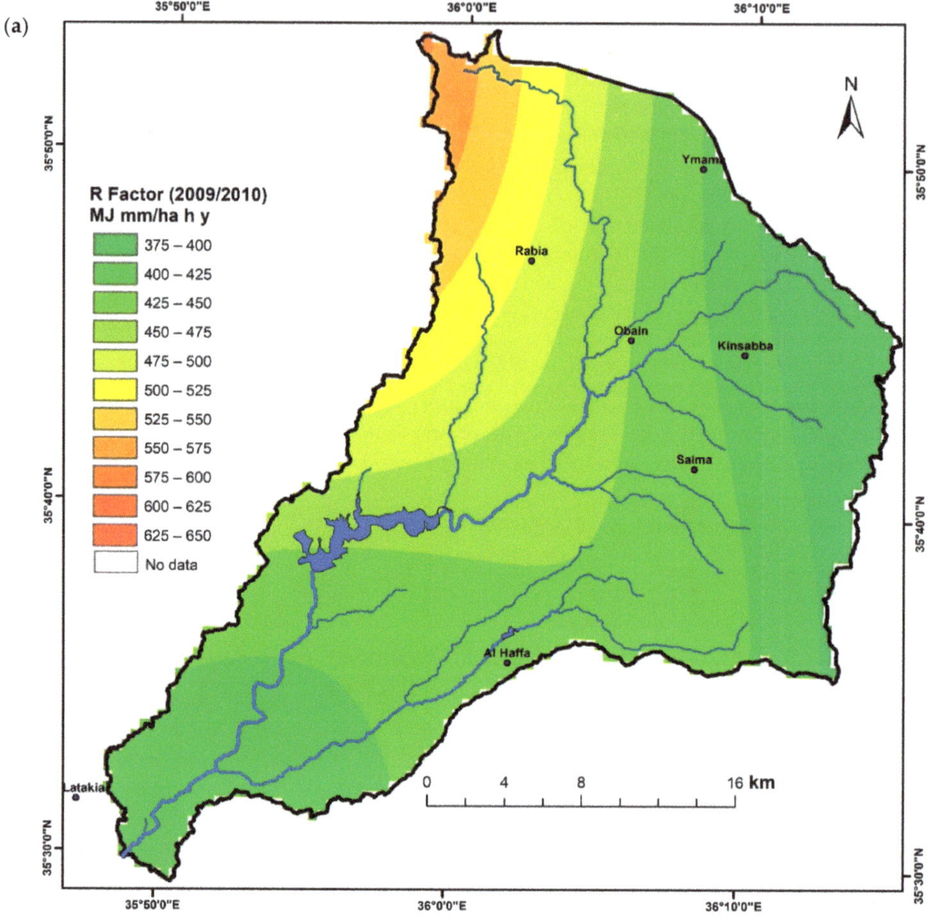

Figure 3. *Cont.*

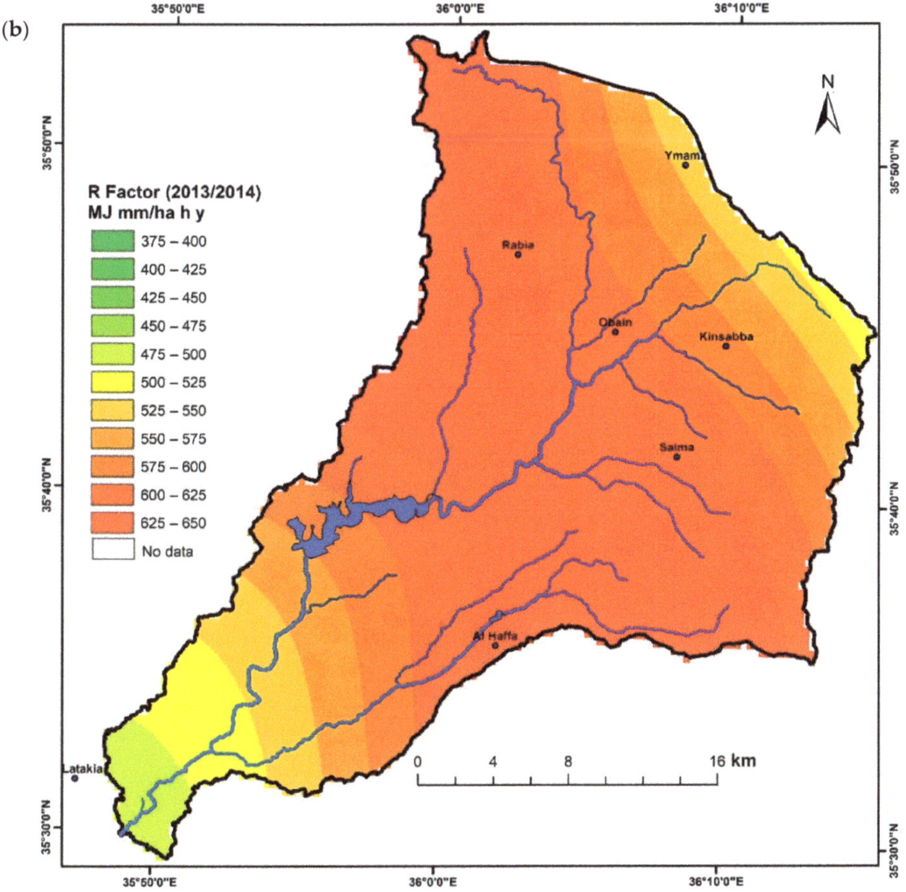

Figure 3. *Cont.*

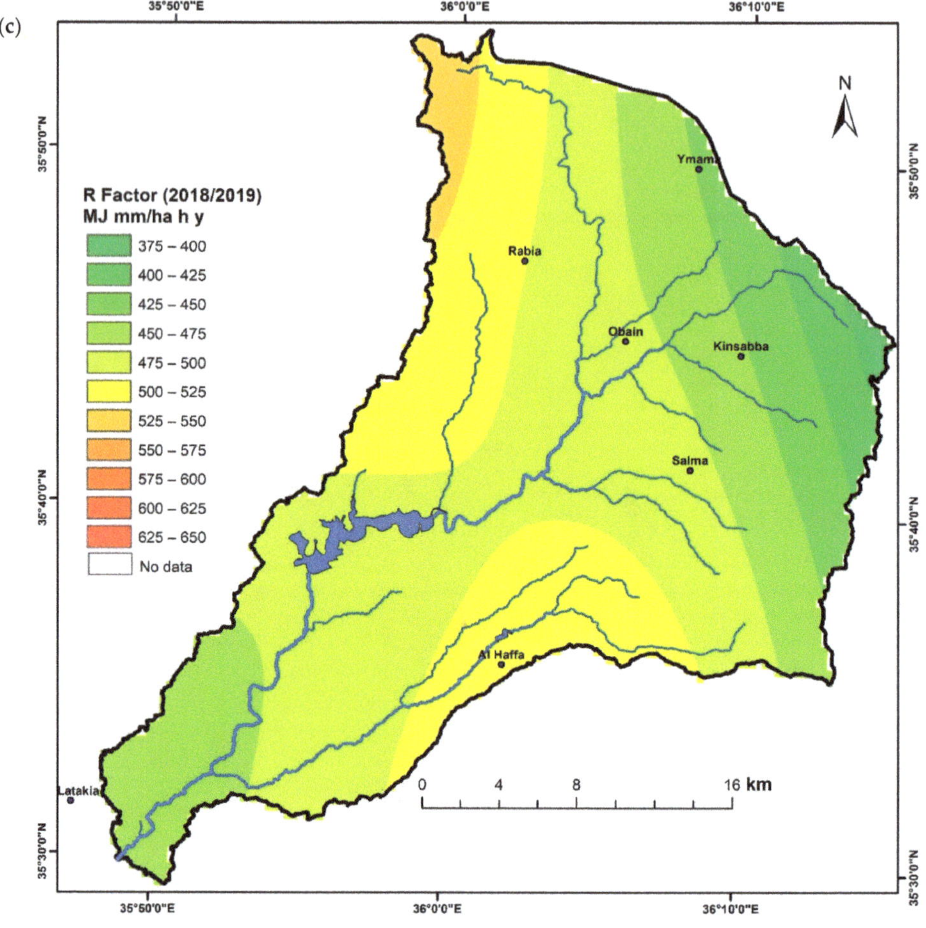

Figure 3. *Cont.*

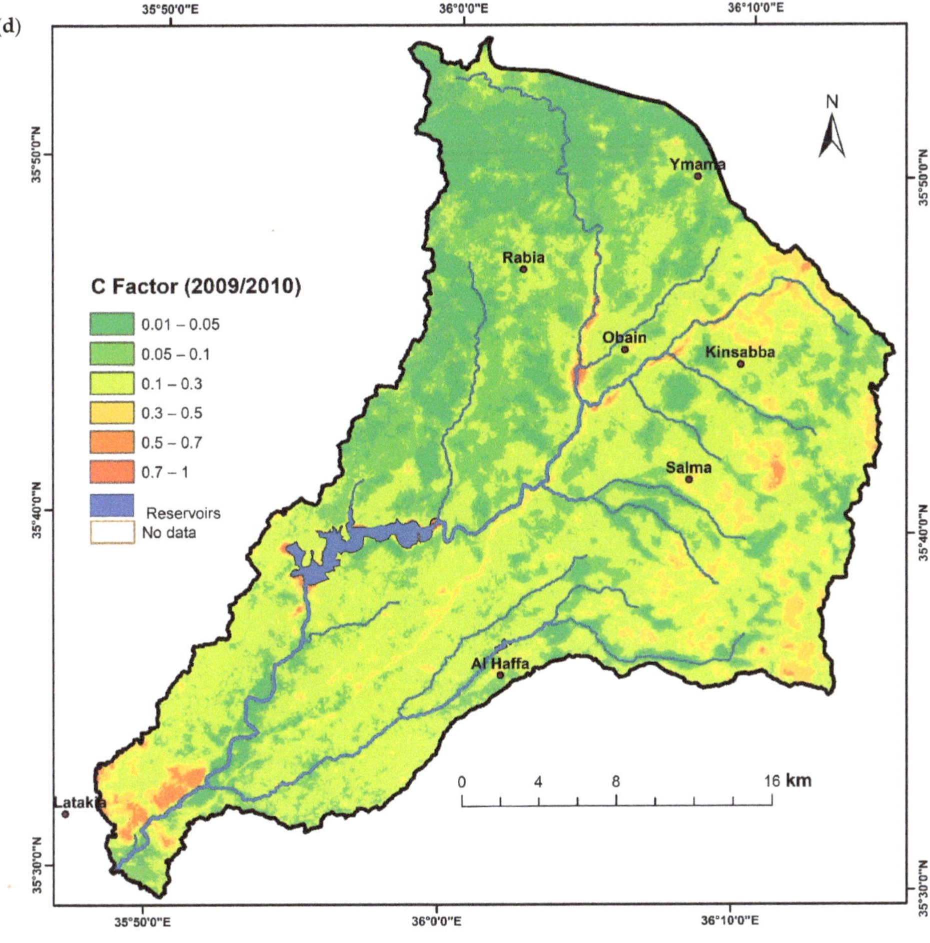

Figure 3. *Cont.*

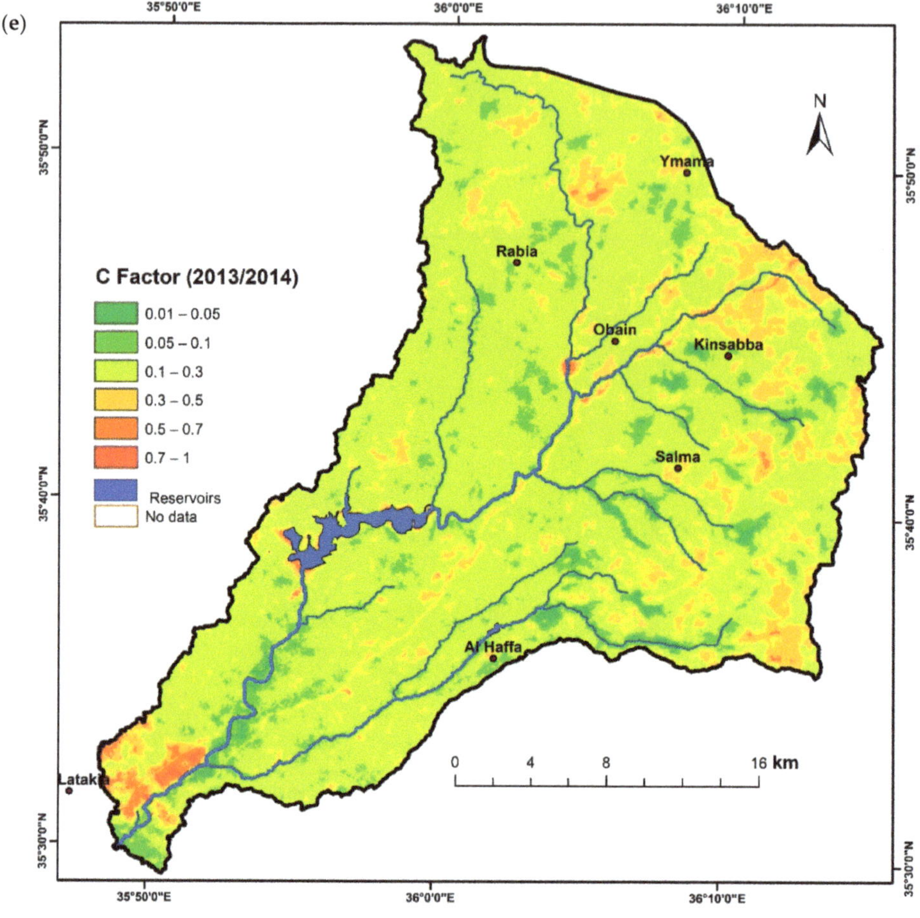

Figure 3. *Cont.*

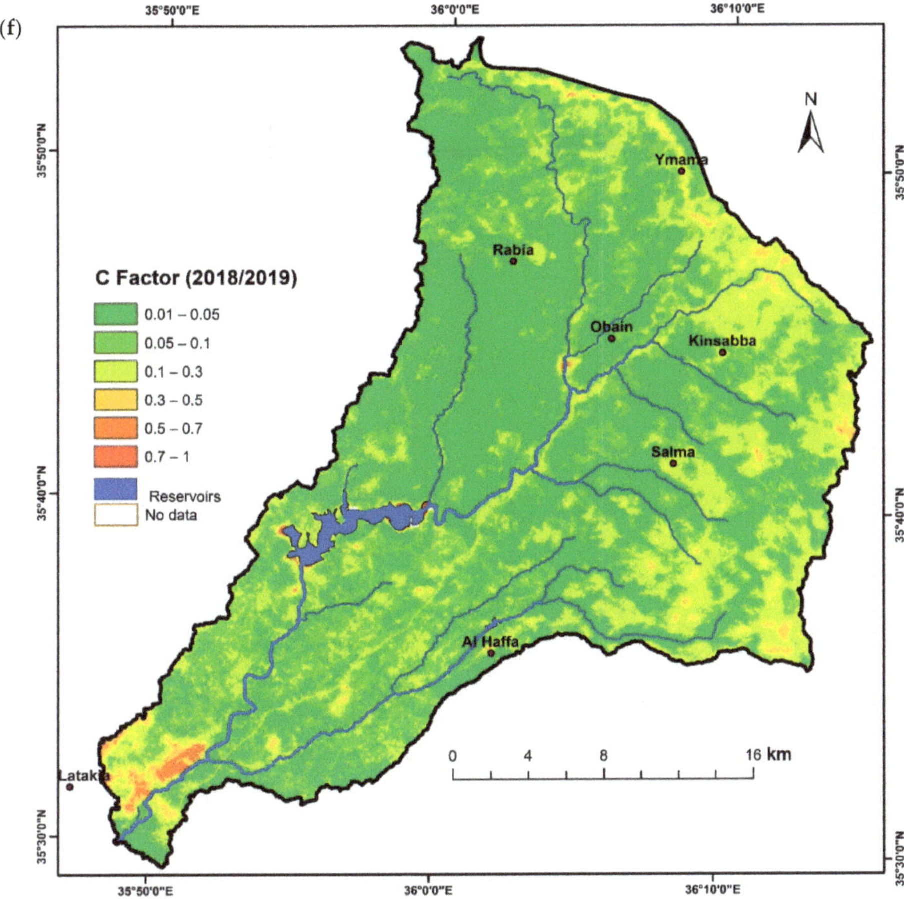

Figure 3. Factor maps of soil erosion modeling of the Northern Al-Kabeer river basin (**a–c**) map of rainfall erosivity factor in 2009/2010, 2013/2014 and 2018/2019, respectively; (**d–f**) cover management factor map in 2009/2010, 2013/2014 and 2018/2019, respectively.

3.2. Estimation and Spatial Distribution of Soil Erosion Rates in the Northern Al-Kabeer River Basin

To facilitate the analysis of the spatial distribution of the soil erosion rates and to promote the visual comparison of the three maps, the basin's soil loss was classified into eight categories [40,61,63] (Table 1). The distribution of soil erosion maps in 2009–210, 2013–2014 and 2018–2019 was produced in ArcGIS 10.7 (Figure 4; Table 1).

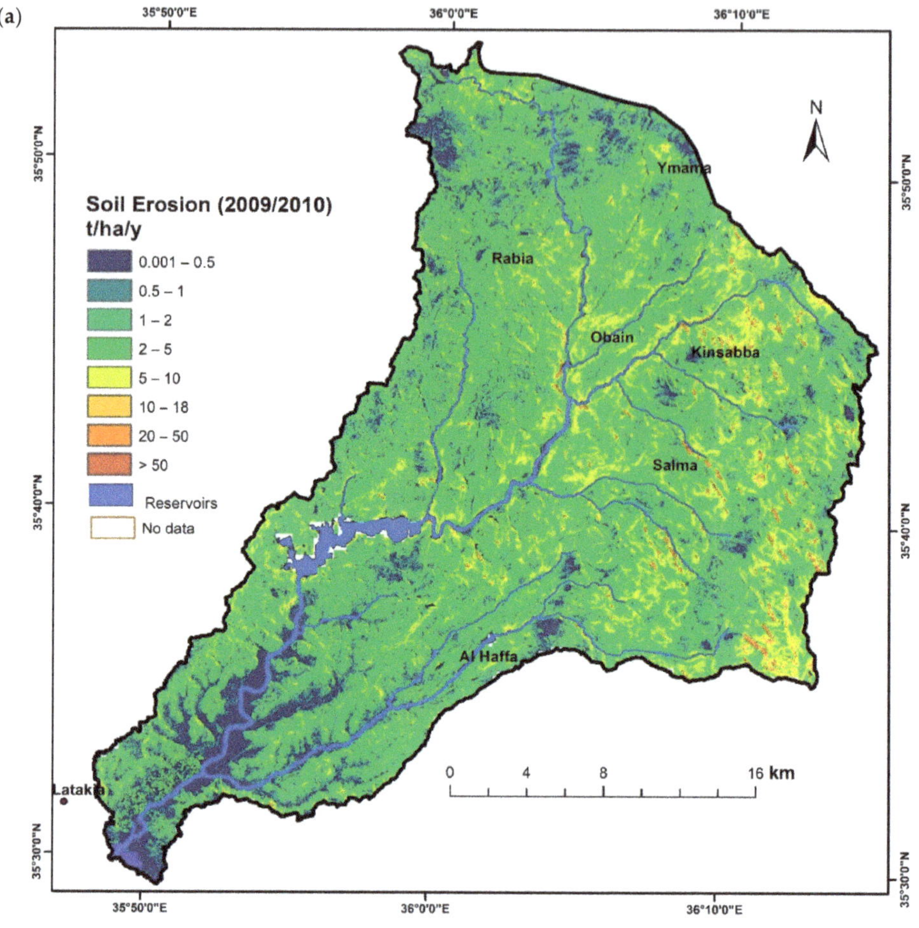

Figure 4. *Cont.*

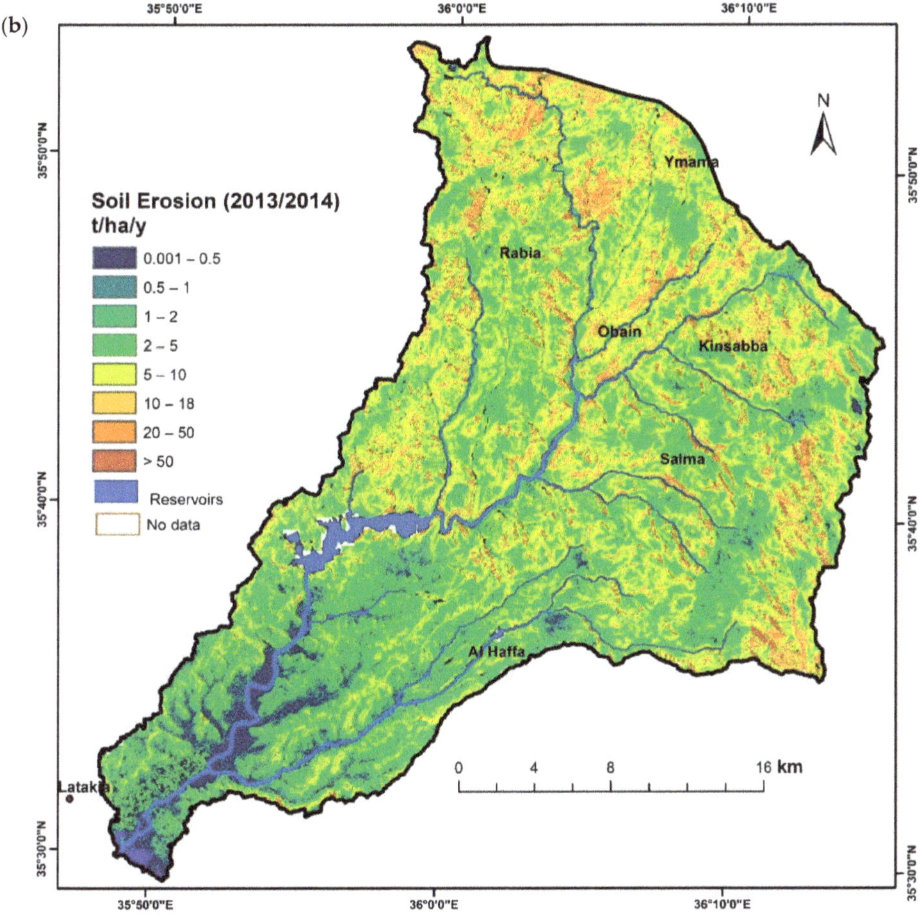

Figure 4. *Cont.*

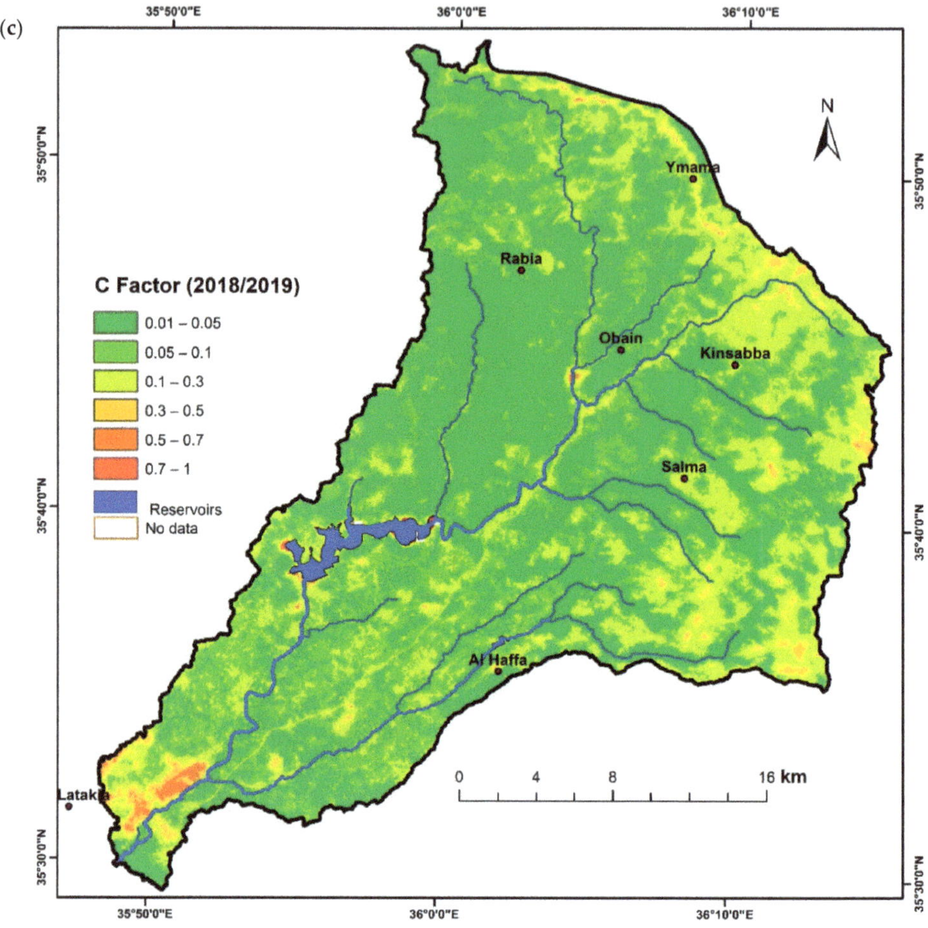

Figure 4. Erosion classification map of the Northern Al-Kabeer watershed in (**a**) 2009/2010, (**b**) 2013/2014 and (**c**) 2018/2019.

Table 1. Soil erosion classes in 2009/2010, 2013/2014 and 2018/2019.

Erosion Classes (t ha^{-1} year^{-1})	2009/2010		2013/2014		2018/2019	
	Area (km^2)	Percent (%)	Area (km^2)	Percent (%)	Area (km^2)	Percent (%)
Very low—VL (0–0.5)	74.5	8.8	42.3	5.0	186.6	22.1
Low—L (0.5–1)	95.9	11.4	36.2	4.3	178.7	21.2
Low medium—LM (1–2)	198.8	23.5	84.1	9.9	217.1	25.7
Medium—M (2–5)	316.5	37.5	238.1	28.2	191	22.6
High Medium—HM (5–10)	113.5	13.4	252.7	29.9	52.3	6.2
High—H (10–20)	31.6	3.7	126.2	14.9	13.5	1.6
Very high—VH (20–50)	12.9	1.5	54.8	6.5	5.1	0.6
Extremely high—EH (>50)	1.4	0.2	10.8	1.3	0.7	0.1

The average rate well exceeds the tolerance limit of 1 t ha^{-1} year^{-1} soil loss for the study areas. In Table 1, it is observed that the soil erosion intensity in most of the study area was classified as low medium to high medium (1–10 t ha^{-1} year^{-1}), which is more than 66% of the river basin. About 24% of the watershed is under the tolerant erosion rate. On the contrary, the distribution of soil losses in the study area of over 10 t ha^{-1} year^{-1} (very high and extremely high soil classes) was lower and accounted for around 10% of the total area. However, the rate of soil erosion remains high compared to other basins in the Mediterranean region. The critically high watershed soil erosion is related to the upstream portion of the river. Removal or alteration of the vegetation, destruction of the forest, fires caused by human activities and shallow depth of the soil above the bedrock significantly increase soil erosion. Overall, the far east and southeast portions of the basin (upstream region and eastern portion of the downstream region of the Northern Al-Kabeer river) is characterized by soil loss, where a shallow depth of the inceptisols (xerofluvents) soil and built up areas are causing erosion (Figure 4). The very low erosion class was found mainly in the upstream region due to land cover by forest, and in the downstream region due to the flat terrain (Figure 4).

From 2009/2010 to 2013/2014, the average annual soil erosion module increased and decreased from 2013/2014 to 2018/2019 with an average of 3.6 t ha^{-1} year^{-1} and a standard deviation of 6.4 t ha^{-1} year^{-1} in 2009/2010; the average was 7.9 t ha^{-1} year^{-1} with a standard deviation of 12 t ha^{-1} year^{-1} in 2013/2014; and the average was 2.2 t ha^{-1} year^{-1} with a standard deviation of 4.3 t ha^{-1} year^{-1} in 2018/2019.

Tables 2–4 show the erosion change matrix between 2009/2010, 2013/2014 and 2018/2019. In the period of 2009/2010 to 2013/2014, the soil erosion area of regions with no changes was 30%. The region shows that the total area under very low erosion, low erosion, low medium erosion and medium erosion in 2009/2010 was nearly 9%, 11%, 24% and 38% of the total area, respectively. Different results were observed for the watershed, where the area under very low erosion, low erosion, low medium erosion and medium erosion was 5%, 4%, 10% and 28%, respectively, in the year 2013/2014. The region of elevated erosion can primarily be attributed to the change from low medium erosion to medium erosion, medium erosion to high medium erosion and high medium erosion to high erosion. The low medium erosion area decreased from 24% to 10%; this 14% transferred to medium erosion and high medium erosion.

Table 2. Change the soil erosion matrix classes from 2009/2010 to 2013/2014 (%).

Soil Erosion Classes		Soil Erosion Classes 2013/2014								Grand Total
		VL	L	LM	M	HM	H	VH	EH	
	VL	4.23	1.80	1.25	1.03	0.36	0.08	0.04	0.02	8.81
	L	0.47	1.77	3.65	3.25	1.63	0.44	0.12	0.03	11.35
	LM	0.19	0.57	4.02	10.73	5.33	2.05	0.55	0.07	23.52
Soil erosion	M	0.08	0.12	0.89	11.95	16.87	5.63	1.73	0.19	37.45
classes 2009/2010	HM	0.02	0.02	0.10	0.91	5.02	5.46	1.68	0.23	13.43
	H	0.01	0.01	0.02	0.21	0.50	1.07	1.69	0.23	3.74
	VH	0.01	0.00	0.01	0.06	0.16	0.19	0.65	0.43	1.53
	EH	0.01	0.00	0.00	0.02	0.02	0.02	0.02	0.08	0.17
	Grand total	5.02	4.30	9.94	28.17	29.89	14.93	6.48	1.28	100.00

From 2013/2014 to 2018/2019, the soil erosion area of the watershed with initially no change decreased to 10%. The areas of very low, low and low medium erosion classes increased, with amplitude reductions of 17%, 17% and 16%, respectively. The areas of medium, high medium, high, very high and extremely high erosion decreased, with amplitudes of 6%, 24%, 13%, 6% and 1%, respectively. The region with decreased erosion was primarily because 10%, 10% and 6% area under medium erosion class was classified as very low, low and low medium erosion classes, respectively. Additionally, 6%, 12%, and 11% transferred from high medium to low, low medium and medium erosion classes, respectively. It is noted that the erosion had decreased significantly in a short time.

Table 3. Change the soil erosion matrix classes from 2013/2014 to 2018/2019 (%).

Soil Erosion Classes		Soil Erosion Classes 2018/2019								Grand Total
		VL	L	LM	M	HM	H	VH	EH	
	VL	4.97	0.05	0.00	0.00	0.00	0.00	0.00	0.00	5.02
	L	3.71	0.54	0.04	0.00	0.00	0.00	0.00	0.00	4.30
	LM	5.34	3.69	0.87	0.06	0.00	0.00	0.00	0.00	9.95
Soil erosion	M	5.94	10.08	9.55	2.48	0.10	0.01	0.00	0.00	28.16
classes 2013/2014	HM	1.87	5.59	11.17	10.32	0.84	0.09	0.00	0.00	29.89
	H	0.23	1.02	3.40	7.28	2.76	0.21	0.03	0.00	14.93
	VH	0.02	0.18	0.65	2.31	2.17	0.93	0.20	0.01	6.47
	EH	0.00	0.00	0.02	0.14	0.32	0.36	0.37	0.07	1.28
	Grand total	22.08	21.15	25.70	22.60	6.19	1.60	0.60	0.08	100.00

From 2009/2010 to 2018/2019, the soil erosion area of the watershed with no alterations increased to 30%. Very low, low and low medium classes showed increasing trends during this time period, in contrast to a decrease in the change rate of high classes. This was due to vegetation regeneration after fires. The area with a decreased rate of change was largely caused by the transition from low medium to low erosion, medium to low medium and high medium to medium erosion classes. Approximately 8% and 14% and 8% of areas transferred from low medium erosion to low erosion, and medium erosion to low medium erosion and low erosion, respectively.

Table 4. Change the soil erosion matrix classes from 2009/2010 to 2018/2019 (%).

Soil Erosion Classes		Soil Erosion Classes 2018/2019								Grand Total
		VL	L	LM	M	HM	H	VH	EH	
	VL	7.22	0.95	0.42	0.16	0.03	0.01	0.01	0.00	8.81
	L	5.76	3.10	1.56	0.75	0.12	0.04	0.01	0.00	11.35
	LM	6.03	7.93	6.06	2.83	0.49	0.13	0.04	0.01	23.52
Soil erosion	M	2.65	8.13	14.23	10.56	1.52	0.27	0.08	0.02	37.45
classes 2009/2010	HM	0.28	0.82	2.89	6.75	2.28	0.30	0.09	0.01	13.43
	H	0.08	0.16	0.41	1.22	1.32	0.44	0.09	0.01	3.74
	VH	0.04	0.05	0.11	0.30	0.40	0.38	0.24	0.01	1.52
	EH	0.02	0.01	0.01	0.03	0.02	0.02	0.05	0.02	0.17
	Grand total	22.08	21.15	25.70	22.60	6.19	1.60	0.60	0.08	100.00

3.3. Impact of Vegetation for Soil Erosion Rates

Figure 5 illustrates spatial distribution of burned areas and front lines 2012–2019 and C factor in burned areas. Extremely high C factors were obtained in areas covered by coniferous forests, which were severely affected by the fire in 2013 and 2014 in southeast basin. These areas are located near the front line in 2014, while the other parts are located in the southeast basin with considerable distance. The main cause is exposed to artillery and missiles. In these areas, the average pre-fire C factor was estimated to be around 0.02–0.1, whereas after the fire this value was estimated at an extremely high value of 0.4–0.6. The cessation of fighting after 2016 contributed positively to the vegetation recovery in most parts of the region. Moreover, the C factor is a parameter critical for burned areas because the density of vegetation cover, which plays a role as a productive agent against soil erosion, is highly significant for the occurrence of wildfires, especially in forested areas.

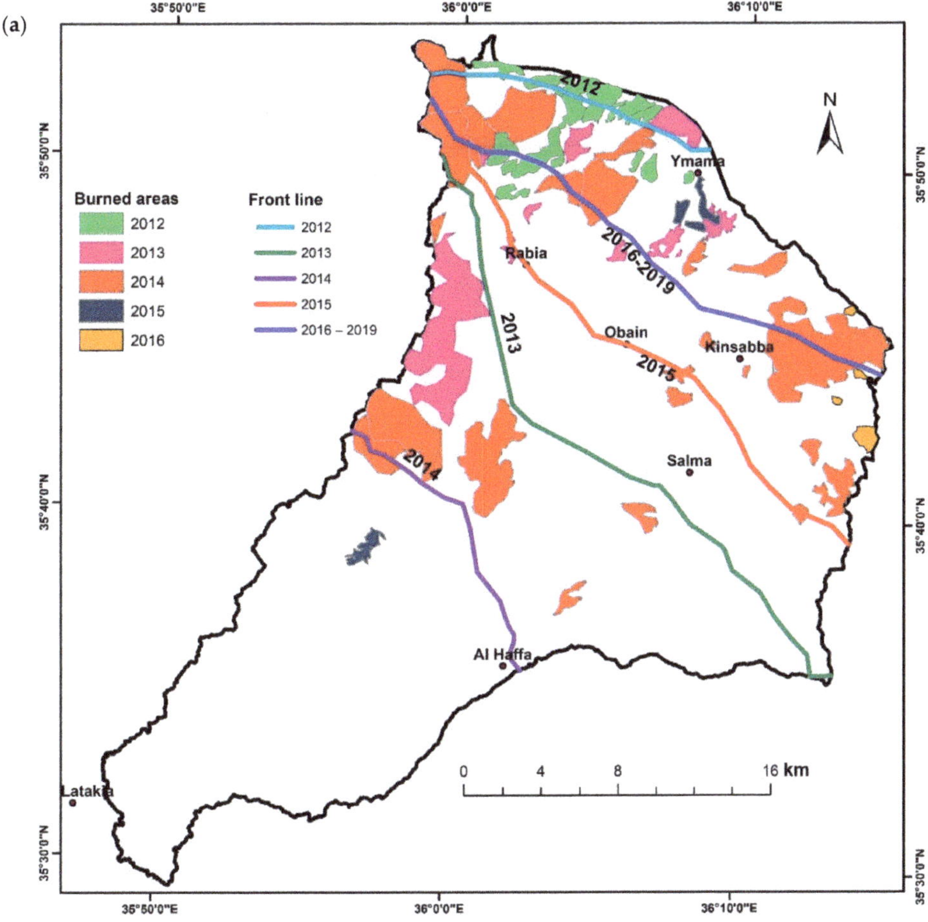

Figure 5. *Cont.*

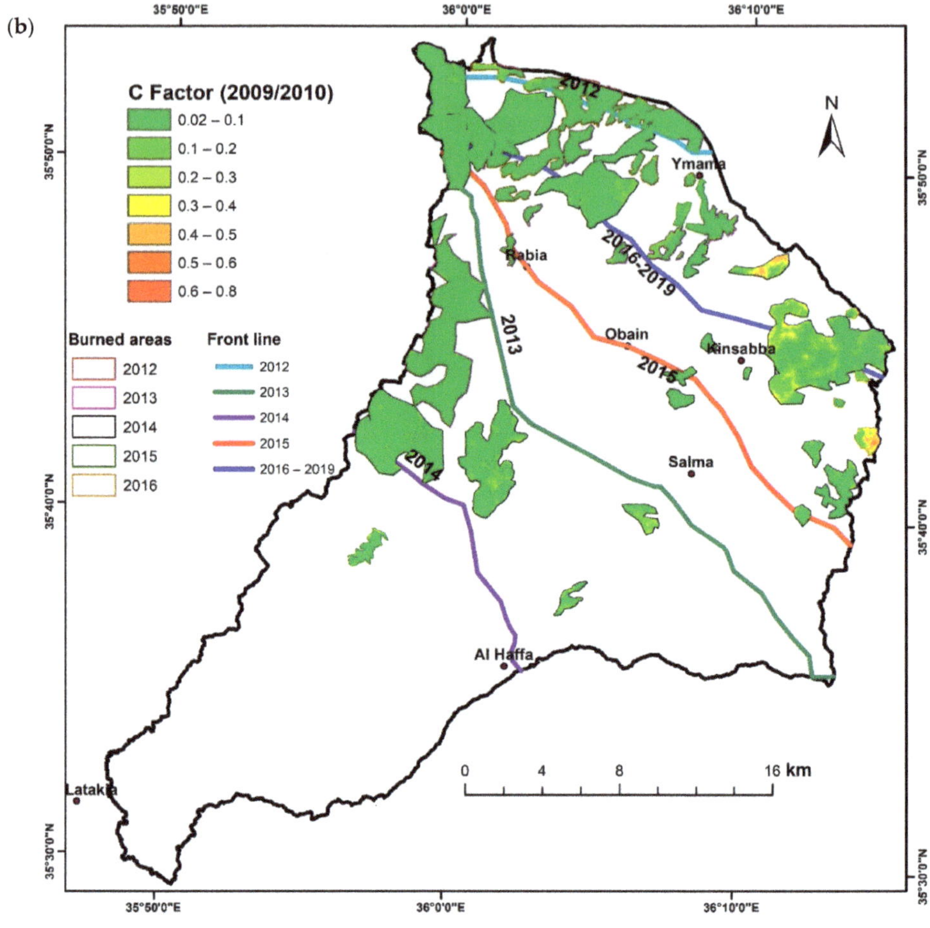

Figure 5. *Cont.*

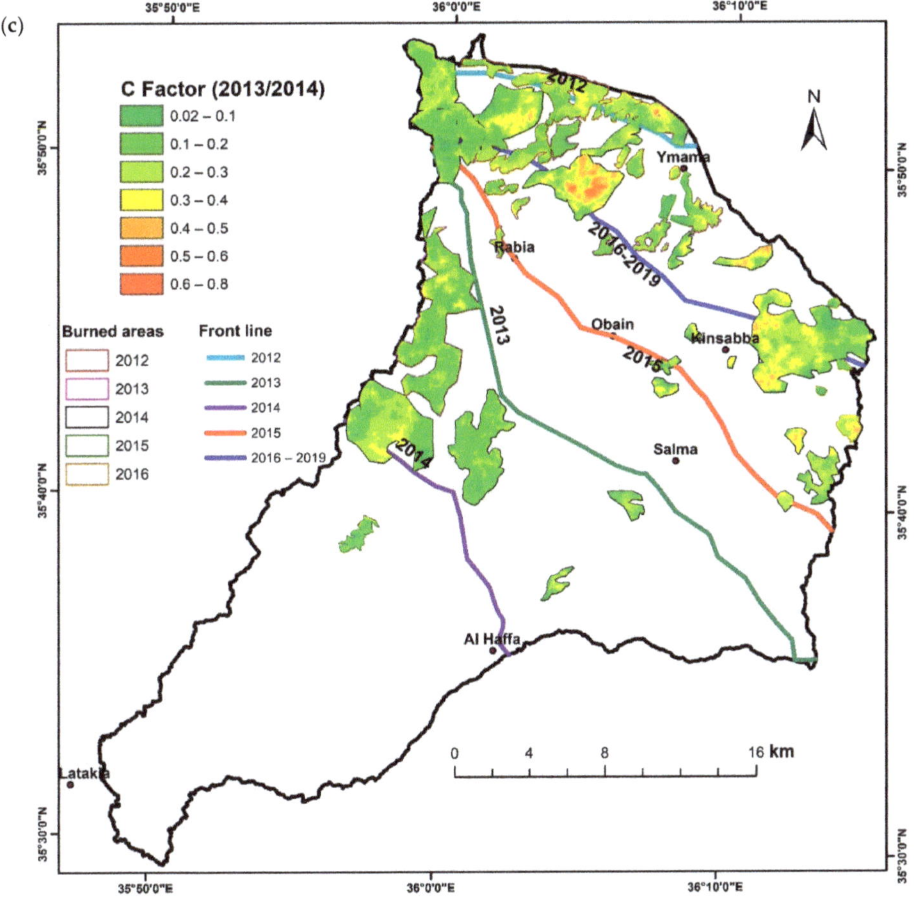

Figure 5. *Cont.*

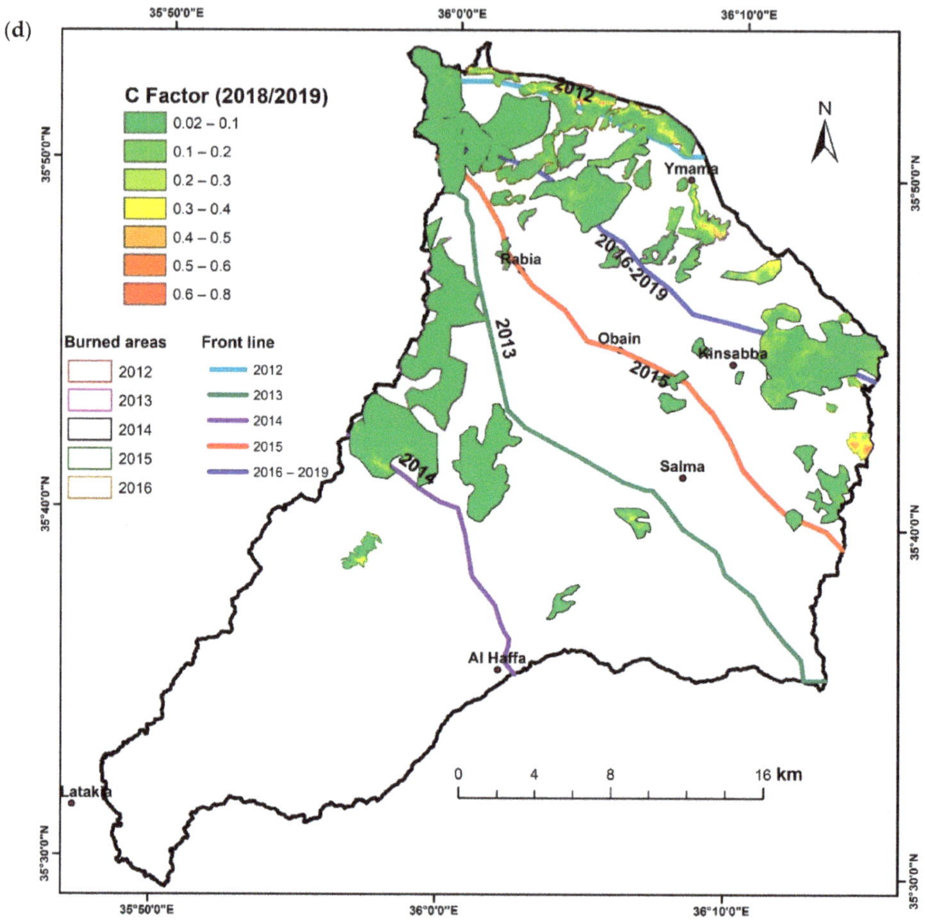

Figure 5. (**a**) Spatial distribution of burned areas and front lines 2012–2019 and the C factor in burned areas in (**b**) 2009/2010, (**c**) 2013/2014 and (**d**) 2018/2019.

Figure 6 shows the impact of factor (C) of the vegetation cover on soil erosion rates in different periods. It displays the 2009/2010 to 2013/2014 time period. Due to the change in vegetation coverage factors, the rate of soil erosion increased in most areas of the basin. The areas with the main increase of soil erosion rates are located in the central and eastern parts of the basin. Coniferous forests as well as transitional forests and scrublands are the dominant forms of land cover, for which the average post-fire soil loss values (by reason of C factor) are 200% to 800% higher than in pre-fire conditions.

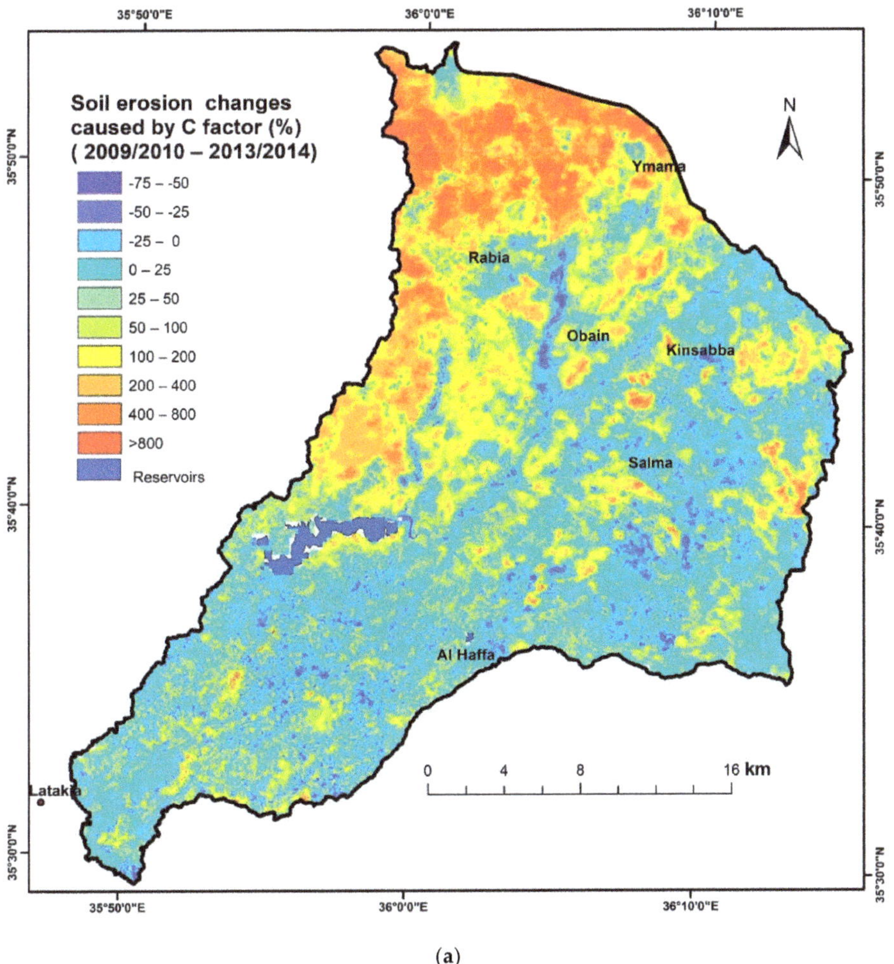

(**a**)

Figure 6. *Cont.*

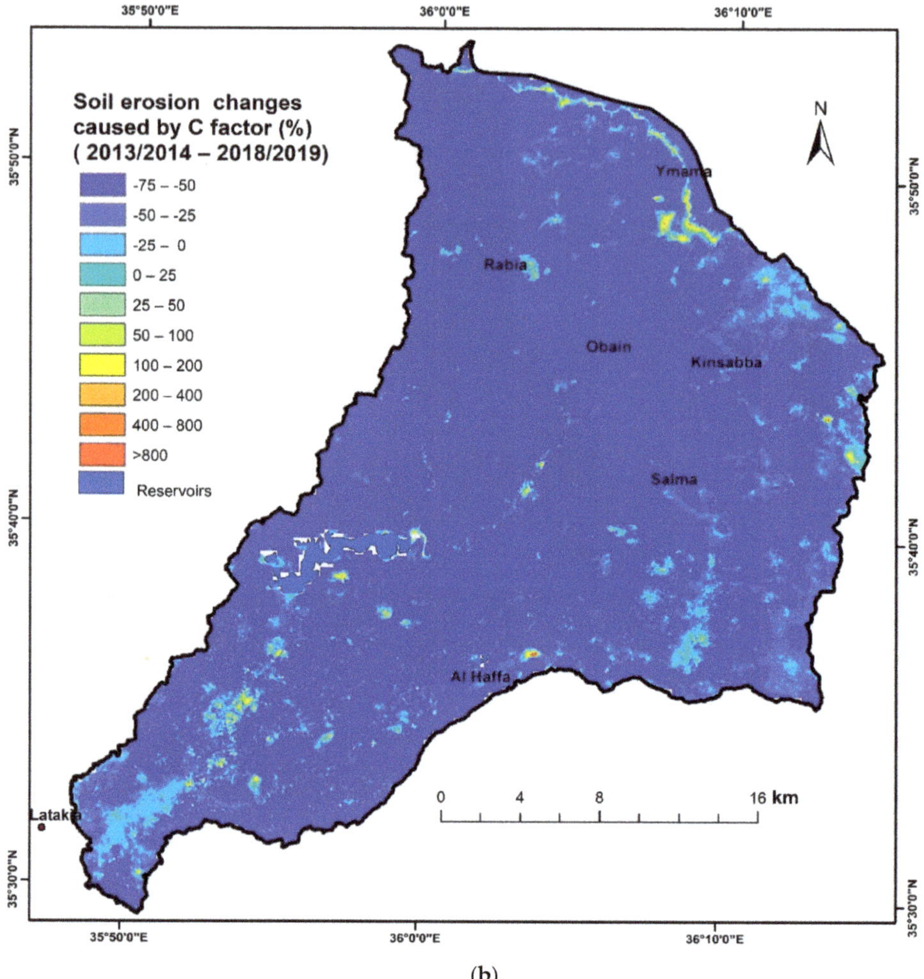

(**b**)

Figure 6. *Cont.*

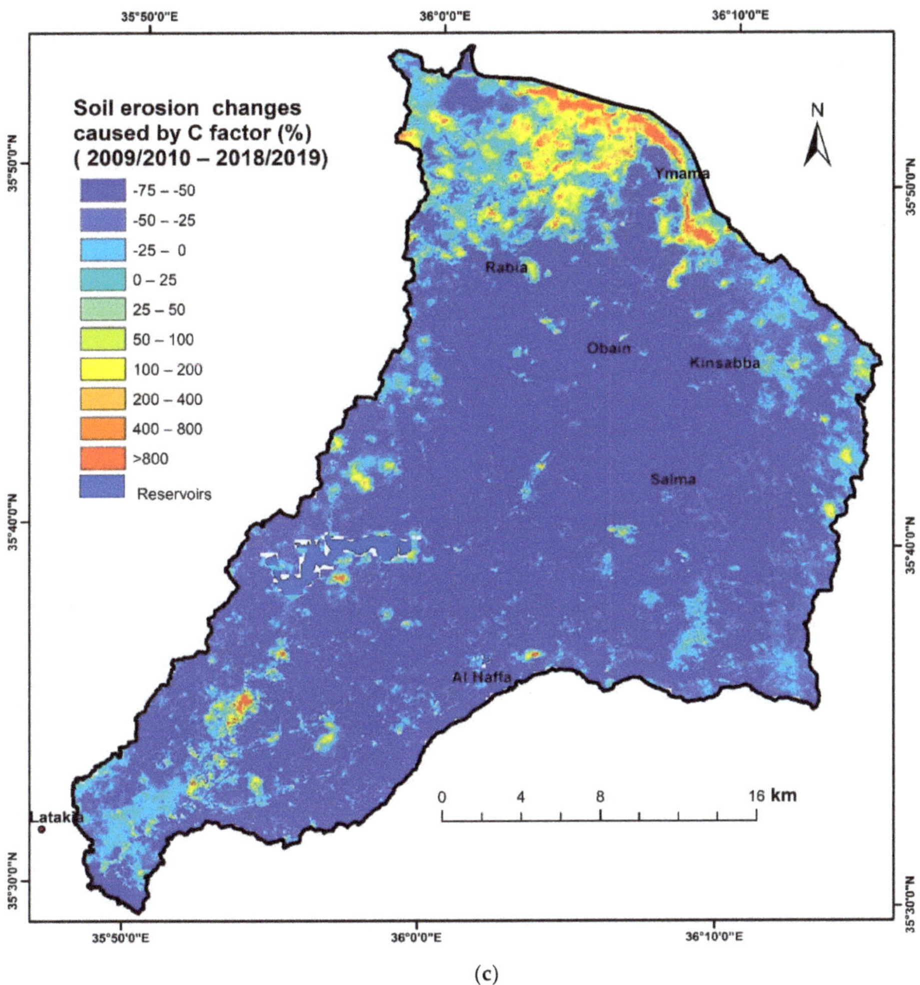

(c)

Figure 6. The effect of the C factor on the rate of soil erosion over different periods. (**a**) 2009/2010–2013/2014 (**b**) 2013/2014–2018/2019 (**c**) 2009/2010–2018/2019.

The decline in vegetation due to forest fires in 2012, 2013 and 2014 were overall due to the civil war, during which hundreds of shells and bombs were fired in this area (Figures 5 and 6). In addition, forest logging operations took place after these fires because of the high prices of oil derivatives, especially in the upper part of the basin near the border area between Syria and Turkey. In comparison, the areas with reduced soil erosion are located in small patches in the western and southwestern regions, where slow slopes and fruit trees dominate.

From 2013/2014 to 2018/2019, the C factor caused a decrease in the average soil erosion rate by 50–100% in most parts of the basin. This period was characterized by a humid climate, in addition to a ceasefire agreed upon after 2016, which enabled the growth and renewal of vegetation cover. The western portion, close to the river mouth, showed no significant change between the two years and is characterized by low precipitation, low slope and low fruit tree coverage.

By comparing 2009/2010 (before war) with 2018/2019 (at the end of the war stage), it can be concluded that the change in C factors slowed down the deterioration trend of soil erosion and reduced the average soil erosion rate in more than half of the basin by about 10–75%.

This area is located relatively far away from the centers of armed conflicts. In contrast, the areas with increased soil erosion by about 60–400% are situated in the northeast and east, with shorter distances to armed conflict centers. These findings can be explained by forest fires, after which the burned forests were turned into agricultural land or refugee camps and road areas.

4. Discussion

Processes of soil erosion and surface degradation have various impacts on the worldwide land use as well as on land condition. As a result, an increasing number of scientists and policy-makers need to deal effectively with soil erosion issues and recognize how to mitigate their impact [2]. The most commonly used soil erosion model, RUSLE, was adopted in the present study. In the most recent 10 years of this study, the average soil erosion of the Northern Al-Kabeer river basin was characterized as mild (1–10 t ha^{-1} year^{-1}), which accounted for more than 66% of the total area. This basin is subject to a mean soil erosion rate of 3.6 t ha^{-1} year^{-1}. Furthermore, only 10% of the area has a tolerable soil loss rate, based on the 1 t ha^{-1} year^{-1} soil tolerance limit for Mediterranean highlands. Many authors consider that the limit of 1 t ha^{-1} year^{-1} has been set as the acceptable soil loss tolerances for environmental protection in the Mediterranean area, when considering the balance between soil formation and erosion [61,64–66]. Mediterranean soils have a low degree of tolerance for soil loss, which is much lower than soils of the temperate humid zone [64,65].

Overall, the average soil loss rate determined by RUSLE over the entire study region (3.6 t ha^{-1} year^{-1}) before the war, approaches the under-average soil erosion rate recorded in other studies for Mediterranean mountain forest areas. In northern Jordan, Alkharabsheh et al. (2013) and Farhan et al. projected an average soil loss rate of 9 t ha^{-1} year^{-1} [37] and 10 t ha^{-1} year^{-1} [38], respectively, whereas Karamesouti et al. (2016) reported an average soil loss rate of 16.9 t ha^{-1} year^{-1} in Athens using the RUSLE model [2]. The average soil loss rate was 7.9 t ha^{-1} year^{-1} during the Syrian civil war. During the period 2013–2014, several fires occurred due to indiscriminate shelling of the area (Figure 7a,b). The method of cutting trees after forest fires (Figure 7c) began to be used for heating and cooking in light of the rising oil prices. A large number of residents cultivated the burned forest land (Figure 8) and planted it with olive trees with the aim to acquire property later. The soil erosion rate decreased to 2.2 t ha^{-1} year^{-1} in 2018/2019, after the ceasefire agreement in 2016 and the transformation of the watershed into a semi-protected area. These changes allowed the vegetation cover to regenerate. The regeneration of vegetation in many watershed places has improved the vegetation cover, as shown in the C Factor map of the current investigation. Mousa et al. (2011) also found an adequate regeneration of most plant species that were in the protected area of west Syria before the fire. This regeneration was either by seed or vegetative propagation, which took place two years after the forest fire. For instance, *Pinus brutia Ten.*, *Quercus cerris L. subsp. pseudocerris* (Boiss.) *Chalabi*, *Quercus infectoria Oliv.*, *Phillyrea media* L., *Pistacia palaestina* (Boiss)., *Rhus cotinus* L. and *Laurus nobilis* L. forests/trees were regenerated after the occurrence of the forest fires [67]. Indeed, the majority of fires occurred in 2012, and we found that most plant species were regenerated, especially in areas that were exposed to light fires. This regeneration greatly reduced erosion compared to 2014. This indicates that the forest is able to regenerate within a short period of time if adequate protection from grazing and logging are put in place, resulting in reduced soil erosion. During the field tour, we found that that the *Pistacia palaestina* (Boiss) (Figure 9) and *Pistacia atlantica* regenerated by vegetative propagation three weeks after the fire. In addition, the vegetation cover and soil properties disturbances, such as the aggregate stability and water repellency, are closely linked to the intensity of the fires.

Figure 7. Photos of burned forests of *Pinus brutia Ten.* in upstream river (**a**,**b**) burned forest by bombing and (**c**) wood Logging after the forest was burned.

Figure 8. Photo of the burned forests changed into farmland in the Syrian-Turkish border area (Yamada), 22 October 2013.

Figure 9. Vegetative regrowth of *Pistacia palestina* (Bioss) three weeks after a fire in Rabia (Autumn 2013).

Most of the basin experienced a significant increase in soil erosion rates during the period of 2013/2014 compared to 2009/2010, with the exception of some fruit tree areas in the direction towards the mouth of the river. This area is not far from the battlefront and where farmers temporarily leave their farming operations and property for security purposes [68]. There was an enhancement in the vegetation cover, owing to the lack of soil tilling, growth of weeds and lack of pruning of fruit trees, as reflected in the NDVI images of the present investigation.

Although soil erosion decreased in almost the entire study area in 2018/2019, soil erosion remained high along the Syrian-Turkish border due to forest fires, the establishment of displacement camps, the construction of roads and the conversion of burned forests into farmland. Many studies (e.g., [69–71]) agreed that a key function for the government is to ensure ownership rights for soil users: lack of secure land tenure is a major impediment to taking erosion-control steps. For many soil conservation initiatives, the need for secure land tenure is particularly significant, as many do not have discernible short-term benefits. Stocking (2003) [71] noted that the greatest harm to the soil occurs where tenure, for example with migrants and refugees, is most unpredictable. Local knowledge is low in such conditions and soil mining is important for survival, at least in the short term.

During 2018/2019, vegetation cover increased by approximately 20–50% in the forest area that burned in 2012 (Figure 5), which led to a decline in soil erosion by about 10–75% (Figure 4). Gyssels et al. (2005) found sheet and rill erosion to be reduced by 75% in coverings of around 30 to 35%, and 90% in coverings of around 60% [34]. The rate of soil erosion decreased with an increase in vegetation cover and root biomass, whereas the soil erosion rate decreased with a rise in vegetation cover and root biomass. The rate of soil erosion further decreased due to forest cover, roots and litter components resulting in soil surface defense against raindrop effect, a decrease in flow velocity caused by roughness and improved infiltration capability [2,10,34,72].

These results reflect the observations made during the field investigation. Areas with extremely severe erosion (i.e., more than $10\ \mathrm{t\ ha^{-1}\ year^{-1}}$) experienced severe erosion in the field, as demonstrated by many gullies and low vegetation cover. The soil loss increased with a rise in slope in burned forests (Figure 10), which has also been recorded [23] for eight different slopes. This indicates that the cumulative soil loss after rainfall increases with the slope gradient due to precipitation intensities and is more pronounced due to higher slopes. Such areas have very steep slopes, showing the importance of using the slope percentage when examining soil erosion. This is consistent with the relationship between erosion and the square of the slope [73]. The variations in the erosion chart (Figure 4) are obviously seen to be identical to those of the LS (Figure 2b) and C factor maps (Figure 3e,f). The areas were presented in the upper part of the basin. In terms of the spatial patterns in China, Hui et al. (2010) also found that the LS and C measurements are highly correlated with those of erosion [73].

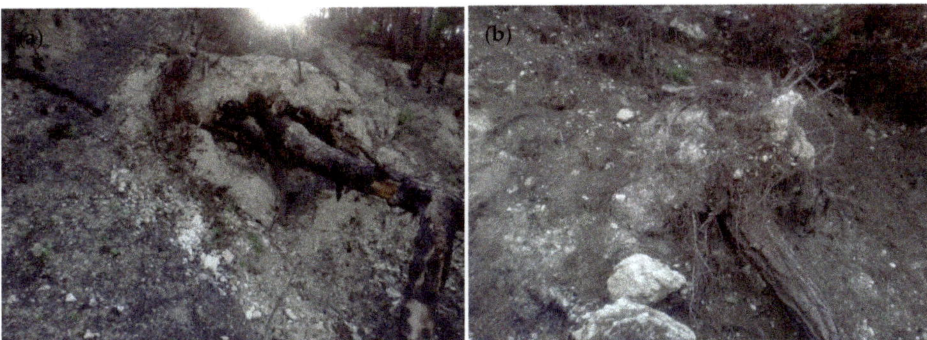

Figure 10. Photos of soil erosion in burned forests at steep slopes in (**a**) Yamada and (**b**) Rabia.

Although the forest loss has decreased in the last three years in most of the study area, it is possible that the forest area will recede again. This recurrence may be due to either renewed battles or an influx of people coming back to the forested areas after the conflict, resulting in unplanned development and settlement. Grime (2019) found that at the end of the armed conflicts in Sri Lanka, Nepal, Peru and the Ivory Coast, on average, there was a 68.08% increase of annual forest loss in the five years following the end of the conflicts. This average is based on analyses of forest-cover data gathered with remote-sensing methods, as compared to the worldwide 7.20% mean [40].

It is imperative, in the reconstruction phase, that environmental systems be given great importance in order to achieve sustainable development. This will require continued research to determine interdisciplinary dialogue of soil erosion science, as well as development and innovation. This could be a political will to create mechanisms that would aid the process of regeneration of war-torn areas, by encouraging the delivery of environmental services. This in turn will minimize the possibility that the region could become the target of potential socio-economic conflicts.

5. Conclusions

This study aimed to determine the impact of land cover change on soil erosion and its spatio-temporal variability in the Northern Al-Kabeer watershed over the past 10 years, using the RUSLE model and GIS. The impact of war on nature in the period of 2012–2014 was significantly negative, mainly as a result of forest fires, which led to a more than 10-fold increase of erosion in steep areas. It is observed that soil erosion increases as we move away from the center of the conflict towards the east and north-east as this region has been exposed to fires, and part of it has been turned into agricultural lands or refugee camps and roads near the Syrian-Turkish border. On the other hand, erosion decreases as we move away from the center of armed conflict towards the west, with this region being covered mostly with fruit trees in addition to some shrubs and forests that were abandoned because of the vicinity to the center of the armed conflict.

After the cease-fire decision in 2016, nature and wildlife benefited from the armed conflict in regions where human pressure decreased in contested areas and the area turned into a semi-nature reserve contaminated with landmines. Despite this improvement, battles may resume again and negatively affect the rest of the forests. Additionally, human interventions in forests usually increase with the end of the war, as happened in Sri Lanka and Colombia, owing to the return of the residents to their homes and their desire to improve their standard of living through the conversion of forests to agricultural lands or tourist facilities. It is apparent from the findings that, even though some inconsistencies and inaccuracies are present, the RUSLE model can be implemented effectively at the watershed scale with limited data requirements. The results of this study make a significant contribution to enhancing our understanding of the effect of forest fires and land cover change on soil erosion, and support the applicability of these models to forecast pre- and post-fire soil erosion rates. However, further research is required to quantitatively expose soil erosion and its influencing factors, and to validate the RUSLE model's success in the field. Analyzed soil loss dynamics can encourage decision-makers and planners to take appropriate forest and soil conservation priority actions, thereby reducing Syria's land loss and land degradation issues as a consequence of high annual fire rates and armed conflicts.

Funding: The article processing charge was funded by the Deanship of Scientific Research, Qassim University.

Acknowledgments: The author would like to thank the Deanship of Scientific Research, Qassim University for funding the publication of this project. I would like to express my sincere gratitude to Dr. Ibrahim Elton from Qassim University for his suggestions and comments. Additionally, the author is grateful to the editors and anonymous reviewers for their constructive comments and suggestions, which have considerably improved the quality of this manuscript.

Conflicts of Interest: The author declares no conflict of interest.

Appendix A. Description of RUSLE Model

Rainfall erosivity factor

Arnoldus (1977) proposed a modified version of the Fournier Index (F) to avoid pitfalls related to the monthly distribution of erosive precipitation over the year. Arnoldus used the F established in erosion-risk areas in North Africa and the Middle East for regions in the USA and Africa [56]. F is the summation of squares of monthly rain fall to its annual rainfall. F is calculated as:

$$F = \sum_{i=1}^{12} \frac{Pi^2}{P} \tag{A1}$$

where pi is the month's average rainfall (mm) with the maximum rainfall and P is the mean annual rainfall (mm).

Arnoldus (1977) computed the equation as $R = 0.264 \times F^{1.5}$, which was used to develop an iso-erodent map for Morocco [74]. Similarly, the regression equations with R and F were obtained by Renard and Freimund (1994) [75] as:

$$R = 0.07397 \times F^{1.847} \tag{A2}$$

Soil Erodibility Factor (K)

The Equation (A3) was used to estimate the erodibility of soil, as suggested by [59]:

$$K = f_{csand} \times f_{cl-si} \times f_{orgc} \times f_{hisand} \tag{A3}$$

where f_{csand} is a low soil erodibility factor for soil with coarse sand and a high amount of soil with a low sand content (Equation (A4)), f_{cl-si} is a factor that provides low soil erodibility with high clay to silt ratios (Equation (A5)), f_{orgc} is a factor that decreases soil erodibility for soil with a high organic carbon content (Equation (A6)) and f_{hisand} is a factor that decreases soil erosion with an exceptionally high sand content (Equation (A7)).

$$f_{csand} = 0.2 + 0.3 \times exp\left[-0.256 \times m_s \times \left(1 - \frac{m_{silt}}{100}\right)\right] \tag{A4}$$

$$f_{cl-si} = \left(\frac{m_{silt}}{m_c + m_{silt}}\right)^{0.3} \tag{A5}$$

$$f_{orgc} = 1 - \frac{0.0256 \times orgC}{orgC + \exp[3.72 - (2.95 \times orgC)]} \tag{A6}$$

$$f_{hisand} = 1 - \frac{0.7 \times \left(1 - \frac{m_s}{100}\right)}{\left(1 - \frac{m_s}{100}\right) + \exp\left[-5.51 + 22.9 \times \left(1 - \frac{m_s}{100}\right)\right]} \tag{A7}$$

where, m_s, m_{silt} and m_c are the percentage of sand, silt and clay, respectively, and $orgC$ is the organic carbon content of the layer (%).

Slope Length and Steepness Factors (*LS*)

The slope length factor (*L*) is provided by the Desmet and Govers (1996) [60], and is enhanced by the USLE estimation technique (Equation (A8)):

$$L_{i,j} \frac{\left(A_{i,j-in} + D^2\right)^{m+1} - A_{i,j-in}^{m+1}}{D^{m+2} \times x_{i,j}^m x_{i,j} \times 22.13^m} \tag{A8}$$

where $L_{i,j}$ is slope length factor for the grid cell with coordinates (i.j); $A_{i,j-in}$ is the flow accumulation or contributing area at the inlet of a grid cell with coordinates measured in m^2; D is grid cell size (meters); $x_{i,j}$ is $sina_{i,j} + cosa_{i,j}$; $a_{i,j}$ is the aspect direction of the grid cell with coordinates; m is a function of the ratio β of the rill to inter-rill erosion. The m varies from 0 to 1 and reaches 0, while the ratio of rill to inter-rill erosion is similar to 0. The exponent m of the following equation was implemented according to the algorithm proposed by McCool et al. (1989) [76] (Equation (A9)):

$$m = \frac{\beta}{1 + \beta} \tag{A9}$$

The β value is derived by the Equation (A10):

$$\beta = \left(\frac{sin\theta}{0.0896}\right) / \left[3(sin\theta)^{0.8} + 0.56\right] \tag{A10}$$

here, θ is the slope angle.

The estimation of the *S*-factor originally proposed by Wischmeier and Smith (1978) was proposed by McCool et al. (1987) in the RUSLE model to achieve an improved representation of the slope steepness factor, taking into account the ratio of rill and inter-rill erosion. McCool et al. (1987) found that soil erosion occurred more rapidly on slopes with a steepness of more than 9%. Therefore, he used one algorithm for slopes < 9% and another for slopes > 9% (Equations (A11) and (A12)):

$$S = 10.8 \times SIN\theta - 0.03 \text{ where slope gradient} < 9\% \tag{A11}$$

$$S = 16.8 \times SIN\theta - 0.5 \text{ where slope gradient} \geq 9\% \tag{A12}$$

References

1. Diodato, N.; Bellocchi, G. MedREM, a rainfall erosivity model for the Mediterranean region. *J. Hydrol.* **2010**, *387*, 119–127. [CrossRef]
2. Karamesouti, M.; Petropoulos, G.P.; Papanikolaou, I.D.; Kairis, O.; Kosmas, K. Erosion rate predictions from PESERA and RUSLE at a Mediterranean site before and after a wildfire: Comparison & implications. *Geoderma* **2016**, *261*, 44–58.
3. Diodato, N. Estimating RUSLE's rainfall factor in the part of Italy with a Mediterranean rainfall regime. *Hydrol. Earth Syst. Sci.* **2004**, *8*, 103–107. [CrossRef]

4. Albergel, J.; Collinet, J.; Zante, P.; Hamrouni, H. Role of Mediterranean forest in soil and water conservation. In *Water for Forests and People in the Mediterranean Region: A Challenging Balance. What Science Can Tell Us*; European Forest Institute: Joensuu, Finland, 2011.

5. Abdelwahab, O.M.M.; Ricci, G.F.; De Girolamo, A.M.; Gentile, F. Modelling soil erosion in a Mediterranean watershed: Comparison between SWAT and AnnAGNPS models. *Environ. Res.* **2018**, *166*, 363–376. [CrossRef] [PubMed]

6. Polykretis, C.; Alexakis, D.D.; Grillakis, M.G.; Manoudakis, S. Assessment of Intra-Annual and Inter-Annual Variabilities of Soil Erosion in Crete Island (Greece) by Incorporating the Dynamic "Nature" of R and C-Factors in RUSLE Modeling. *Remote Sens.* **2020**, *12*, 2439. [CrossRef]

7. Hategekimana, Y.; Allam, M.; Meng, Q.; Nie, Y.; Mohamed, E. Quantification of soil losses along the coastal protected areas in Kenya. *Land* **2020**, *9*, 137. [CrossRef]

8. Woldemariam, G.; Iguala, A.; Tekalign, S.; Reddy, R. Spatial Modeling of Soil Erosion Risk and Its Implication for Conservation Planning: The Case of the Gobele Watershed, East Hararghe Zone, Ethiopia. *Land* **2018**, *7*, 25. [CrossRef]

9. Eisenberg, J.; Muvundja, F.A. Quantification of Erosion in Selected Catchment Areas of the Ruzizi River (DRC) Using the (R)USLE Model. *Land* **2020**, *9*, 125. [CrossRef]

10. Food and Agricultural Organization (FAO). *Soil Erosion*; FAO: Rome, Italy, 2019.

11. Food and Agricultural Organization (FAO). *Forests and Climate Change in the Near East Region*; Forests and Climate Change Working Papers 9; FAO: Rome, Italy, 2010.

12. Habib, L.; Ibrahim, W. STATUS OF SOIL RESOURCE IN SYRIA. In Proceedings of the Assessment of Existing Soil Information System, Damascus, 2007, Tunis, Tunisia, 26–31 May 2007.

13. Mohammed, S.A.; Alkerdi, A.; Nagy, J.; Harsányi, E. Syrian crisis repercussions on the agricultural sector: Case study of wheat, cotton and olives. *Reg. Sci. Policy Pract.* **2020**, *12*, 519–537. [CrossRef]

14. Selby, J.; Dahi, O.S.; Fröhlich, C.; Hulme, M. Climate change and the Syrian civil war revisited. *Political Geogr.* **2017**, *60*, 232–244. [CrossRef]

15. Global Forest Watch Syria. Available online: https://www.globalforestwatch.org/dashboards/country/SYR/ (accessed on 7 June 2020).

16. Nackoney, J.; Molinario, G.; Potapov, P.; Turubanova, S.; Hansen, M.C.; Furuichi, T. Impacts of civil conflict on primary forest habitat in northern Democratic Republic of the Congo, 1990–2010. *Biol. Conserv.* **2014**, *170*, 321–328. [CrossRef]

17. Ordway, E.M. Political shifts and changing forests: Effects of armed conflict on forest conservation in Rwanda. *Glob. Ecol. Conserv.* **2015**, *3*, 448–460. [CrossRef]

18. Machlis, G.E.; Hanson, T. Warfare Ecology. *NATO Sci. Peace Secur. Ser. C Environ. Secur.* **2011**, *113*, 33–40.

19. Kim, K.C. Preserving biodiversity in Korea's demilitarized zone. *Science* **1997**, *278*, 242–243. [CrossRef]

20. Baumann, M.; Kuemmerle, T. The impacts of warfare and armed conflict on land systems. *J. Land Use Sci.* **2016**, *11*, 672–688. [CrossRef]

21. Hecht, S.B.; Saatchi, S.S. Globalization and Forest Resurgence: Changes in Forest Cover in El Salvador. *Bioscience* **2007**, *57*, 663–672. [CrossRef]

22. Al-Ali, Y.; Shater, Z.; Kheder, R. Studying the Effect of Forest Fire on Soil Erosion and Loss of Some Mineral Elements in the Forest of Ein Al-Jaouz/Tartous. *Tishreen Univ. J. Res. Sci. Stud. Biol. Sci. Ser.* **2014**, *36*, 277–290.

23. Kabibo, I.A.-D.; Bou-Issa, A.; Ibrahim, J. Studying the Effect of Soil Erosion for Eight Different Systems with Different Slopes in the Coastal Area under Forests, Burned Forest and Planted Soil System. *Tishreen Univ. J. Res. Sci. Stud. Biol. Sci. Ser.* **2017**, *1*, 25–38.

24. Mohammed, S.; Abdo, H.G.; Szabo, S.; Pham, Q.B.; Holb, I.J.; Linh, N.T.T.; Anh, D.T.; Alsafadi, K.; Mokhtar, A.; Kbibo, I.; et al. Estimating Human Impacts on Soil Erosion Considering Different Hillslope Inclinations and Land Uses in the Coastal Region of Syria. *Water* **2020**, *12*, 2786. [CrossRef]

25. Mohammed, S.; Alsafadi, K.; Talukdar, S.; Kiwan, S.; Hennawi, S.; Alshihabi, O.; Sharaf, M.; Harsanyie, E. Estimation of soil erosion risk in southern part of Syria by using RUSLE integrating geo informatics approach. *Remote Sens. Appl. Soc. Environ.* **2020**, *20*, 100375. [CrossRef]

26. Almohamad, H.; Knaack, A.L.; Habib, B.M. Assessing Spatial Equity and Accessibility of Public Green Spaces in Aleppo City, Syria. *Forests* **2018**, *9*, 706. [CrossRef]

27. Almohamad, H.; Dittmann, A. Oil in Syria between Terrorism and Dictatorship. *Soc. Sci.* **2016**, *5*, 20. [CrossRef]

28. Renard, B.K.G.; Foster, G.R.; Weesies, G.A.; Porter, J.I. Revised universal soil loss equation (Rusle). *J. Soil Water Conserv.* **1991**, *46*, 30–33.

29. Marti, N.; Goci, M. Changes in Soil Erosion Intensity Caused by Land Use and Demographic Changes in the Jablanica River Basin, Serbia. *Agriculture* **2020**, *10*, 345.

30. Licciardello, F.; Govers, G.; Cerdan, O.; Kirkby, M.J.; Vacca, A.; Kwaad, F.J.P.M. Evaluation of the PESERA model in two contrasting environments. *Earth Surf. Process. Landf.* **2009**, *34*, 629–640. [CrossRef]

31. Kirkby, M.J.; Irvine, B.J.; Jones, R.J.A.; Govers, G.; Boer, M.; Cerdan, O.; Daroussin, J.; Gobin, A.; Grimm, M.; Le Bissonnais, Y.; et al. The PESERA coarse scale erosion model for Europe. I.—Model rationale and implementation. *Eur. J. Soil Sci.* **2008**, *59*, 1293–1306. [CrossRef]

32. Baigorria, G.A.; Romero, C.C. Assessment of erosion hotspots in a watershed: Integrating the WEPP model and GIS in a case study in the Peruvian Andes. *Environ. Model. Softw.* **2007**, *22*, 1175–1183. [CrossRef]

33. Laflen, J.M.; Lane, L.J.; Foster, G. WEPP: A new generation of erosion prediction technology. *J. Soil Water Conserv.* **1991**, *46*, 34–38.

34. Gyssels, G.; Poesen, J.; Bochet, E.; Li, Y. Impact of plant roots on the resistance of soils to erosion by water: A review. *Prog. Phys. Geogr.* **2005**, *29*, 189–217. [CrossRef]

35. Arabameri, A.; Cerda, A.; Tiefenbacher, J.P. Spatial pattern analysis and prediction of gully erosion using novel hybrid model of entropy-weight of evidence. *Water* **2019**, *11*, 1129. [CrossRef]

36. Angulo-Martínez, M.; Beguería, S. Estimating rainfall erosivity from daily precipitation records: A comparison among methods using data from the Ebro Basin (NE Spain). *J. Hydrol.* **2009**, *379*, 111–121. [CrossRef]

37. Farhan, Y.; Zregat, D.; Farhan, I. Spatial Estimation of Soil Erosion Risk Using RUSLE Approach, RS, and GIS Techniques: A Case Study of Kufranja Watershed, Northern Jordan. *J. Water Resour. Prot.* **2013**, *5*, 1247–1261. [CrossRef]

38. Alkharabsheh, M.M.; Alexandridis, T.K.; Bilas, G.; Misopolinos, N.; Silleos, N. Impact of Land Cover Change on Soil Erosion Hazard in Northern Jordan Using Remote Sensing and GIS. *Procedia Environ. Sci.* **2013**, *19*, 912–921. [CrossRef]

39. Vinson, J.A.; Barrett, S.M.; Aust, W.M.; Bolding, M.C. Suitability of soil erosion models for the evaluation of bladed skid trail BMPs in the Southern appalachians. *Forests* **2017**, *8*, 482. [CrossRef]

40. Uddin, K.; Matin, M.A.; Maharjan, S. Assessment of land cover change and its impact on changes in soil erosion risk in Nepal. *Sustainability* **2018**, *10*, 4715. [CrossRef]

41. Das, T. *Estimation of Annual Average Soil Loss and Preparation of Spatially Distributed Soil Loss Map: A Case Study of Dhansiri River Basin*; Indian Institute of Technology Guwahati: Guwahati, India, 2017; p. 38.

42. Ferreira, V.; Panagopoulos, T. Seasonality of soil erosion under Mediterranean conditions at the Alqueva dam watershed. *Environ. Manag.* **2014**, *54*, 67–83. [CrossRef] [PubMed]

43. Alewell, C.; Borrelli, P.; Meusburger, K.; Panagos, P. Using the USLE: Chances, challenges and limitations of soil erosion modelling. *Int. Soil Water Conserv. Res.* **2019**, *7*, 203–225. [CrossRef]

44. Abu Hammad, A.; Lundekvam, H.; Børresen, T. Adaptation of RUSLE in the eastern part of the Mediterranean region. *Environ. Manag.* **2004**, *34*, 829–841. [CrossRef]

45. Aiello, A.; Adamo, M.; Canora, F. Remote sensing and GIS to assess soil erosion with RUSLE3D and USPED at river basin scale in southern Italy. *Catena* **2015**, *131*, 174–185. [CrossRef]

46. Hasan, I.A.; Hammad, M.; Ahmad, M.; Boubou, M.; Yazaji, Z. Studying of Sedimentation in Reservoir of 16 November Dam. *Tishreen Univ. J. Res. Sci. Stud. Eng. Sci. Ser.* **2019**, *41*, 23–43.

47. Niu, L.; Shao, Q. Soil conservation service spatiotemporal variability and its driving mechanism on the Guizhou Plateau, China. *Remote Sens.* **2020**, *12*, 2187. [CrossRef]

48. Aslam, B.; Maqsoom, A.; Shahzaib; Kazmi, Z.A.; Sodangi, M.; Anwar, F.; Bakri, M.H.; Faisal Tufail, R.; Farooq, D. Effects of Landscape Changes on Soil Erosion in the Built Environment: Application of Geospatial-Based RUSLE Technique. *Sustainability* **2020**, *12*, 5898. [CrossRef]

49. Barakat, M.; Mahfoud, I.; Kwyes, A.A. Study of soil erosion risk in the basin of Northern Al-Kabeer river at Lattakia-Syria using remote sensing and GIS techniques. *Mesop. J.* **2014**, *29*, 29–44.

50. Ghazal, A. *Landscape Ecological, Phytosociological and Geobotanical Study of Eu-Mediterranean in West of Syria*; University of Hohenheim: Stuttgart, Germany, 2008.

51. Renard, K.G.; Foster, G.R.; Weesies, G.A.; McCool, D.K.; Yoder, D.C. *Predicting Soil Erosion by Water: A Guide to Conservation Planning with the Revised Universal Soil Loss Equation (RUSLE)*; United States Government Printing: Washington, DC, USA, 1997.

52. Biswas, S.S.; Pani, P. Estimation of soil erosion using RUSLE and GIS techniques: A case study of Barakar River basin, Jharkhand, India. *Model. Earth Syst. Environ.* **2015**, *1*, 42. [CrossRef]

53. Karamage, F.; Shao, H.; Chen, X.; Ndayisaba, F.; Nahayo, L.; Kayiranga, A.; Omifolaji, J.K.; Liu, T.; Zhang, C. Deforestation effects on soil erosion in the Lake Kivu Basin, D.R. Congo-Rwanda. *Forests* **2016**, *7*, 281. [CrossRef]

54. Koirala, P.; Thakuri, S.; Joshi, S.; Chauhan, R. Estimation of Soil Erosion in Nepal using a RUSLE modeling and geospatial tool. *Geosciences* **2019**, *9*, 147. [CrossRef]

55. Barakat, M.; Ileen Mahfood, A.A.A.-K. Study of the soil water erosion in the basin of 16 Tishreen Dam in the province of Lattakia using Geographic Information System techniques (GIS). *Tishreen Univ. J. Res. Sci. Stud.* **2013**, *35*, 85–104.

56. Sabri, E.; Boukdir, A.; Mabrouki, M.; Romaric, V.; Mbaki, E. *Predicting Soil Erosion and Sediment Yield in Oued El Abid Watershed, Morocco*; ATINER's Conference Paper Series GEO2016–2091; Athens Institute for Education and Research ATINER: Athens, Greece, 2016; p. 23.

57. Yavuz, M.; Tufekcioglu, M. Estimating Surface Soil Losses in the Mountainous Semi-Arid Watershed using RUSLE and Geospatial Technologies. *Fresenius Environ. Bull.* **2019**, *28*, 2589–2598.

58. Irvem, A.; Topaloğlu, F.; Uygur, V. Estimating spatial distribution of soil loss over Seyhan River Basin in Turkey. *J. Hydrol.* **2007**, *336*, 30–37. [CrossRef]

59. Sharpley, A.N.; Williams, J.R. EPIC: The erosion-productivity impact calculator. *U.S. Dep. Agric. Tech. Bull.* **1990**, *1768*, 235.

60. Desmet, P.J.J.; Govers, G. A GIS procedure for automatically calculating the USLE LS factor on topographically complex landscape units. *J. Soil Water Conserv.* **1996**, *51*, 427–433.

61. Van der Knijff, J.M.F.; Jones, R.J.A.; Montanarella, L. *Soil Erosion Risk Assessment Italy*; European Soil Bureau, European Commission: Brussels, Belgium, 1999.

62. Wang, Z.; Su, Y. Assessment of Soil Erosion in the Qinba Mountains of the Southern Shaanxi Province in China Using the RUSLE Model. *Sustainability* **2020**, *12*, 1733. [CrossRef]

63. Panagos, P.; Borrelli, P.; Poesen, J.; Ballabio, C.; Lugato, E.; Meusburger, K.; Montanarella, L.; Alewell, C. The new assessment of soil loss by water erosion in Europe. *Environ. Sci. Policy* **2015**, *54*, 438–447. [CrossRef]

64. Poesen, J.W.A.; Hooke, J.M. Erosion, flooding and channel management in Mediterranean environments of southern Europe. *Prog. Phys. Geogr.* **1997**, *21*, 157–199. [CrossRef]

65. García-Ruiz, J.M.; Nadal-Romero, E.; Lana-Renault, N.; Beguería, S. Erosion in Mediterranean landscapes: Changes and future challenges. *Geomorphology* **2013**, *198*, 20–36. [CrossRef]

66. Panagos, P.; Borrelli, P.; Meusburger, K. A new European slope length and steepness factor (LS-factor) for modeling soil erosion by water. *Geosciences* **2015**, *5*, 117–126. [CrossRef]

67. Mousa, F.H.; Sarhan Laykah, M.A. Studying the Effect of Fire on Regeneration of Natural Vegetation in Kassab. *Damascus Univ. J. Basic Sci.* **2011**, *27*, 115–136.

68. Jaafar, H.H.; Zurayk, R.; King, C.; Ahmad, F.; Al-Outa, R. Impact of the Syrian conflict on irrigated agriculture in the Orontes Basin. *Int. J. Water Resour. Dev.* **2015**, *31*, 436–449. [CrossRef]

69. Juerges, N.; Hansjürgens, B. Soil governance in the transition towards a sustainable bioeconomy—A review. *J. Clean. Prod.* **2018**, *170*, 1628–1639. [CrossRef]

70. Shiferaw, B.A.; Okello, J.; Reddy, R.V. Adoption and adaptation of natural resource management innovations in smallholder agriculture: Reflections on key lessons and best practices. *Environ. Dev. Sustain.* **2009**, *11*, 601–619. [CrossRef]

71. Stocking, M.A. Tropical Soils and Food Security: The Next 50 Years. *Science* **2003**, *302*, 1356–1359. [CrossRef] [PubMed]

72. Raclot, D.; Le Bissonnais, Y.; Annabi, M.; Sabir, M.; Smetanova, A. Main Issues for Preserving Mediterranean Soil Resources From Water Erosion Under Global Change. *Land Degrad. Dev.* **2018**, *29*, 789–799. [CrossRef]

73. Hui, L.; Xiaoling, C.; Lim, K.J.; Xiaobin, C.; Sagong, M. Assessment of soil erosion and sediment yield in Liao watershed, Jiangxi Province, China, Using USLE, GIS, and RS. *J. Earth Sci.* **2010**, *21*, 941–953. [CrossRef]

74. Arnoldus, H.M.J. Methodology Used to Determine the Maximum Potential Range Average Annual Soil Loss to Sheet and Rill Erosion in Morocco. Assessing Soil Degradation. *FAO Soils Bull.* **1977**, *34*, 39–48.

75. Renard, K.G.; Freimund, J.R. Using monthly precipitation data to estimate the R-factor in the revised USLE. *J. Hydrol.* **1994**, *157*, 287–306. [CrossRef]
76. Oliveira, J.A.; Dominguez, J.M.L.; Nearing, M.A.; Oliveira, P.T.S. A GIS-Based procedure for automatically calculating soil loss from the universal soil loss Equation: GISus-M. *Appl. Eng. Agric.* **2015**, *31*, 907–917.

Article

Comparison of the Applicability of Different Soil Erosion Models to Predict Soil Erodibility Factor and Event Soil Losses on Loess Slopes in Hungary

Boglárka Keller [1], Csaba Centeri [1,*], Judit Alexandra Szabó [2], Zoltán Szalai [2,3] and Gergely Jakab [2,3]

[1] Institute of Wildlife Management and Nature Conservation, Szent István Campus, Hungarian University of Agriculture and Life Sciences, 2100 Gödöllő, Hungary; bogi87@gmail.com
[2] Geographical Institute, Research Centre for Astronomy and Earth Sciences, 1112 Budapest, Hungary; szabojuditalexandra@gmail.com (J.A.S.); szalai.zoltan@csfk.mta.hu (Z.S.); jakab.gergely@csfk.org (G.J.)
[3] Department of Environmental and Landscape Geography, ELTE University, 1117 Budapest, Hungary
* Correspondence: Centeri.Csaba@uni-mate.hu

Abstract: Climate change induces more extreme precipitation events, which increase the amount of soil loss. There are continuous requests from the decision-makers in the European Union to provide data on soil loss; the question is, which ones should we use? The paper presents the results of USLE (Universal Soil Loss Equation), RUSLE (Revised USLE), USLE-M (USLE-Modified) and EPIC (Erosion-Productivity Impact Calculator) modelling, based on rainfall simulations performed in the Koppány Valley, Hungary. Soil losses were measured during low-, moderate- and high-intensity rainfalls on cultivated soils formed on loess. The soil erodibility values were calculated by the equations of the applied soil erosion models and ranged from 0.0028 to 0.0087 t ha h ha^{-1} MJ^{-1} mm^{-1} for the USLE-related models. EPIC produced larger values. The coefficient of determination resulted in an acceptable correlation between the measured and calculated values only in the case of USLE-M. Based on other statistical indicators (e.g., NSEI, RMSE, PBIAS and relative error), RUSLE, USLE and USLE-M resulted in the best performance. Overall, regardless of being non-physically based models, USLE-type models seem to produce accurate soil erodibility values, thus modelling outputs.

Keywords: rainfall simulation; field measurement; water erosion model; USLE; event scale

Citation: Keller, B.; Centeri, C.; Szabó, J.A.; Szalai, Z.; Jakab, G. Comparison of the Applicability of Different Soil Erosion Models to Predict Soil Erodibility Factor and Event Soil Losses on Loess Slopes in Hungary. *Water* **2021**, *13*, 3517. https://doi.org/10.3390/w13243517

Academic Editor: Maria Mimikou

Received: 31 October 2021
Accepted: 7 December 2021
Published: 9 December 2021

Publisher's Note: MDPI stays neutral with regard to jurisdictional claims in published maps and institutional affiliations.

1. Introduction

Water erosion is a worldwide problem that causes several economic and environmental impacts [1] due to deteriorating soil functions [2,3]. The negative impacts can occur on-site by reduced production and off-site by sediment transportation. Sediments can deliver pollutants into surface waters or onto other lands, threatening sustainable land use [4]. As Miller et al. [5] announced, nearly 90 percent of the total soil loss of nitrogen and phosphorus [6] can be delivered with the soil loss and later accumulated within the sedimented area. Recognizing the severity of accelerated soil erosion processes, a soil loss tolerance value was introduced [7,8] in the USA for soil conservation planning in the 1940s. Later, Mannering [9] and Skidmore [10] suggested the involvement of establishing tolerable rates into the regulation of farming. Morgan [11] defined the tolerable amount of soil loss as an amount that does not considerably decrease soil fertility; thus it is a key factor of sustainable soil management. The concept of tolerable soil loss was applied in soil conservation planning, as well as soil loss evaluation and mapping. The annual and long-term average soil loss can serve as the basis of soil erosion predictions and nonpoint source pollution control. With climate change, the occurrence of rare but more erosive rainfall events is going to increase [12]; therefore, studying the effect of these events on soil degradation is becoming more essential and crucial [13], because these events cause the larger part of annual soil erosion. Therefore, many scientists [14–17] emphasize that

141

the conservation strategies should take into account large storms rather than average weather conditions.

With rising competition for the limited soil resources, soil erosion prediction has become an essential part of soil conservation planning and practices, as there is a need for simple and cost-effective soil erosion- and soil erodibility measurements [18–22]. There are several types of soil erosion models; the new ones require a lot of input parameters. This can be the reason why USLE (Universal Soil Loss Equation) [8] is still one of the most often used models. Compared to other models, USLE is still the most cited of all soil erosion models. USLE users do not need to calculate event rainfalls for the calculation of a yearly average [23–25]. Still, it has been recognised that there is a need to predict soil losses on a shorter time scale, which resulted in many studies with rainfall simulation measurements. Some studies [26–29] warn us that these simulations reflect single rainfall events which over- or underestimate average soil loss [30–34]. This phenomenon is common when the EI30 index is used in estimating erosion on bare fallow areas [35,36]. Boardman [37] stated that the application of USLE can assume unconcern in event-driven erosion, whereas it is unsatisfactory for estimating soil loss at event scale. Foster et al. [38] recommended that the erosivity factor, including rainfall amount, rainfall intensity and runoff amount, should be better than the single-storm erosion index (EI30) by Wischmeier and Smith [8]. On the other hand, Kinnel et al. [39] stated that USLE and similar models, such as RUSLE, can predict the event soil loss for runoff-producing events when we assume that the event soil loss is directly related to the product of event storm kinetic energy (E) and the maximum 30 min rainfall intensity (I30) [40]. This way, both USLE and RUSLE can model short-term soil losses. To provide better predictions, several similar but modified models were built, such as MUSLE [41], RUSLE [42], WEPP [43,44], USLE-M [45], etc.; while USLE and RUSLE are detachment limited models that hypothesise that soil loss is not affected by the runoff capacity of detached soil particles, USLE-M studies the runoff, operating, in this way, as a transport-limited model.

In practical soil conservation measures, the most important element of soil erosion modelling is soil erodibility or "K-factor", as it became known by the work of Wischmeier and Smith [8]. This factor is defined as the susceptibility of a soil to erode, representing the effect of soil properties on soil loss and the resistance against surface water runoff [42]. Unfortunately, the utilization of soil erodibility is not consistent in the literature [22,46,47], as it follows many inaccuracies and uncertainties within the soil erosion modelling soil erodibility determinations [48–54] and soil loss estimations [55]. Cassol et al. [56] also drew attention to the fact that the USLE K-factors can differ from annual estimation values in the case they originate from individual rainstorm events. Although there is a well-known standard procedure for the USLE model [57] (22.13 m long with 9% uniform slope, natural and continuously tilled fallow field plots, long-term measurements), there are still several studies with different circumstances trying to simplify the observations [49]. Many pieces of literature exist, providing non-satisfying but detailed determination procedures; as a result, some uncertainty concerning the accuracy of the reported K-values exists. This means that different estimation methods behave differently and their results are relevant just under the given circumstances. The K-factor has high variability and mostly depends on the method of the experimental setup, the main soil properties and various and numerous input parameters [49]. Besides, in event-based, mostly rain-simulated erosional studies, erodibility is regarded as a dynamic variable due to the main exogenic forces [54], especially the rainfall and erosional processes driven by rain splash. To serve the purpose of simplifying the long-term measurements based on physical and chemical properties, several nomographs were created in the models (USLE, RUSLE and EPIC). At the same time, El-Swaify and Dangler [58] reported that erodibility estimation is quite risky, based on soil classification only. Based on the modelling research studies performed so far, we can conclude that there is a huge need for soil erodibility measurements [59].

Accordingly, the problem of how the soil erodibility values measured by event rainfall simulation can be comparable arises. However, to date, no comparison has been made relatively to the ability of the USLE, RUSLE and USLE-M models, nor of USLE and EPIC nomographs, in soil erodibility determination and soil loss estimations produced by individual high-intensity simulated rainfall events on different slope gradients and under different soil moisture conditions on soils formed on loess parent rock in the Koppány River Valley. Therefore, the specific objectives of this paper are the following: (a) to statistically test and compare the erodibility factors calculated by the USLE, RUSLE and USLE-M [35,36,45] models and the calculated values by the USLE and EPIC nomograph equations [60] based solely on soil physical and chemical properties; (b) to determine the difference between each predicted and measured soil loss based on high-intensity rainfall simulations; (c) to determine how accurate the event rainfall simulation results are, compared to other similar long term estimations; and (d) to establish how the different antecedent soil moisture content can affect the erodibility values and soil loss estimation results.

2. Materials and Methods

2.1. Study Site and Soil Properties

The concerned area is situated in the north-eastern part of the South-Transdanubian Region, on the east side of Somogy county, in Hungary. Gerézdpuszta is situated in the Koppány valley, adjacent to the floodplain of the Koppány River (Figure 1). This landscape is characterized by loess-covered asymmetric hilly areas, where the northern hillsides are short and steep, while the southern areas are longer and have gentle slopes [61]. Due to loess-like deposits, the area is prone to soil erosion; however, most of the arable lands are under conventional tillage where large-scale farming is typical. The less eroded hillslopes are characterized by Cambisols, while, in cultivated areas, Regosols and Colluviums can be found [62]. The climate is moderately warm and wet. The annual average temperature is between 10.0 °C and 10.2 °C and the average annual precipitation is between 605 mm and 700 mm. Most of the hillsides are characterized by agricultural cultivation and almost half of that is situated on slopes steeper than 12%; further, usually the farmers do not use any soil protection measures.

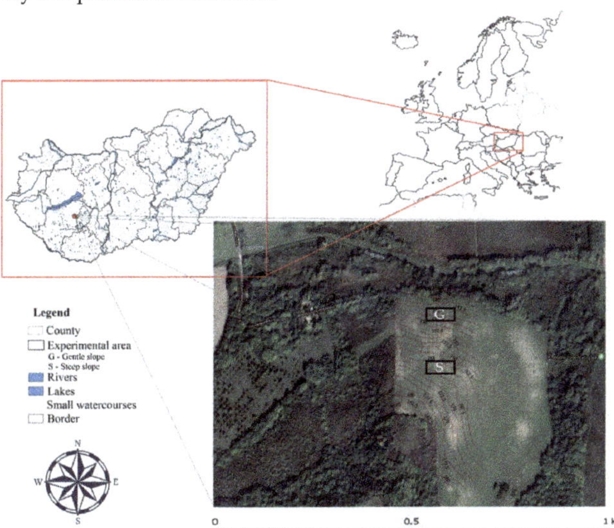

Figure 1. The situation of the study site and the experimental area within the field.

The in situ experiments were performed on freshly tilled Regosol, on gentle slopes (6.7%, 8.0% and 7.7%) and on steep slopes (18.32%, 17.26% and 17.5%), in 3 repetitions. The soil properties of the 0–30 cm layer are reported in Table 1.

Table 1. Soil properties of the examined soils in the 0–30 cm layer.

					Coarse Sand	Fine Sand	Sand (mm)	Silt	Clay	Soil Permeability	Soil Structure Class
	Plasticity	pH_{H_2O}	$CaCO_3$ (%)	SOC (%)	>0.25	0.25–0.0	0.05–0.02	0.02–0.002	<0.002	mm h^{-1}	
Gentle slope	41	8.06	2.7	1.37	0.9	29	22	20.2	28	55.43	fine
Steep slope	44	8.22	13	1.39	1.1	26.5	19.2	23.8	29.3		granular

The texture is sandy clay loam. The gentle slope section has low $CaCO_3$ content, while the steep slope has high $CaCO_3$ content; the soil organic carbon content is low in both sections. These characteristics indicate that the area has been severely prone to soil water erosion for a long time.

Plasticity indicates the amount of distilled water needed to reach the state of plasticity (or saturation). Soils are considered loam with values from 37 to 42 and clayey loam with values from 43 to 49 (accuracy is ± 2). The pH (H_2O) was measured with a WTW inolab pH7310 pH measuring device. SOC was measured with the wet combustion method. The particle sizes were measured with a Fritsch Laser analyser. Soil permeability was measured with a double-ring infiltrometer in the field. Soil structure was evaluated by experts on the field before the rainfall simulations.

2.2. Rainfall Simulator and Experimental Design

The Shower Power (SP) 02 is a portable field unit applicable on a plot scale with alternating nozzles which were designed and constructed by the Geographical Institute, Research Centre for Astronomy and Earth Sciences, Hungarian Academy of Sciences, in 2014. It is a portable field scale device with a measuring plot of 3×2 m; however—to exclude the border effects—the irrigated plot size was 12 m^2. The drop-forming unit was an alternating axis equipped with two 80,100 Veejet nozzles placed next to each other with a distance of 2 m and 3 m in height. This way the overlapping between the nozzles was perfect, resulting in a uniform spatial intensity and drop-size distribution. The intensity can vary in the range 30–100 mm h^{-1}, depending on the axis alternation frequency. Using the V-jet nozzles, the rainfall energy was 77% of natural rainfall when the pressure and rainfall intensity of the simulator was constantly held at 0.41 KPa and 65 mm/h intensity rate [63]. The average kinetic energy of the simulator was 24 J m^{-2} mm^{-1}. The plot was fenced using metal sheets pushed into the soil to a depth of 15 cm. At the bottom of the plot, a pair of collector triangles gathered the runoff. During the rainfall simulation, the starting and ending time of simulation, ponding and runoff time were recorded. The volume of runoff was continuously measured by two scaled dishes as a function of time and the whole runoff was collected separately into buckets. Time and temporary runoff subsample values were measured and stored in controlling software. The subsample measurements and sediment concentration determination were made from the proportion of the total soil loss over the simulation. In the laboratory, the total amount of soil loss was oven-dried for 48 h at 60 °C and soil loss was measured afterward. Altogether, 18 simulations were conducted (Table 2). Further details in Appendix A.

Table 2. Details of 18 rainfall simulations. The name of each treatment originates from the following: F—field experiment; P—Gerézdpuszta, the location of the experiments; G—gentle slope, or S—steep slope.

Treatment ID	s	Measured Rainfall Intensity	Rainfall	Rainfall Duration	Soil Moisture Content
	(%)	(mm h^{-1})	(mm)	(s)	
FPG1	6.7	15.06	19.18	4584	dry
FPG2	6.7	56.24	16.81	1076	moist
FPG3	6.7	84.68	10.63	452	wet
FPG4	8	22.65	28.96	4603	dry
FPG5	8	70.19	13.53	694	moist
FPG6	8	86.12	10.41	435	wet
FPG7	7.7	26.32	35.10	4801	dry
FPG8	7.7	56.92	15.19	961	moist
FPG9	7.7	80.44	12.4	555	wet
FPS1	18.32	20.87	26.64	4595	dry
FPS2	18.32	66.78	16.49	889	moist
FPS3	18.32	103.48	11.67	406	wet
FPS4	17.26	24.42	31.10	4585	dry
FPS5	17.26	49.89	10.53	760	moist
FPS6	17.26	63.16	8.05	459	wet
FPS7	17.6	18.68	19.9	4627	dry
FPS8	17.6	44.05	13.86	1121	moist
FPS9	17.6	76.69	11.18	525	wet

Three rainfall simulation treatment rates were sequentially applied: (1) dry antecedent moisture content before the application of the low rainfall intensity rate (30 mm/h^{-1}); (2) moist antecedent moisture after the low rainfall intensity rate (30 mm/h^{-1}) and before the medium rainfall intensity (60 mm/h^{-1}) (60 min after the previous treatment); and (3) wet antecedent moisture content after the medium rainfall intensity rate (60 mm/h^{-1}) (after the ponds disappeared from the surface). In the first 8 experiments, the selected slope section was gentle, while the following 8 experiments were prepared on a steep slope. The real amount of rainfall was measured with the help of 0.5 L small buckets placed around the plot borders. The measured soil loss from the dry, moist and wet rainfall simulation treatments was used in this study. The total sum of the rainfall simulation experiments was denoted later as the pooled dataset.

2.3. Soil Erosion Modelling

The evaluation of various soil erosion models was performed by comparing the measured and predicted soil losses for each rainfall simulation for the applied slopes and rainfall intensities. Since rainfall simulations were based on a specific design of rainfall events, the models had to be used based on the single rainfall event inputs. The event-based predictions were possible using the procedures for calculating individual design storm rainfall erosivity values provided by the rainfall simulation measurements.

K-factor values were determined by five direct methods using a rainfall simulator and by two indirect methods (based on soil physical properties).

The basis of the erodibility, K-factor calculation is provided by one of the several estimation methods, the widely spread and well-known USLE equation by Wischmeier and Smith [8]. The Revised Universal Soil Loss Equation (RUSLE) [64] computes with the same formula, as follows:

$$A = R \cdot K \cdot L \cdot S \cdot C \cdot P \tag{1}$$

where A (t ha^{-1} y^{-1}) is the annual soil loss, R (MJ mm ha^{-1} h^{-1} y^{-1}) is the rainfall erosivity factor, K (t ha h ha^{-1} MJ^{-1} mm^{-1}) is the soil erodibility factor, L is the slope length (dimensionless), S is the slope gradient factor (dimensionless), C is the cropping cover management factor (dimensionless) and P is the agricultural (or support) practice factor (dimensionless). This way the soil erodibility can be calculated as $K = A\,R^{-1}\,L^{-1}\,S^{-1}$.

A, the annual soil loss (t $\times$ ha^{-1} $\times$ y^{-1}), is originated from the measured runoff in the field on a 6 m^2 plot area.

R, the rainfall erosivity index, is based on the equation by Foster et al. [38],

$$R = EI_{30} = EI_{max30} \tag{2}$$

where E is the kinetic energy (MJ ha^{-1} mm^{-1}) and I_{30} is the rainfall intensity (mm h^{-1}). Kinetic energy can be obtained from the following expression:

$$E = 0.119 + 0.0873\, log_{10}I \tag{3}$$

LS, the slope length and gradient factor in USLE, is based on the equation by Presbitero [65],

$$LS = (\lambda\, 22.13^{-1})^m\, (65.41\, \sin^2\theta + 4.56 \sin\theta + 0.0654) \tag{4}$$

where λ is the slope length; θ is the slope angle ($^\circ$); m is equivalent to 0.5 for s > 5%, 0.4 for 3% < s $\leq$ 5%, 0.3 for 1% < s $\leq$ 3% and 0.2 for s $\leq$ 1% slope gradient.

m, the exponent in RUSLE, based on Foster et al. [38], is defined as

$$m = \beta/(1 + \beta) \tag{5}$$

where β is a function of slope, moderately susceptibility for rill and inter-rill erosion [42], i.e.,

$$\beta = (\sin\theta/0.0896)/3(\sin\theta)^{0.8} + 0.56 \tag{6}$$

where λ is the slope length, m is a variable length-slope exponent, β is a factor that varies with slope gradient and θ is the slope angle ($^\circ$) for slopes 1–30°.

S, in RUSLE, based on McCool et al. [42], where the slopes are shorter than 4.6 m, is

$$S = 3.0 (\sin\theta)0.8 + 0.56 \tag{7}$$

USLE-M can be used to evaluate parameters from single runoff events, using a combination of storm erosivity and runoff ratio, with $Q_R EI_{30}$ term [35,36,45,66] as the erosivity index.

$$A_e = Q_R\, EI_{30}\, K_{UM}\, LS\, C_{UM}\, P_{UM} \tag{8}$$

where Q_R (dimensionless) is the event runoff coefficient; EI_{30} is the event erosivity factor. which has changed from the original EI_{30} to $Q_R EI_{30}$ factor; K_{UM} is the soil erodibility factor (t ha h ha^{-1} MJ^{-1} mm^{-1}); L is the USLE slope-length factor; S is the USLE slope-steepness factor; C_{UM} is the crop factor; and P_{UM} is crop management and support practice factor. The subscript U represents the factor associated with the USLE which has to be modified (subscript M).

The K-factor was also calculated with the following USLE nomograph's equation [67]:

$$K = 2.77 \times M^{1,14} \times (10^{-7}) \times (12 - SOM) + 4.28 \times (10^{-3}) \times (SS - 2) + 3.29 \times (10^{-3}) \times (PP - 3) \tag{9}$$

where M is (0.002–0.1 mm grain-size fraction in %) $\times$ (below 100–0.002 mm grain-size fraction in %); SOM is the soil organic matter in %; SS is the soil structure class (1 = very fine granular; 2 = fine granular; 3 = medium or coarse granular; 4 = blocky, platy, or massive); and PP is the profile permeability class (1 = rapid (150 $\leq$ mm h^{-1}); 2 = from moderate to rapid (50–150 mm h^{-1}); 3 = moderate (12–50 mm h^{-1}); 4 = from slow to moderate (5–15 mm h^{-1}); 5 = slow (1–5 mm h^{-1}); 6 = very slow (<1 mm h^{-1})).

The EPIC model k-value estimation was performed with the following formula by Williams et al. [41]:

$$K = \{0.2 + 0.3 \exp[-0.0256\,\text{SAN}(1 - 0.01\,\text{SAN})]\}\,(\text{SIL}/\text{CLA} + \text{SIL})^{0.3}\,\{1.0 - [0.25\,\text{C}/[\text{C} + \exp(3.72 - 2.95\,\text{C})]]\}$$
$$\{1.0 - [[0.7\,\text{SN1}]/[\text{SN1} + \exp(-5.51 + 22.9\,\text{SN1})]\} \times 0.1317 \tag{10}$$

where SAN refers to the content of sand (0.1 mm size < 2 mm) in %; SIL refers to the silt particle content (0.002 mm size < 0.1 mm) in %; CLA refers to the clay particle size (<0.002 mm) in %; C refers to the organic carbon content in %; and SN1 = $-$SAN/100. The unit of the soil erodibility k-value is t ha h ha^{-1} MJ^{-1} mm^{-1}.

The event soil loss and runoff values, erosivity indices and other input parameters of different soil loss estimation methods were calculated by adding the measured variables at the event scale. These latter gave the basis of K-factor determinations which were inserted into the selected models. Soil loss predictions in the case of different models were obtained by the substitution of the mean, measured soil- and model-specific K-factors.

2.4. Statistical Evaluation of the Model Performance

The Pearson correlation coefficient (r), coefficient of determination (r^2), relative error (*RE*) and Nash–Sutcliffe model efficiency index (*NSEI*) were adopted to evaluate the models' performance.

The Pearson correlation coefficient (r) and the coefficient of determination (r^2) describe the degree of collinearity of predicted and measured data. The correlation coefficient ranges between -1 and 1 and indicates the linear relationship between measured and calculated data. R^2 ranges from 0 to 1, where the higher values indicate less error variance and above 0.5 can be considered acceptable.

The *NSEI* determines the relative magnitude of the residual variance compared to the measured data variance. The *NSEI* indicates how well the plot of measured versus predicted data fits the 1:1 line. It is calculated with the following relationship (Equation (11)):

$$NSEI = 1 - \frac{\sum_{i=1}^{N}\left(A_{e,measured} - A_{e,calculated}\right)^2}{\sum_{i=1}^{N}\left(A_{e,measured} - A_{e,mean}\right)^2} \tag{11}$$

where $A_{e,measured}$ is the measured soil loss, $A_{e,calculated}$ is the soil loss value calculated by the models and $A_{e,mean}$ is the mean of the measured values. When the model accounts for all the variation in the measurements ($A_{e,calculated} = A_{e,measured}$), the index has a value of 1.0; NSEI ranges between $-\infty$ and 1.0. Values between 0.0 and 1.0 mean acceptable levels of performance. Zero indicates that the model predictions are as accurate as of the mean of the measured values. A negative *NSEI* indicates that the measured mean is better than the predictions. Based on Ahmad et al.'s [68] study on erosion prediction evaluation, the model performance criteria is *NSEI* > 0.4.

Some of the commonly used error indices were applied. The relative error (RE) and the root-mean-square error (RMSE) were calculated to indicate the error between calculated and measured models. In the case of these error indices, the value of 0 indicates a perfect fit.

$$RE = [\text{K-calculated} - \text{K-measured}]/\text{K-measured as percent} \tag{12}$$

$$RMSE = \sqrt{\frac{\sum_{i=1}^{n}(y_i - \hat{y})^2}{n}} \tag{13}$$

The percent bias (PBIAS) measures the average tendency of the predicted data to be larger or smaller than their measured counterparts [69]. The optimal value is 0.0, with low-magnitude values indicating an accurate model simulation, while positive values indicate under-estimation of bias and negative values indicate over-estimation of bias (Equation (14)). The PBIAS can be considered very good if PBIAS < $\pm$ 15%; good, if $15\% \leq \text{PBIAS} < \pm 30\%$; satisfactory, if $30\% \leq \text{PBIAS} < \pm 55\%$; and unsatisfactory, if $\text{PBIAS} \geq \pm 55\%$.

$$PBIAS = \left[\frac{\sum_{i=1}^{n} \left(y_i^{measured} - y_i^{predicted} \right) * (100)}{\sum_{i=1}^{n} \left(y_i^{measured} \right)} \right] \tag{14}$$

The Mann–Whitney U test was used to determine whether the estimates from the models were significantly different from the measured data ($p < 0.05$ means significant difference). The applied software was GraphPad Prism.

3. Results

3.1. Evaluation of the Erodibility Estimated by the Applied Models

The first evaluation of erodibility values was performed by comparing the calculated K-results based on the measured parameters by USLE, RUSLE and USLE-M and by the EPIC and USLE nomographs (Figure 2). In the separated, calculated results, in the case of the USLE and RUSLE models, the wet runs resulted in higher K-values than the dry runs, while USLE-M indicated higher erodibility with dry runs. These results owe to the high variance within the measured hydrological parameters. There was a higher difference among different slope categories with USLE-M calculations resulting in higher values for steep slopes, while the USLE and RUSLE models showed lower variation. The pooled data indicate higher variability with a higher difference between the K-values calculated by each model. The K-factors obtained by the USLE nomograph, in most cases, were under the other calculated values, while the K-factors resulting from the EPIC nomograph were above the USLE-type models. In the case of all models, the variability within the measurements increased with the increase in soil moisture content and slope degree.

The mean values ranged from 0.0028 to 0.0087 for the USLE-type models (Table 3).

Table 3. Different soil erosion models and their calculated K-values from the measured parameters of each rainfall simulation (K_U—USLE; K_{RU}—RUSLE; K_{U-M}—ULSE-M; K_{EPIC}—EPIC nomograph equation; K_{NOMO}—USLE nomograph equation).

	Pooled Data				
	K_U	K_{RU}	K_{U-M}	K_{EPIC}	K_{NOMO}
MEAN	0.0041	0.0028	0.0087		
SD	0.0030	0.0024	0.0101		
Min	0.0001	0.0000	0.0014		
Max	0.0096	0.0095	0.0398	0.0112	0.0018
Median	0.0039	0.0026	0.0049		
CV%	73.5113	86.3314	116.1762		
CV rel	18.3778	21.5828	29.0440		

Comparing the calculated values based on the measured rainfall simulation parameters with the results of soil texture-based nomographs, the EPIC model returned a 0.0112 erodibility value, while USLE returned 0.0018. According to the coefficient of variations (CV), the most variable results occurred in the case of the USLE-M (CV = 116.17%) model's calculations.

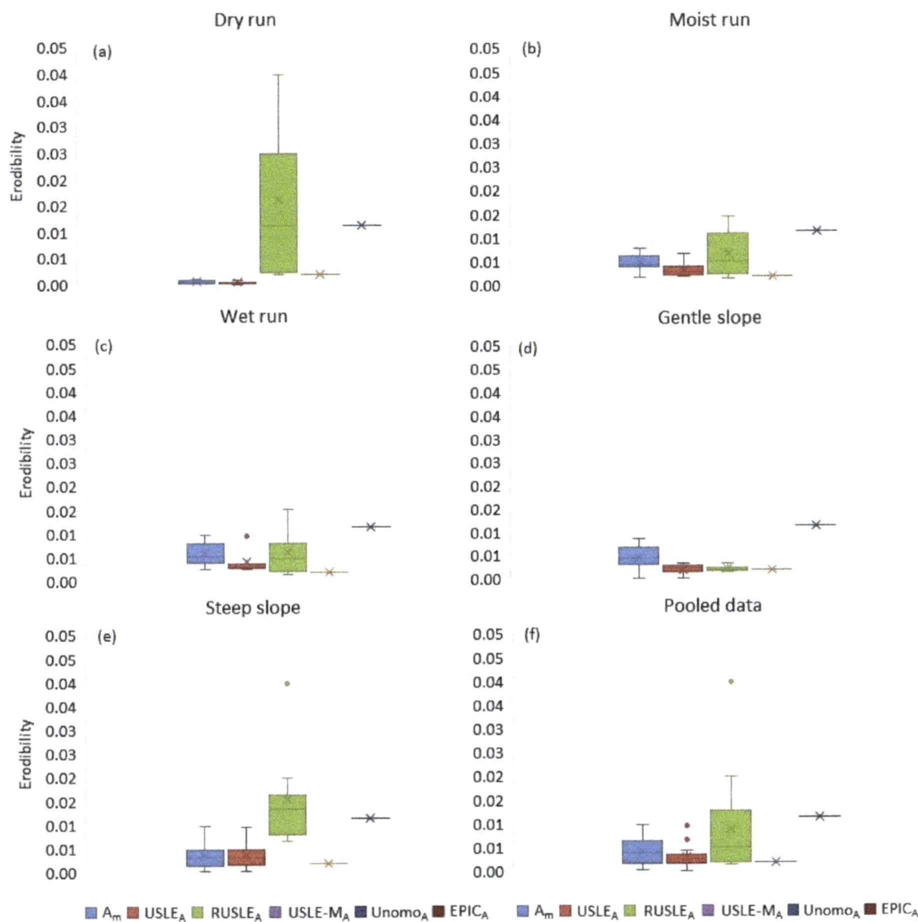

Figure 2. Comparison of calculated erodibility values by USLE, RUSLE and USLE-M (t ha h ha^{-1} MJ^{-1} mm^{-1}), as well as USLE and EPIC nomograph equations (t ha h ha^{-1} MJ^{-1} mm^{-1}) according to (**a**) dry runs, (**b**) moist runs, (**c**) wet runs, (**d**) gentle slope, (**e**) steep slope and (**f**) pooled data.

3.2. Evaluation of Event Soil Loss Estimation

Soil erosion measured results confirmed that the higher rainfall intensities with higher soil moisture content (moist and wet runs) generated more runoff and soil loss during each event containing higher suspended sediment concentration (Table 4).

Table 4. Erosion resulting from individual rainfall simulations (mean values in brackets).

Moisture	Rainfall Intensity (mm h^{-1})	Runoff (mm h^{-1})	Suspended Sediment Concentration (g L^{-1})	Total Soil Loss (g m^{-2} h^{-1})
dry	20.87–26.32	0.2–0.9 (0.5)	3.6–10.3 (7.04)	2.9–3.4 (3.21)
moist	44.05–70.19	14.1–22.8 (19.2)	9.01–19.25 (12.23)	150.5–438.5 (237.88)
wet	63.16–103.48	32.1–48.1 (37.5)	11.26–26.19 (16.53)	406.3–1061.5 (633.69)

The highest rates of erosion were recorded in storms of 63.16–103.48 mm h^{-1}, where the total soil loss ranged from 406.3 g m^{-2} h^{-1} to 1061.5 g m^{-2} h^{-1}.

Model evaluations were first performed by comparing the measured soil loss values with the predicted ones estimated by the USLE, RUSLE, USLE-M, EPIC and USLE nomographs (Figure 3).

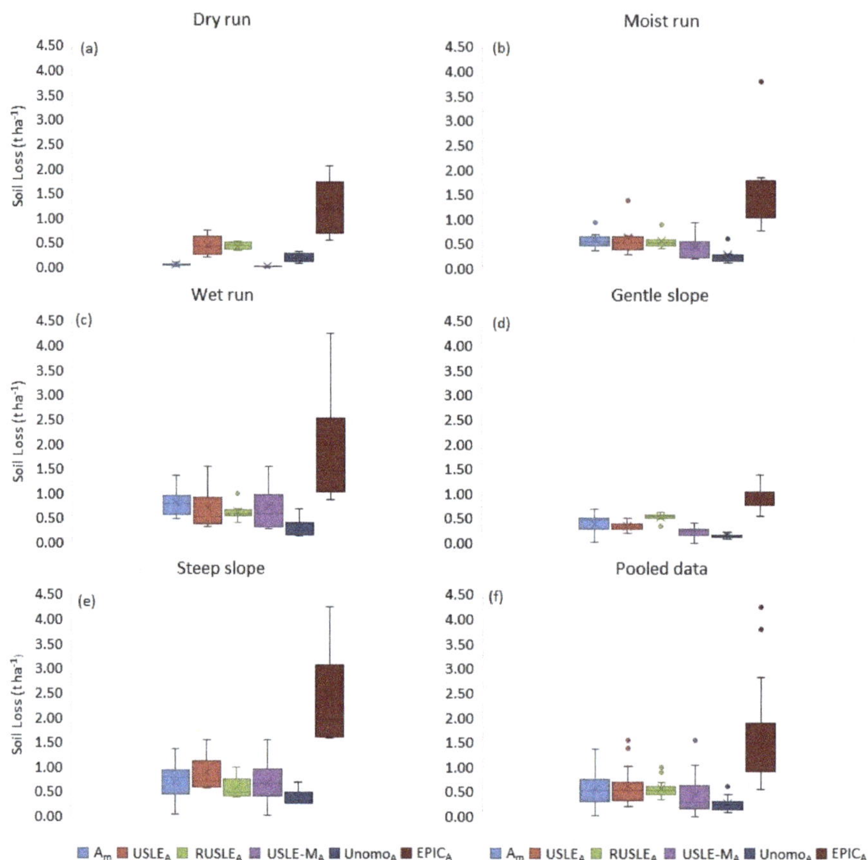

Figure 3. Comparison of measured soil loss (A_m) (t ha^{-1}) and the estimation results by the USLE, RUSLE, USLE-M (t ha^{-1}), USLE (Unomo$_A$) and EPIC nomograph (EPIC$_A$) equations according to (**a**) dry runs, (**b**) moist runs, (**c**) wet runs, (**d**) gentle slope, (**e**) steep slope and (**f**) pooled data.

Generally, for dry runs, the highest variability in soil loss estimation occurred in the case of USLE, followed by the EPIC nomograph estimation. The closest results were derived from USLE-M estimations. USLE, RUSLE and the nomographs overestimated the measured amounts. Based on moist runs, soil losses estimated by USLE, RUSLE and USLE-M were close to the measured ones, while the EPIC nomograph resulted in nearly threefold higher values (also with higher variability). The nomograph underestimated the measured soil loss. The wet rainfall simulation results indicate higher variance within the data which resulted in a higher discrepancy among the estimations of USLE and USLE-M, as well as the EPIC nomograph. The USLE nomograph underestimated the average soil loss, while EPIC overestimated it. Correspondence could be found among the USLE, RUSLE and USLE-M models. Separating the data by slope steepness, then, in the case of the gentle slope, high variability occurred among the results in the attendance of many outlier values. The estimation of the EPIC nomograph indicated almost threefold greater loss than the measured one. The nearest soil loss values were those of USLE and RUSLE.

In the case of the steep slope, the closest estimation was obtained by USLE-M, then RUSLE and USLE, while other models indicated two- and threefold amounts of soil loss. Based on all pooled data, the best estimations were obtained by RUSLE, USLE, then USLE-M, while the EPIC nomograph overestimated the soil loss against the measured losses and the USLE nomograph underestimated it. At the same time, soil loss estimations by RUSLE, USLE and the USLE nomograph had the lowest variability.

The coefficient of determination shows an acceptable correlation between the measured soil loss and the predicted ones by USLE-M (Table 5).

Table 5. Soil loss calculations by different soil loss estimation models (A_U—USLE; A_{RU}—RUSLE; A_{U-M}—USLE-M; A_{nomo}—USLE nomograph; A_{EPIC}—EPIC nomograph) and their Pearson correlation coefficient (r), coefficient of determination (r^2), Nash–Sutcliffe coefficient (NSEI), root-mean-square error (RMSE), relative error (RE), percent bias (PBIAS), relative difference (R_{diff}) and *p*-value (Mann–Whitney U test), used to test models' performance.

	A_U	A_{RU}	A_{U-M}	A_{nomo}	A_{EPIC}
r	0.3988	0.2825	0.7706	0.3988	0.3988
r^2	0.1591	0.0798	0.5938	0.1591	0.1591
NSEI	−0.2744	0.0431	−1.7041	−0.3745	−15.5122
RMSE	0.1732	0.1300	0.3674	0.1868	2.2437
RE	0.1340	0.0359	−0.1779	−0.4970	2.1296
PB	−13.3993	−3.5878	−50.9319	49.7031	−212.9582
p-value	0.7520	0.8672	0.3414	0.0615	<0.0001

The model efficiency of almost all models—except RUSLE—was negative in the simulations. The negative NSEI values indicated that the mean measured soil loss from the field was a better representation of soil erosion for the concerned simulations than the estimations of specific soil erosion models. Based on the NSEI, the RUSLE model resulted in the best prediction, followed by the USLE and USLE-M, while the nomograph EPIC estimation seemed the worst predictor. RMSE values indicated that the best model performance was obtained in the case of the RUSLE, USLE and USLE nomographs. Based on the relative error (RE), a satisfactory relationship occurred between the measured and predicted datasets in the case of RUSLE, USLE and the USLE-M, which is the third with a negative value owing to higher measured soil loss. PBIAS (PB) with negative values indicated over-estimation of bias almost in all cases, except the USLE nomograph, which under-estimated the observed soil loss (Figure 4) with satisfactory estimation (49%); in the case of USLE (−13%), RUSLE (−3%) and USLE-M (−50%), we obtained good and satisfactory estimations. The estimates from the models departed significantly from the measured data ($p < 0.05$) with the EPIC model.

The accuracy of model predictions and the tendencies of the models to over- or underpredict soil losses are graphically visualized in Figure 4a–e.

The 1:1 line represents the perfect match between the measured and predicted values. The data points fitting or positioned closer to the 1:1 line indicate accurate predictions, while the values situated above the 1:1 line reveal overestimation and the ones situated below refer to the data that underestimated the amount of soil losses. USLE and RUSLE overestimated soil loss by an average of 13.4% and 3.6%, respectively (Figure 5a,b), while the EPIC model showed the worse performance, with an average overestimation of 212.95% (Figure 5d,e). RUSLE model predictions were sensitive for the measured values, when this increased above 0.6 t/ha; then, the model predictions indicated underestimations. USLE-M and the USLE nomograph (Figure 5c,e) underestimated the measured values by an average rate of 17.7% and 49.7%, respectively. However, except for some outlier values, the USLE-M showed the best performance and fit among the existing models which were not sensitive to the changes within the measured values.

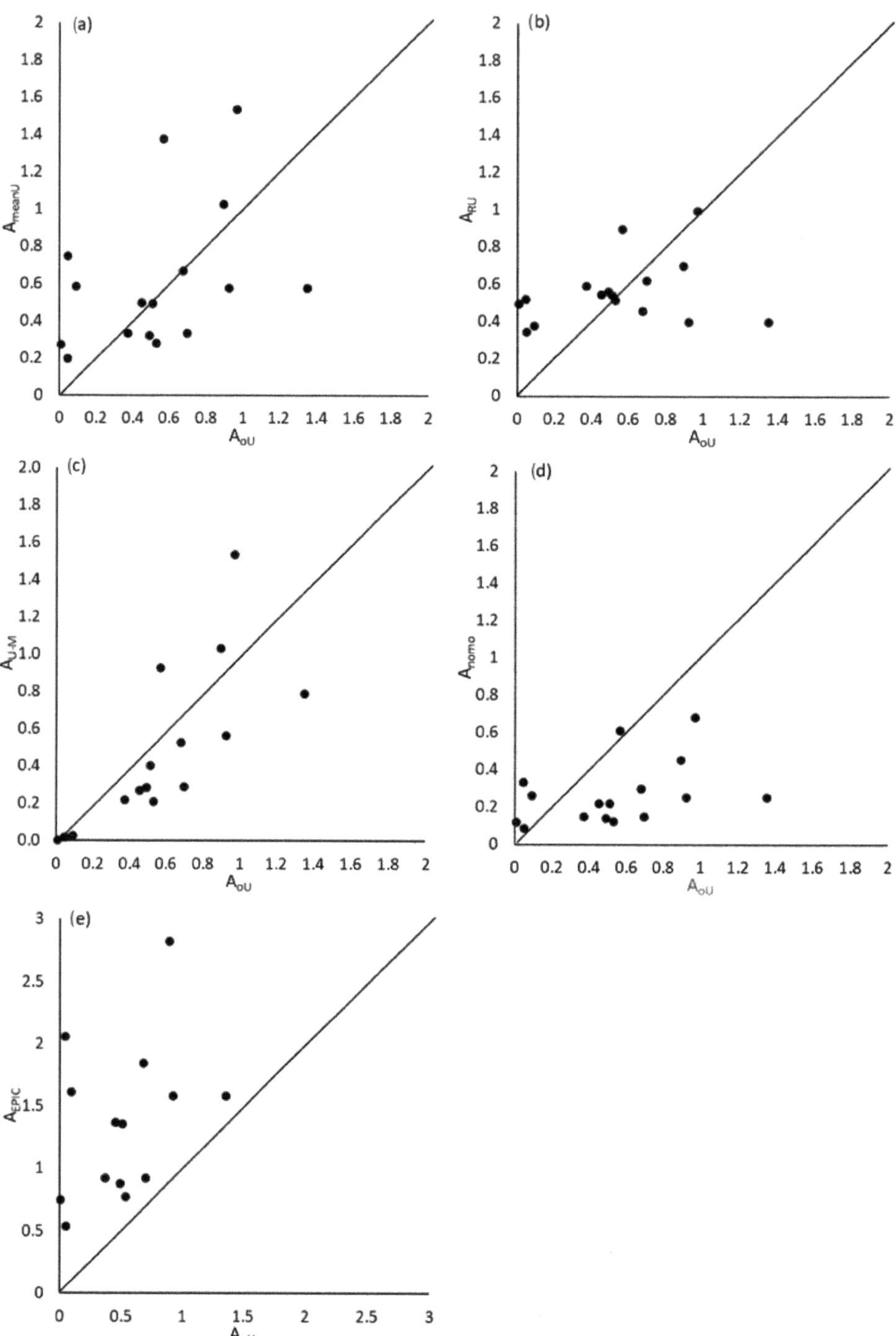

Figure 4. Comparison between the measured event soil loss values and the predicted ones by (**a**) USLE (**b**) RUSLE, (**c**) USLE-M, (**d**) USLE nomograph equation and (**e**) EPIC nomograph equation.

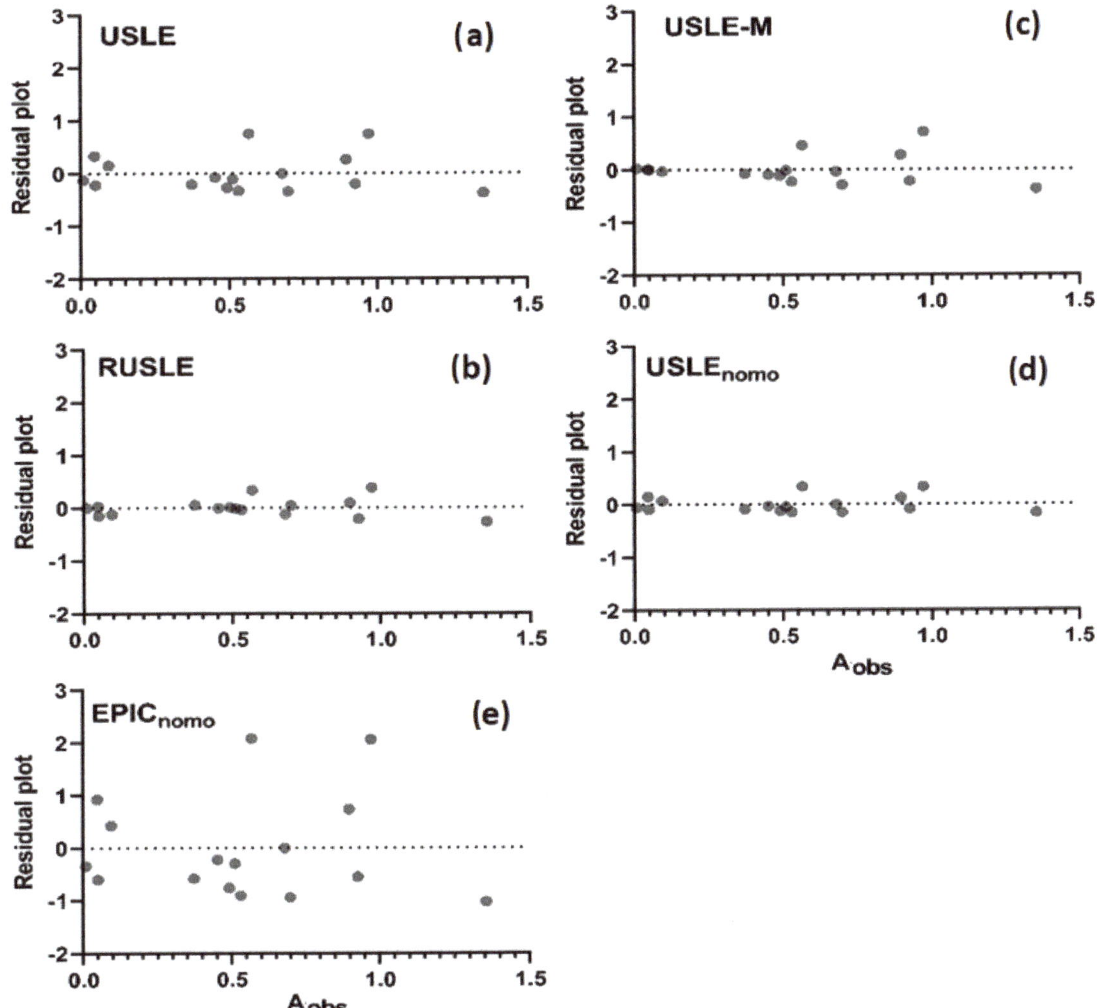

Figure 5. Predicted soil loss (from dry, moist and wet rainfall simulation treatments) plotted against residual values (measured–predicted soil loss): (**a**) USLE (**b**) RUSLE, (**c**) USLE-M, (**d**) USLE nomograph equation, (**e**) EPIC nomograph equation.

In the case of RUSLE and the USLE nomograph, high-density of points are close to the origin. In the case of USLE and USLE-M, the scattering from the origin increases but these are still independent and normally distributed, while, for the EPIC nomograph, the residuals and their deviation increase with the increase in measured soil loss; therefore, these are not independent.

4. Discussion

USLE and RUSLE resulted in higher soil erodibility values in the case of wet runs, while, surprisingly, USLE-M resulted in higher K-values for dry runs. This is not consistent with some studies where increased soil moisture content resulted in decreased sediment concentration, thus decreased soil erodibility values [59,70–72]. At the same time, we can conclude that the variability of the calculated erodibility values increased in parallel with the increase in soil moisture content and slope degree.

The effect of slope gradient on soil loss and soil erodibility is well known. In this study, USLE-M indicated higher erodibility in steep slopes with higher variation than USLE- and RUSLE-based erodibilities. Kinnel et al. [3] reported, in their study, that values of both USLE-M and RUSLE varied with the slope gradient, which is parallel with our findings. In the case of gentle slopes, the event soil loss estimation's results showed high variance, but the best estimations were obtained by RUSLE and USLE-M for both gentle and steep slopes. These results indicate that, in the case of event soil loss measurements, many influencing factors should be taken into account in order to conclude with a correct interpretation.

The mean of the calculated soil erodibility values obtained by different soil erosion models ranged from 0.0028 to 0.0087 for USLE-type models. The previously reported K-factor, according to Regosol in Hungarian literature (derived from in situ rainfall simulation experiments), is 0.038 ($t \times ha \times h \times ha^{-1} \times MJ^{-1} \times mm^{-1}$) [73], which is significantly higher than our calculated K-factors. This difference could be derived from the difference in physical properties of the investigated Regosol, from the different initial soil moisture content and the applied replicates. The different rainfall duration and intensities were eliminated during the calculations. Generally, the EPIC nomograph estimated larger erodibility factors, while USLE estimated smaller ones. The observation of the EPIC nomograph's overestimation is in agreement with the findings by Zhang et al. [74] and Wang et al. [59], who reported gross overprediction of soil loss in the case of nomograph estimators.

The event soil loss evaluation confirmed that the initial soil moisture content and rainfall intensity has great importance in runoff and soil loss generation [75,76]; therefore, it is important to emphasize this parameter when we publish results and draw conclusions from these results, regarding the accuracy of models. For the dry run, when the amount of soil loss and the rainfall intensity was the lowest, USLE-M obtained the best performance. In the case of moist antecedent soil moisture content, where the soil loss and intensity increased, the USLE-type models predicted soil loss rates close to the measured ones. When the antecedent soil moisture content and intensity were higher, the estimated soil loss values were more variable, where the best performance was again obtained by the USLE-type models.

Based on the whole dataset and the statistical evaluation, the coefficient of determination resulted in an acceptable correlation between the measured and predicted values only in the case of USLE-M. Based on other statistical indicators, such as NSEI, RMSE, PBIAS and relative error, RUSLE and USLE, followed by USLE-M, resulted in the best performance. PBIAS indicated overestimation of bias in almost all cases, except the USLE nomograph. Similar results could be inferred by examining the graphical visualisation of predicted and measured values fitted to the 1:1 line. The lowest overestimation occurred with the application of RUSLE (3.6%) and USLE (13.4%), while a slight underestimation was obtained in the case of USLE-M (17.7%). At the same time, RUSLE indicated sensitivity in prediction, whereas, above 0.6 t/ha, event soil loss underestimation occurred in the estimations, an observation similar to the findings by Kinnel et al. [39] and Nearing [77]. While the USLE nomograph resulted in constant underestimation, EPIC overestimated the soil losses. Risse et al. [29], Tiwari et al. [78] and Kinnel et al. [36] reported that USLE overestimates the small annual and event soil losses while underestimating the high annual and event soil losses. In this study, these findings do not occur, except for a general overprediction of soil losses. Di Stefano et al. [79] found that both USLE and USLE-M tended to overestimate low event soil losses, while, in our study, the opposite was observed. These observations are partly due to the size of the plot, which was a unit in their study, and the values of the measured soil losses originated from natural rainfall and not from simulated precipitation. Based on the residuals, the ranking is similar to the above-mentioned results concerning the USLE-type models. Kinnel et al. [39] emphasized that USLE and RUSLE can predict event soil losses, while USLE-M can predict event soil losses better than RUSLE. These statements are just in partial agreement with our findings, whereas the capability of USLE and RUSLE to predict event soil losses is confirmed; however, USLE-M did not perform better than USLE or RUSLE. The nomographs are mostly based on solely soil physical parameters; therefore, in this study, we confirmed that they are unsatisfactory for this kind of prediction and, applied alone, cannot be used to estimate individual soil loss events [58]. It seems that the more rainfall simulation data are analysed, the more we know about soil erodibility; however, due to the complex nature of soil, the data gathered and analysed so far did not provide enough spatial coverage and in-depth knowledge to cover all needs. Similar research studies have been conducted in the past decades, dealing with both estimations and evaluations of the K-factor under natural and artificial rainfall conditions [80]. Furthermore, similar experiments have been conducted in different settings, e.g., tropical climate [81], that also supported the idea that more experiments and evaluation are needed.

5. Conclusions

Using the measured event soil losses from six runoff plots under different antecedent soil moisture content and rainfall intensities in loess covered Koppány River Valley, this paper presents a set of measured K-values and soil loss data by a set of soil erosion modelling methods, namely, USLE, RUSLE and USLE-M, as well as USLE and EPIC nomographs. The differences in the measured and predicted soil losses were evaluated in the case of different initial moisture content, rainfall intensities and slope gradients. The results can serve as values for event rainfall simulations for areas without natural runoff plots. The main conclusions are as follows:

(1) The K-values in Hungary for the loess-covered Koppány River Valley ranged from 0.0028 to 0.087 t ha h ha^{-1} MJ^{-1} mm^{-1}. The calculated erodibility results vary with initial soil moisture content and slope degree. Regarding the pooled data, the CV% indicated less relative variability in the case of USLE and RUSLE, followed by USLE-M.

(2) In respect to the event soil loss estimation, the results indicate high variance with the increase in initial soil moisture content and slope steepness. Based on the pooled data, the best estimations were obtained by RUSLE, USLE and USLE-M.

(3) Both soil erodibility and soil loss estimation draw attention to the importance of the effect of soil moisture conditions.

(4) Investigating the measured and estimated soil losses statistically, we found that RUSLE resulted in the best performance in event soil loss estimation. At the same time, RUSLE indicated sensitivity for the measured values.

(5) Within these experiments, the USLE-type models' capability for estimating event soil losses was confirmed. At the same time, we cannot emphasize enough that the non-standardized circumstances involving many affecting factors could result in high variance among the results.

(6) This study points out the high variability of soil erodibility factors and soil loss estimation values in the case of event rainfall simulations. Besides, it draws attention to the importance of further studies on this topic, because rainfall simulation is necessary for the determination of each soil type's erodibility and short or long time soil loss estimation. That is why we need to develop an accurate, standardized method for event soil loss measurements, where the observable parameters are not solely the type of soil and other soil properties because these methods cannot give satisfactory results.

(7) This study focused on soils formed on loess parent material. These soils cover a wide range of geographical regions. Thus, this study can provide useful information for many areas that struggle with similar water erosion problems, suggesting using erosion modelling as a tool for finding a solution.

Author Contributions: For Conceptualization, B.K. and C.C.; methodology, B.K., J.A.S. and C.C.; formal analysis, J.A.S.; investigation, C.C., G.J., Z.S. and J.A.S.; writing—original draft preparation, B.K. and C.C.; writing—review and editing, Z.S. and G.J., visualization, B.K.; supervision, C.C. All authors have read and agreed to the published version of the manuscript.

Funding: This research received no external funding.

Institutional Review Board Statement: Not applicable.

Informed Consent Statement: Not applicable.

Data Availability Statement: Not applicable.

Acknowledgments: The authors are grateful to the staff of Geographical Institute, Research Centre for Astronomy and Earth Sciences, Hungarian Academy of Sciences, for providing the rainfall simulator equipment for our measurements and to the laboratory of ELTE University for enabling the laboratory measurements. In addition, we thank all the colleagues who took part in the field measurements: Pálma Anna Varga-Kertész, Balázs Bolf, Bernadett Bolf, Igor Dekemati, Zsófia Dobó, Marianna Ringer, Norbert Keller. Besides, we are grateful to the field's owner, who contributed to the in situ measurements and water supply.

Conflicts of Interest: The authors declare no conflict of interest. The funders had no role in the design of the study; in the collection, analyses, or interpretation of data; in the writing of the manuscript; or in the decision to publish the results.

Appendix A

Table A1. Summary of the different individual rainfall, runoff and soil loss measurements.

Treatment	s	Soil Moisture Content	RI_e	P_e	EI_{30}	Q_e	Q_R	$Q_R EI_{30}$	A_{obs}
	%		mm h^{-1}	mm	MJ mm ha^{-1} h^{-1}	mm		MJ mm ha^{-1} h^{-1}	t ha^{-1}
FPG1	6.7	moist	56.24	16.81	256.93	4.22	0.25	64.53	0.45
FPG2	6.7	wet	84.68	10.63	254.79	4.04	0.38	96.90	0.51
FPG3	8	dry	22.65	28.96	155.66	1.20	0.04	6.43	0.05
FPG4	8	moist	70.19	13.53	266.09	4.12	0.30	81.06	0.37
FPG5	8	wet	86.12	10.41	253.61	4.36	0.42	106.22	0.49
FPG6	7.7	dry	26.32	35.1	224.49	0.21	0.01	1.35	0.01
FPG7	7.7	moist	56.92	15.19	235.44	5.28	0.35	81.79	0.53
FPG8	7.7	wet	80.44	12.4	282.30	4.95	0.40	112.68	0.70
FPS1	18.32	dry	20.87	26.64	130.20	0.53	0.02	2.61	0.09
FPS2	18.32	moist	66.78	16.49	306.47	5.17	0.31	96.08	0.57
FPS3	18.32	wet	103.48	11.67	341.76	5.42	0.46	158.82	0.97
FPS4	17.26	dry	24.42	31.10	182.39	0.40	0.01	2.33	0.05
FPS5	17.26	moist	49.89	10.53	140.42	4.81	0.46	64.11	0.93
FPS6	17.26	wet	63.16	8.05	140.47	5.17	0.64	90.12	1.35
FPS7	17.6	moist	44.05	13.86	158.61	5.03	0.37	58.11	0.68
FPS8	17.6	wet	76.69	11.18	243.18	5.23	0.47	113.51	0.90

s (%), slope gradient; RI_e (mm h^{-1}), event rainfall intensity; P_e (mm), event rainfall amount; EI_{30} (MJ mm ha^{-1} h^{-1}), rainfall erosivity index; Q_e (mm), event runoff amount; Q_R (dimensionless), event runoff coefficient; A_{obs} (t ha^{-1}), event soil loss per unit area; μ, mean value; CV, coefficient of variations.

References

1. Wang, B.; Zheng, F.; Guan, Y. Improved USLE-K factor prediction: A case study on water erosion areas in China. *Int. Soil Water Conserv. Res.* **2016**, *4*, 168–176. [CrossRef]
2. Pimentel, D.; Harvey, C.; Resosudarmo, P.; Sinclair, K.; Kurz, D.; McNair, M. Environmental and economic costs of soil erosion and conservation benefits. *Science* **1995**, *267*, 1117–1123. [CrossRef] [PubMed]
3. Pimentel, D. Soil erosion: A food and environmental threat Environment. *Dev. Sustain.* **2006**, *8*, 119–137. [CrossRef]
4. Lal, R. Enhancing crop yield in the developing countries through restoration of soil organic carbon pool in agricultural lands. *Land Degrad. Dev.* **2006**, *17*, 197–209. [CrossRef]

5. Miller, G.A.; Amemiya, M.; Jolly, R.W.; Melvin, S.W.; Nowak, P.J. *Soil Erosion and the Iowa Soil 2000 Program*; Iowa State University: Ames, IA, USA, 1988.

6. Sharpley, A.N.; McDowell, R.W.; Kleinman, P.J.A. Phosphorus loss from land to water: Integrating agricultural and environmental management. *Plant Soil* **2001**, *237*, 287–307. [CrossRef]

7. Smith, D.D. Interpretation of soil conservation data for field use. *Agric. Eng.* **1941**, *22*, 173–175.

8. Wischmeier, W.H.; Smith, D.D. Predicting rainfall erosion losses. In *A Guide to Conservation Planning. USDA Agriculture Handbook*; USDA: Blacksburg, VA, USA, 1978.

9. Mannering, J.V. The use of soil loss tolerances as a strategy for soil conservation. In *Soil Conservation: Problems and Prospects*; Morgan, R.P.C., Ed.; John Wiley and Sons: New York, NY, USA, 1981; pp. 337–349.

10. Skidmore, E.L. *Soil Loss Tolerance [Chapter 8], Determinants of Soil Loss Tolerance*; ASA, SSSA: Madison, WI, USA, 1982; pp. 87–93. [CrossRef]

11. Morgan, R.P.C. *Soil Erosion and Conservation*; Blackwell Publishing: Hoboken, NJ, USA, 2005; p. 299. [CrossRef]

12. IPCC. Climate Change and Land. Intergovernmental Panel on Climate Change. 2019. Available online: https://www.ipcc.ch/site/assets/uploads/2019/08/4.-SPM_Approved_Microsite_FINAL.pdf (accessed on 6 September 2021).

13. Edwards, W.M.; Owens, L.B. Large Storm Effects on Total Soil Erosion. *J. Soil Water Conserv.* **1991**, *46*, 75–78. Available online: https://www.jswconline.org/content/46/1/75.short (accessed on 28 October 2021).

14. Bagarello, V.; Di Stefano, C.; Ferro, V.; Pampalone, V. Statistical distribution of soil loss and sediment yield at Sparacia experimental area, Sicily. *Catena* **2010**, *82*, 45–52. [CrossRef]

15. Bagarello, V.; Di Stefano, C.; Ferro, V.; Pampalone, V. Using plot loss distribution for soil conservation design. *Catena* **2011**, *86*, 172–177. [CrossRef]

16. Bagarello, V.; Di Stefano, C.; Ferro, V.; Kinnell, P.I.A.; Pampalone, V.; Porto, P.; Todisco, F. Predicting soil loss on moderate scope using an empirical model for sediment concentration. *J. Hydrol.* **2011**, *400*, 267–273. [CrossRef]

17. Strohmeier, S.; Laaha, G.; Holzmann, H.; Klik, A. Magnitude and occurrence probability of soil loss: A risk analytical approach for the plot scale for two sites in lower Austria. *Land Degrad. Dev.* **2016**, *27*, 43–51. [CrossRef]

18. Lang, K.; Prunty, L.; Schroeder, S.; Disrud, L. Interrill erosion as an index of mined land soil erodibility. *Trans. ASAE* **1984**, *99*, 109.

19. Ali, S.A.; Hamelmal, H. Estimation of soil erosion using USLE and GIS in Awassa Catchment, Rift valley, Central Ethiopia. *Geoderma Reg.* **2016**, *7*, 159–166. [CrossRef]

20. Pacheco, F.A.L.; Varandas, S.G.P.; Sanches Fernandes, L.F.; Valle, R.F., Jr. Soil losses in rural watersheds with environmental land use conflicts. *Sci. Total Environ.* **2014**, *485–486*, 110–120. [CrossRef]

21. Bagarello, V.; Di Stefano, C.; Ferro, V.; Giordano, G.; Iovino, M.; Pampalone, V. Estimating the USLE Soil Erodibility Factor in Sicily, South Italy. *Appl. Eng. Agric.* **2012**, *28*, 199–206. [CrossRef]

22. Borselli, L.; Cassi, P.; Sanchis, P. Soil Erodibility Assessment for Applications at Watershed Scale. In *Manual of Methods for Soil and Land Evaluation*; Springer: New York, NY, USA, 2009; pp. 98–117. [CrossRef]

23. González-Hidalgo, J.C.; Batalla, R.J.; Cerdà, A.; de Luis, M. Contribution of the largest events to suspended sediment transport across the USA. *Land Degrad. Dev.* **2010**, *21*, 83–91. [CrossRef]

24. González-Hidalgo, J.C.; Batalla, R.J.; Cerdà, A.; de Luis, M. A regional analysis of the effects of largest events on soil erosion. *Catena* **2012**, *95*, 85–90. [CrossRef]

25. Larson, W.E.; Lindstrom, M.J.; Schumacher, T.E. The role of severe storms in soil erosion: A problem needing consideration. *J. Soil Water Conserv.* **1997**, *52*, 90–95.

26. Iserloh, T.; Fister, W.; Seeger, M.; Willger, H.; Ries, J. A small portable rainfall simulator for reproducible experiments on soil erosion. *Soil Tillage Res.* **2012**, *124*, 131–137. [CrossRef]

27. Iserloh, T.; Ries, J.; Arnáez, J.; Boix-Fayos, C.; Butzen, V.; Cerdà, A.; Echeverría, M.; Fernandez-Galvez, J.; Fister, W.; Geißler, C.; et al. European small portable rainfall simulators: A comparison of rainfall characteristics. *Catena* **2013**, *110*, 100–112. [CrossRef]

28. Iserloh, T.; Pegoraro, D.; Schlösser, A.; Thesing, H.; Seeger, M.; Ries, J.B. Rainfall Simulation Experiments: Influence of Water Temperature, Water Quality and Plot Design on Soil Erosion and Runoff. *Geophys. Res. Abstr.* **2015**, *17*, EGU2015-5817. Available online: https://meetingorganizer.copernicus.org/EGU2015/EGU2015-5817.pdf (accessed on 28 October 2021).

29. Risse, L.M.; Nearing, M.A.; Nicks, A.D.; Laflen, J.M. Error assessment in the Universal Soil Loss Equation. *Soil Sci. Soc. Am. J.* **1993**, *57*, 825–833. [CrossRef]

30. Todisco, F.; Mannocchi, F.; Vergni, L.; Vinci, A. Plot Scale Measurements of Rainfall Erosion Losses in Central Italy. In Proceedings of the Role of Hydrology in Water Resources Management Symposium, Capri, Italy, 13–16 October 2008; Available online: http://www.scopus.com/inward/record.url?eid=2-s2.0-79551547410&partnerID=MN8TOARS (accessed on 28 October 2021).

31. Todisco, F. The internal structure of erosive and non-erosive storm events for interpretation of erosive processes and rainfall simulation. *J. Hydrol.* **2014**, *519*, 3651–3663. [CrossRef]

32. Todisco, F.; Brocca, L.; Termite, L.F.; Wagner, W. Use of satellite and modeled soil moisture data for predicting event soil loss at plot scale. *Hydrol. Earth Syst. Sci.* **2015**, *19*, 3845–3856. [CrossRef]

33. Bagarello, V.; Di Piazza, G.V.; Ferro, V.; Giordano, G. Predicting unit plot soil loss in Sicily, south Italy. *Hydrol. Process.* **2008**, *22*, 586–595. [CrossRef]

34. Bagarello, V.; Di Stefano, C.; Ferro, V.; Pampalone, V. Comparing theoretically supported rainfall-runoff erosivity factors at the Sparacia [South Italy] experimental site. *Hydrol. Process.* **2018**, *32*, 507–515. [CrossRef]

35. Kinnell, P. Runoff dependent erosivity and slope length factors suitable for modeling annual erosion using the Universal Soil Loss Equation. *Hydrol. Process.* **2007**, *21*, 2681–2689. [CrossRef]
36. Kinnell, P.I.A. Event soil loss, runoff and the Universal Soil Loss Equation family of models: A review. *J. Hydrol.* **2010**, *385*, 384–397. [CrossRef]
37. Boardman, J. Soil erosion science: Reflections on the limitations of current approaches. *Catena* **2006**, *68*, 2–3. [CrossRef]
38. Foster, G.R.; Moldenhauer, W.C.; Wischmeier, W.H. Transferability of U.S. technology for prediction and control of erosion in the tropics. In *Soil Erosion and Conservation in the Tropics*; ASA: Madison, WI, USA, 1981; pp. 135–149. [CrossRef]
39. Kinnell, P.; Wang, J.; Zheng, F. Comparison of the abilities of WEPP and the USLE-M to predict event soil loss on steep loessial slopes in China. *Catena* **2018**, *171*, 99–106. [CrossRef]
40. Wischmeier, W.C.; Smith, D.D. Rainfall energy and its relationship to soil loss. *Trans. Am. Geophys. Union* **1958**, *39*, 285–291. [CrossRef]
41. Williams, J.R. Sediment yield prediction with universal equation using runoff energy factor. In *Present and Prospective Technology for Predicting Sediment Yields and Sources*; USDA: Blacksburg, VA, USA, 1975.
42. Renard, K.G.; Foster, G.R.; Weesies, G.A.; McCool, D.K.; Yoder, D.C. Predicting soil erosion by water: A guide to conservation planning with the Revised Universal Soil Loss Equation [RUSLE]. In *USDA Agriculture Handbook*; USDA: Blacksburg, VA, USA, 1997.
43. Flanagan, D.C.; Nearing, M.A. *USDA-Water Erosion Prediction Project: Hillslope Profile and Watershed Model Documentation*; NSERL Report no.10; USDA-ARS National Soil Erosion Research Laboratory: West Lafayette, IN, USA, 1995.
44. Nearing, M.A.; Foster, G.R.; Lane, L.J.; Finckner, S.C. A process-based soil erosion model for USDA—Water Erosion Prediction Project Technology. *Trans. Am. Soc. Agric. Eng.* **1989**, *32*, 1587–1593. [CrossRef]
45. Kinnell, P.I.A.; Risse, L.M. USLE-M: Empirical modeling rainfall erosion through runoff and sediment concentration. *Soil Sci. Soc. Am. J.* **1998**, *62*, 1667–1672. [CrossRef]
46. Bryan, R.B.; Govers, G.; Poesen, J. The concept of soil erodibility and some problems of assessment and application. *Catena* **1989**, *16*, 393–412. [CrossRef]
47. Bryan, R.B. Soil erodibility and processes of water erosion on hillslope. *Geomorphology* **2000**, *32*, 385–415. [CrossRef]
48. Wang, G.; Gertner, G.; Liu, X.; Anderson, A. Uncertainty assessment of soil erodibility factor for revised universal soil loss equation. *Catena* **2001**, *46*, 1–14. [CrossRef]
49. Torri, D.; Poesen, J.; Borselli, L. Predictability and uncertainty of the soil erodibility factor using a global dataset. *Catena* **1997**, *31*, 1–22. [CrossRef]
50. Jamshidi, R.; Dragovich, D.; Webb, A.A. Catchment scale geostatistical simulation and uncertainty of soil erodibility using sequential Gaussian simulation. *Environ. Earth Sci.* **2014**, *71*, 4965–4976. [CrossRef]
51. Buttafuoco, G.; Conforti, M.; Aucelli, P.P.C.; Robustelli, G.; Scarciglia, F. Assessing spatial uncertainty in mapping soil erodibility factor using geostatistical stochastic simulation. *Environ. Earth Sci.* **2012**, *66*, 1111–1125. [CrossRef]
52. Kinnell, P.I.A. Applying the QREI30 index within the USLE modelling environment. *Hydrol. Process.* **2013**, *28*, 591–598. [CrossRef]
53. Kinnell, P.I.A. Accounting for the influence of runoff on event soil erodibilities associated with the *EI 30* index in RUSLE2. *Hydrol. Process.* **2015**, *29*, 1397–1405. [CrossRef]
54. Govers, G.; Takken, I.; Helming, K. Soil roughness and overland flow. *Agronomie* **2000**, *20*, 131–146. [CrossRef]
55. Nearing, M.A. Soil erosion and conservation. In *Environmental Modelling: Finding Simplicity in Complexity*, 2nd ed.; Wainwright, J., Mulligan, M., Eds.; John Wiley & Sons, Ltd.: Hoboken, NJ, USA, 2013; pp. 365–378. [CrossRef]
56. Cassol, E.A.; Silva, T.S.; Eltz, F.L.F.; Levien, R. Soil Erodibility under Natural Rainfall Conditions as the K Factor of the Universal Soil Loss Equation and Application of the Nomograph for a Subtropical Ultisol. *Rev. Bras. Ciênc. Solo* **2018**, *42*, 1–12. [CrossRef]
57. Römkens, M.J.M. The soil erodibility factor: A perspective. In *Soil Erosion and Conservation*; El-Swaify, S.A., Moldenhauer, W.C., Lo, A., Eds.; Soil and Water Conservation Society: Ankeny, IA, USA, 1985; pp. 445–461.
58. El-Swaify, S.A.; Dangler, E.W. Rainfall erosion in the tropics: A state of art. In *Determinants of Soil Loss Tolerance*; Krebs, D.M., Ed.; American Society of Agronomy: Madison, WI, USA, 1982; pp. 1–25.
59. Wang, B.; Zheng, F.; Römkens, J.M. Comparison of soil erodibility factors in USLE, RUSLE2, EPIC and Dg models based on a Chinese soil erodibility database. *Acta Agric. Scand. Sect. B Soil Plant Sci.* **2013**, *63*, 69–79. [CrossRef]
60. Williams, J.R.; Sharply, A.N. EPIC-Erosion Productivity Impact Calculator, I. In *Model Documentation*; US Department of Agriculture Technical Bulletin: Washington, DC, USA, 1990; p. 1768.
61. Dövényi, Z. *Inventory of Microregions in Hungary*; MTAFKI: Budapest, Hungary, 2010.
62. Szabó, B.; Centeri, C.; Szalai, Z.; Jakab, G.; Szabó, J. Comparison of soil erosion dynamics under extensive and intensive cultivation based on basic soil parameters. *Növénytermelés* **2015**, *64*, 23–26. [CrossRef]
63. Szabó, J.A.; Centeri, C.; Keller, B.; Hatvani, I.G.; Szalai, Z.; Dobos, E.; Jakab, G. The use of various rainfall simulators in the determination of the driving forces of changes in sediment concentration and clay enrichment. *Water* **2020**, *12*, 2856. [CrossRef]
64. Renard, K.G.; Foster, G.R.; Weesies, G.A.; Porter, J.P. RUSLE: Revised Universal Soil Loss Equation. *J. Soil Water Conserv.* **1991**, *46*, 30–33.
65. Presbitero, A.L. Soil Erosion Studies on Steep Slopes of Humid-Tropic Philippines. In *School of Environmental Studies*; Nathan Campus, Griffith University: Queensland, Australia, 2003.

66. Kinnell, P.I.A. Runoff ratio as a factor in the empirical modeling of soil erosion by individual rainstorms. *Aus. J. Soil Res.* **1997**, *35*, 1–14. [CrossRef]
67. Rosewell, C.J.; Edwards, K. *SOILOSS—A Program to Assist in the Selection of Management Practices to Reduce Erosion, Tech. Handbook No. 11*; Soil Conservation Service of New South Wales: Sydney, Australia, 1998; p. 71.
68. Ahmad, H.M.N.; Sinclair, A.; Jamieson, R.; Madani, A.; Hebb, B.; Havard, P.; Yiridoe, E.K. Modeling sediment and nitrogen export from a rural watershed in Eastern Canada Using the soil and water Assessment Tool. *J. Environ. Qual.* **2011**, *40*, 1182–1194. [CrossRef]
69. Moriasi, D.; Arnold, J.; Van Liew, M.; Bingner, R.; Harmel, R.D.; Veith, T. Model Evaluation Guidelines for Systematic Quantification of Accuracy in Watershed Simulations. *Trans. ASABE* **2007**, *50*, 885–900. [CrossRef]
70. Defersha, M.B.; Quraishi, S.; Melesse, A. The effect of slope steepness and antecedent moisture content on interrill erosion, runoff and sediment size distribution in the highlands of Ethiopia. *Hydrol. Earth Syst. Sci.* **2011**, *15*, 2367–2375. [CrossRef]
71. Vermang, J.; Demeyer, V.; Cornelis, W.; Gabriels, D. Aggregate Stability and Erosion Response to Antecedent Water Content of a Loess Soil. *Soil Sci. Soc. Am. J.* **2009**, *73*, 718–726. [CrossRef]
72. Ziadat, F.M.; Taimeh, A.Y. Effect of rainfall intensity, slope, land use and antecedent soil moisture on soil erosion in an arid environment. *Land Degrad. Dev.* **2013**, *24*, 582–590. [CrossRef]
73. Centeri, C. In situ soil erodibility values versus calculations. In *25 Years of Assessment of Erosion—International Symposium*; Gabriels, D., Cornelis, W., Eds.; International Center for Eremology—University of Ghent: Ghent, Belgium, 2003; pp. 135–140.
74. Zhang, K.L.; Shu, A.P.; Xu, X.L.; Yang, Q.K.; Yu, B. Soil erodibility and its estimation for agricultural soils in China. *J. Arid. Environ.* **2008**, *72*, 1002–1011. [CrossRef]
75. Bronstert, A.; Bárdossy, A. The role of spatial variability of soil moisture for modelling surface runoff generation at the small catchment scale. *Hydrol. Earth Syst. Sci.* **1999**, *3*, 505–516. [CrossRef]
76. Zehe, E.; Blöschl, G. Predictability of hydrologic response at the plot and catchment scales: Role of initial conditions. *Water Resour. Res.* **2004**, *40*, W10202. [CrossRef]
77. Nearing, M.A. Why soil erosion models over-predict small soil losses and under-predict large soil losses. *Catena* **1998**, *32*, 15–22. [CrossRef]
78. Tiwari, A.K.; Risse, L.M.; Nearing, M.A. Evaluation of WEPP and its comparison with USLE and RUSLE. *Trans. ASAE* **2000**, *43*, 1129–1135. [CrossRef]
79. Di Stefano, C.; Ferro, V.; Pampalone, V. Applying the USLE Family of Models at the Sparacia (South Italy) Experimental Site. *Land Degrad. Dev.* **2016**, *28*, 994–1004. [CrossRef]
80. Singh, M.J.; Khera, K.L. Nomographic estimation and evaluation of soil erodibility under simulated and natural rainfall conditions. *Land Degrad. Dev.* **2009**, *20*, 471–480. [CrossRef]
81. Marques, V.S.; Ceddia, M.B.; Antunes, M.A.H.; Carvalho, D.F.; Anache, J.A.A.; Rodrigues, D.B.B.; Oliveira, P.T.S. USLE K-Factor method selection for a tropical catchment. *Sustainability* **2019**, *11*, 1840. [CrossRef]

Article

Effects of Infiltration Amounts on Preferential Flow Characteristics and Solute Transport in the Protection Forest Soil of Southwestern China

Mingfeng Li [1], Jingjing Yao [2], Ru Yan [1] and Jinhua Cheng [1,*]

[1] Jinyun Forest Ecosystem Research Station, School of Soil and Water Conservation, Beijing Forestry University, Beijing 100083, China; m17710660921@163.com (M.L.); yanru2016@163.com (R.Y.)
[2] Environmental Protection Research Institute of Light Industry, Research Centre for Urban Environment, Beijing Academy of Science and Technology, Beijing 100095, China; yaojing1989_lucky@163.com
* Correspondence: jinhua_cheng@126.com

Citation: Li, M.; Yao, J.; Yan, R.; Cheng, J. Effects of Infiltration Amounts on Preferential Flow Characteristics and Solute Transport in the Protection Forest Soil of Southwestern China. *Water* **2021**, *13*, 1301. https://doi.org/10.3390/w13091301

Academic Editors: Domenico Cicchella and Viktor O. Polyakov

Received: 18 February 2021
Accepted: 3 May 2021
Published: 6 May 2021

Abstract: Preferential flow has an important role as it strongly influences solute transport in forest soil. The quick passage of water and solutes through preferential flow paths without soil absorption results in considerable water loss and groundwater pollution. However, preferential flow and solute transport under different infiltration volumes in southwestern China remain unclear. Three plots, named P20, P40 and P60, were subjected to precipitation amounts of 20, 40 and 60 mm, respectively, to investigate preferential flow and solute transport characteristics via field multiple-tracer experiments. Stained soils were collected to measure Br^- and NO_3^- concentrations. This study demonstrated that precipitation could promote dye tracer infiltration into deep soils. The dye tracer reached the maximum depth of 40 cm in P60. Dye coverage generally reduced with greater depth, and sharp reductions were observed at the boundary of matrix flow and preferential flow. Dye coverage peaked at the soil depth of 15 cm in P40. This result demonstrated that lateral infiltration was enhanced. The long and narrow dye coverage pattern observed in P60 indicated the occurrence of macropore flow. Br^- and NO_3^- were found at each soil depth where preferential flow had moved. Increasing precipitation amounts increased Br^- and NO_3^- concentration and promoted solute movement into deep soil layers. Solute concentration peaked at near the end of the preferential flow path and when preferential flow underwent lateral movement. These results indicated that the infiltration volume and transport capacity of preferential flow had important effects on the distribution of Br^- and NO_3^- concentrations. The results of this study could help expand our understanding of the effects of preferential flow on solute transport and provide some suggestions for protection forest management in southwestern China.

Keywords: multiple-tracer experiments; precipitation amounts; preferential flow; solute transport; protection forest

1. Introduction

Preferential flow, considered as a typical soil water movement process in unsaturated soils, has an important influence on water infiltration and solute transport [1–3]. It is a rapid and nonequilibrium process that accounts for 11% to 54% of the entire water flow [4,5]. Preferential flow can cause water and solutes to rapidly bypass most of the soil matrix through preferential flow paths, which are formed by cracks, biological activities, root channels, erosive actions, soil shrinkage cracking and freezing–thawing phenomena [6–8]. Preferential flow thus drastically decreases the utilisation rate of water and nutrients and positively increases the contamination risks of groundwater [9].

Preferential flow has received considerable attention because it is an important factor that considerably influences water and nutrient transport in rainfall events. It is induced by numerous factors, including soil texture and structure; channels formed by biological

161

activities, such as roots and earthworms; crop management; antecedent soil water; rainfall intensity and precipitation [6,10]. Soil types and structure have a complex effect on preferential flow because of their spatial heterogeneities, which can directly change the hydraulic properties, quantities and distribution of soil macropores. Soil clay content could change the soil pore structure and thus affect the type of preferential flow occurring in the soil [11]. The greater the clay content in the soil, the more obvious the macropore structure and the more favourable the formation of preferential flow [12,13]. Biological activities create complex channel systems that can serve as preferential flow paths, further influencing the lateral and vertical movements of preferential flow [14]. The role of antecedent soil water in preferential flow may differ under different soil and macropore conditions [15]. The effect of rainfall intensity on preferential flow has been well discussed, and several studies have shown that the proportion of macropore flow usually increases with the increase in rainfall intensity when it is higher than the infiltration rate [6,10]. However, preferential flow and its influence on water and solute transport under different precipitation amounts, which can affect soil moisture and soil water repellency during rainfall events, have received limited attention.

Vidon and Cuadra [10] showed that the proportion of the contribution of macropore flow to total flow flux is positively correlated with precipitation amounts. However, the spatial changes shown by preferential flow under different precipitation amounts have not been fully described and quantitatively evaluated. Such information is crucial for understanding the mechanism of preferential flow in different rainfall events. Moreover, identifying the role of preferential flow in solute transport under different precipitation volumes can help reduce groundwater pollution.

Tracing experiments are increasingly used to study preferential flow in the field and laboratory [16,17]. The food-grade dye Brilliant Blue is widely applied to describe preferential flow because of its advantages of low cost, high water solubility, limited toxicity, stability and distinct visibility [15]. However, it cannot describe solute transport during different steps of infiltration. Therefore, multiple-tracer experiments have been gradually performed to characterise preferential flow and solute transport given the similarity of their advantages with those of experiments with Brilliant Blue FCF [1,3]. In these experiments, image processing is used to describe the distribution of preferential flow, and the concentrations of tracer ions are measured to characterise solute transport during infiltration.

The Three Gorges reservoir area is located in the combination of the middle and upper reaches of the Yangtze River, with complex topography, large spatial variation of natural resources and strong anthropogenic interference. It is a sensitive ecological area and an important functional area for soil and water conservation in China. The protection forest optimisation project of nearly 270,000 ha has been implemented to solve the problems of unreasonable spatial distribution and structure of forest species, poor quality of forest stands and low ecological protection effectiveness in the Three Gorges reservoir area of the Yangtze River. However, fertilisers and pesticides are widely applied in the process of optimising the construction of protective forests, which has an impact on the quality of water resources in the Three Gorges reservoir area. High mean annual precipitation reaching 1031 mm increases groundwater contamination risks in this region [15]. More than 80% of rainfall is received during April to October, with the region receiving the highest amount of rainfall during June to August. The utilisation rate of water, fertilisers and pesticides may be seriously influenced by preferential flow during these months. In addition, groundwater contamination is promoted by the increase in the amount of solutes that infiltrate deep soil together with preferential flow. Thus, determining preferential flow characteristics and their effect on solute transport under different infiltration amounts is highly desirable. The objective of the present study was to (1) characterise the distribution of preferential flow and (2) determine solute transport with preferential flow under different infiltration amounts via multiple-tracer experiments.

2. Materials and Methods

Field Experiments

Field multiple-tracer experiments were conducted in protection forest soil in Simian Mountain (28°36′ N, 106°23′ E), Chongqing Province, Three Gorges Area, China. This area is located at the end of the Three Gorges reservoir area where it has a subtropical humid monsoon climate with 17–19 °C annual mean temperature and 1000–1250 mm mean annual precipitation received mainly from April to October. According to the Chinese soil taxonomy, the soil type of the study area is yellow earth, which belongs to the class of ferralsols, the subclass of wet-warm ferralsols. To improve the protective forest system, people have been optimising the species and spatial configuration of the protection forests in this region from 2000 to the present. Our study selected the protection forests which were mainly dominated by *Cunninghamia lanceolate* and *Quercus acutissima*, and the understory companion plant species include *Itea oblonga*, *Eurga loquaiana*, *Plagiogyria distinctissima* and *Aster ageratoides*. In the first 3 years, pest and disease control were conducted in the protection forest regularly. The pesticides (Use 0.75~1% Bordeaux, 75% chlorothalonil wettable powder, or 65% Dyson zinc) were applied during July to August once a week for 3 consecutive weeks, with 50 to 70 kg of the pesticides per mu. The type of fertiliser, the amount of fertiliser applied, and the time of fertiliser application depend on the actual forest growth.

Undisturbed soil cores with volumes of 100 cm^3 and soil samples with weights of approximately 500 g were collected at the depth intervals of 0–10, 10–20, 20–30, 30–40, 40–50 and 50–60 for the measurement of soil properties, including soil texture, bulk density, total porosity and organic matter content, in the laboratory. Soil characteristics are listed in Table 1.

Table 1. Soil characteristics of the studied field.

Depth (cm)	Sand Content (%) (2–0.02 mm)	Silt Content (%) (0.02–0.002 mm)	Clay Content (%) (≤0.002 mm)	Bulk Density (g·cm^3)	Total Porosity (%)	Organic Matter Content (%)
0–10	66.33 ± 6.03	28.76 ± 1.03	4.91 ± 0.09	1.05 ± 0.07	60.30 ± 8.83	4.78 ± 0.23
10–20	59.34 ± 3.34	30.67 ± 0.78	9.99 ± 0.23	1.13 ± 0.03	58.15 ± 3.08	4.33 ± 0.12
20–30	45.32 ± 2.92	42.19 ± 2.26	12.49 ± 0.43	1.22 ± 0.06	50.95 ± 1.98	3.21 ± 0.13
30–40	44.32 ± 3.00	43.18 ± 2.23	12.50 ± 0.48	1.30 ± 0.02	52.20 ± 1.89	2.56 ± 0.13
40–50	43.21 ± 4.25	46.03 ± 1.30	10.76 ± 0.92	1.29 ± 0.03	49.36 ± 1.23	2.01 ± 0.12
50–60	40.45 ± 3.74	49.06 ± 3.24	10.49 ± 0.45	1.27 ± 0.05	49.71 ± 0.89	1.54 ± 0.11

Note: Data in the table are the average value ± standard deviation.

In this study, three areas with similar site conditions were selected for the replication of the field multiple-tracer experiments. In each area, three different infiltration levels of 20, 40 and 60 mm were established to simulate three different infiltration amounts, which were designated as P20, P40 and P60, respectively. Each level had two plots, which were treated with different solutions. A total of 18 plots were established for the six treatments, with each type of treatment having three replicates (3 replicates × 6 plots each = 18 plots total). For each plot, two rectangular iron frames, namely, an inner frame with dimensions of 60 cm (length) × 60 cm (width) × 50 cm (height) and an outer frame with dimensions of 80 cm × 80 cm × 50 cm, were concentrically embedded into the soil to a depth of 30 cm after the experimental surface had been cleaned and smoothed. After embedding, the soil within 5 cm was compacted by using a wooden hammer to prevent dye tracer infiltration along the frames. Due to the relatively high initial water content of the soil in the study area, the accumulation of water will rapidly form on the soil surface layer. Therefore, the solution was quickly applied to the soil surface of the inner frame to simulate instantaneous ponding infiltration. In accordance with double-ring infiltration, the same depth of freshwater was simultaneously applied to the soil surface of the outer frame at each step to force the solution to infiltrate into the inner frame fully. The details of the field multiple-tracer experiments are shown in Figure 1 and Table 2. These details include the layout of the plots, the solute of the solution, the volume of the solution and the time consumed for solution infiltration. The solution from step 2 (3) should be added only when the solution from step 1 (2) is completely infiltrated to avoid the mixing of different solutions. The plots were covered with plastic tarpaulin to prevent evaporation and rainfall. After 24 h from the beginning of the experiment, five vertical slices were excavated from the core region of the dye area to avoid excavating disturbed soil edges. Each vertical slice was spaced 10 cm apart. A fine mist of a solution of starch (50 g L^{-1}) and $Fe_2(SO_4)_3$ (20 g L^{-1}) was uniformly sprayed onto the excavated soil surface immediately after a slice was exposed. The solution was an iodide indicator solution and showed a distinct purple colour when it encountered iodine molecules. After sufficient reaction for approximately 10 min, a digital camera (2560 × 1960) was used to photograph the dye-stained patterns of the horizontal and vertical slices. Colour images were classified as stained or unstained areas in accordance with the procedure of Yao et al. [15]. In this study, each 1 mm × 1 mm area of the original slice was represented by one pixel. After photography, soil samples weighing approximately 100 g were collected from the stained area at a site 5 mm below each excavated soil surface to remove the solution of starch and $Fe_2(SO_4)_3$. These soil samples were collected to measure the concentrations of Br^- and NO_3^-. The initial concentrations of Br^- in the soil were ignored because they were significantly lower than the applied concentrations. The concentration of NO_3^- in plot 1 was considered as the initial concentration of NO_3^- in the soil.

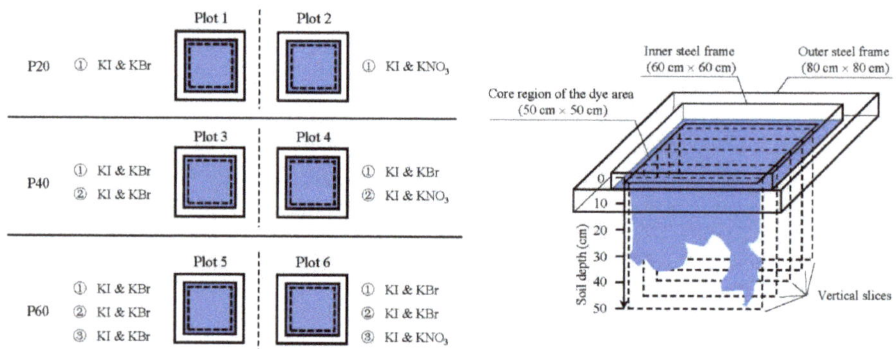

Figure 1. Diagram of the experimental set-up for multiple-tracer experiments.

Table 2. Periods of each step for P20, P40 and P60.

Level	Plots	Sequence of Matrix Solution Application	Total Applied Solution Amount (mm)	Period of Mixture Solution Infiltration (min)
P20	Plot 1	KI + KBr	20	13
	Plot 2	KI + KNO₃	20	15
P40	Plot 3	KI + KBr → KI + KBr	40	17 → 44
	Plot 4	KI + KBr → KI + KNO₃	40	20 → 39
P60	Plot 5	KI + KBr → KI + KBr → KI + KBr	60	14 → 38 → 35
	Plot 6	KI + KBr → KI + KBr → KI + KNO₃	60	15 → 45 → 40

Note: KI + KBr: Solution of KI (20 g L^{-1}) and KBr (10 g L^{-1}); KI + KNO₃: Solution of KI (20 g L^{-1}) and KNO₃ (10 g L^{-1}).

3. Results and Discussion

3.1. Spatial Variation Characteristics of Preferential Flow

We compared and calculated the data extracted from vertical sections under different infiltration amounts. The dye coverage distributions and the images of dye flow patterns in vertical sections are shown in Figure 2. The nonuniform or dissimilar distributions of dye areas under different infiltration amounts suggested that different preferential flow patterns had developed. As shown in Figure 2, most of the vertical slices were stained with a dye coverage of more than 80% in the topsoil, indicating that the topsoil experienced matrix flow [15,18]. Meanwhile, the matrix flow depths for P20 (Figure 2a), P40 (Figure 2b) and P60 (Figure 2c) were 1.6, 7 and 9.3 cm. These results were similar to the conclusion of a study in plantation forests in the southwest karst region [19], which has shown that the depth of matrix flow increased with the increase in infiltration amounts. The increase in infiltration water corresponded to a longer rainfall period, increasing the potential energy of water supply, and the increase in water potential gradient increased the uniform infiltration of soil water. Additionally, the increase in infiltration water increased the water content in the lower soil layer, slowed down the rate of water infiltration, and promoted the lateral movement of water in the top soil layer to form matrix flow [20].

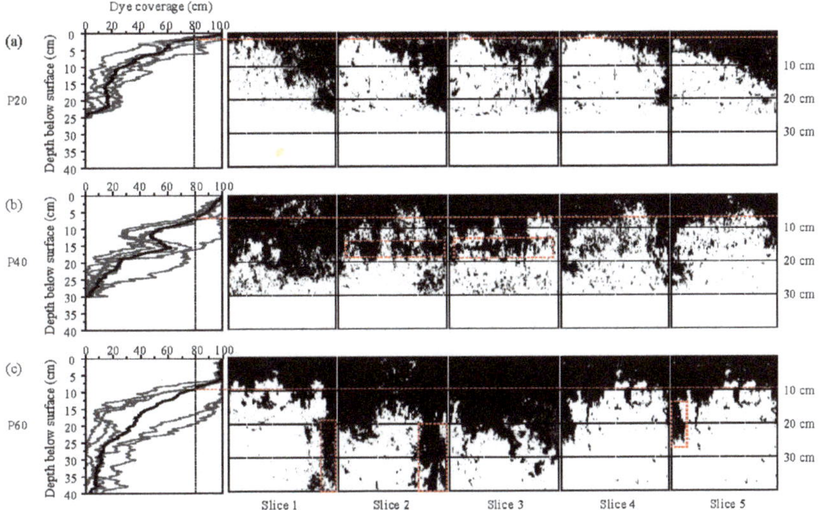

Figure 2. Images showing vertical slices from the dye-stained soil profiles and average dye coverage of vertical slices with depth under different water amounts: (**a**) 20, (**b**) 40 and (**c**) 60 mm.

Due to the increased depth of matrix flow, it was also expected that dye would infiltrate deeper under the higher infiltration amounts. Indeed, our results showed that the maximum infiltration depths for P20, P40 and P60 were 25, 30 and 40 cm, respectively. The consistency of these results with the findings of several other researchers [20,21] indicated the extensive variation range of the movement of the water in the soil under high infiltration amounts. However, our results showed that the matrix flow depth and the maximum infiltration depth occurred in deeper soil layers compared to the study in the plantation forests in the southwest karst region. This could be due to the fact that the sand content and total soil porosity in the soils of our study area (66.33%; 60.30%) were higher than those of the above study area (27.59%; 37.91%). Previous studies have reported that water movement in clay soils was dominated by preferential flow with less matrix flow compared to loamy soils, as the clay soils were fine-textured and well-structured and were more conducive to the presence of macropores. Additionally, Yan [22] also found that the matrix flow depth was 12 cm and the maximum infiltration depth was 28 cm in sandy soils with a sand content of about 80%. Therefore, in future studies of preferential flow, the depth of occurrence of preferential flow and the maximum depth of infiltration could be predicted based on soil conditions and infiltration amounts.

In our study, infiltration amounts had a substantial effect on the spatial heterogeneities of preferential flow despite the similar properties of the three plots (Figure 2). Reductions in the patterns of dye coverage showed different trends amongst P20, P40 and P60. Compared with that in P40 and P60, which showed considerably different patterns amongst vertical slices with the reduction in depth, the pattern of decrement in P20 showed no evident change, indicating small spatial heterogeneity (Figure 2a). This result may be attributed to the small depth of matrix flow that lacked sufficient head pressure to allow the formation of preferential flow path networks. Thus, the dye tracer infiltrated into deep soil layers only through a small part of preferential flow paths under low infiltration. In P40, unexpected increases in dye coverage were detected as the soil depth increased to 15 cm in P40, demonstrating that high amounts of the dye tracer had infiltrated into these layers (Figure 2b). Therefore, lateral preferential flow became increasingly evident with the increase in infiltration amount because the infiltration amount exceeded the capacity for the vertical transport of the preferential flow to block preferential flow paths. This phenomenon could also be associated with preferential flow path characteristics, resulting in the high lateral movement of the dye tracer. Root channels are considered as primary preferential flow paths in forest soils [9,23]. An investigation into the root systems in the study area revealed that the total root length of the root with root diameter <1 mm in the 10–20 cm soil layers was greater than that in the other soil layers except topsoil (0–5 cm) (Table 3). Consequently, in this region, the large number of intricate root distributions affected the connectivity of priority flow paths and thus increased lateral preferential flow along root channels. The dye coverage distribution in P60 was consistent with that in P20. Specifically, it lacked a peak and decreased gradually with soil depth. The dye coverage in the vertical slices from P60 had large standard deviations. Three of the five dyed coverage distribution curves showed sharp reductions in the 10–15 cm soil layers and an obvious trailing phenomenon at the soil depths of 15–40 cm, indicating the existence of penetrating macropore flow (Figure 2c). This result was largely explained by the fact that the increasing amount of infiltration increased the head pressure of the preferential flow. This effect then prompted the formation of the preferential flow network and increased the connectivity of the preferential flow network. In this case, the transport patterns of preferential flow could be divided into two types under the influence of spatial heterogeneity caused by soil texture and roots and biological activities [24,25]. In one case, preferential flow paths were connected vertically to form macropores, and water was transported rapidly through macropores to deep soil layers, thus bypassing most of the soil matrix. By contrast, in the absence of macropores, the unstable wetting front continued to expand because the infiltration rate was lower than the saturated hydraulic conductivity of the soil. This result, in turn, led to the finger-like preferential flow pattern of water and solutes. At the same time, as the infiltration amount was increased, the flow pattern transformed from one dominated by preferential flow to one dominated by matrix flow. Furthermore, macropore flow and finger-like flow appeared as illustrated in Figure 2. The appearance of these flows provided support for the above explanation.

Table 3. Root length (cm).

Soil Depth (cm)	Root Diameter				
	<1 mm	1–3 mm	3–5 mm	5–10 mm	>10 mm
0–5	2549 ± 322	1284 ± 211	499 ± 142	125 ± 23	19 ± 12
5–10	1504 ± 97	528 ± 200	139 ± 53	54 ± 8	20 ± 11
10–15	2058 ± 88	426 ± 44	149 ± 40	138 ± 22	38 ± 10
15–20	2068 ± 282	394 ± 86	189 ± 40	104 ± 58	23 ± 10
20–25	720 ± 182	476 ± 112	111 ± 62	87 ± 54	38 ± 13
25–30	565 ± 300	212 ± 0	243 ± 13	44 ± 0	0 ± 0
30–35	373 ± 58	396 ± 58	158 ± 31	27 ± 1	0 ± 0
35–40	366 ± 21	122 ± 6	48 ± 6	0 ± 0	0 ± 0

Note: In each soil layer, the root length under different root diameter ranges is the sum of the root length of that root diameter, and the volume of each soil layer is 125 cm^3.

Combining these results, the soil in the study area was dominated by sandy loam, and when the infiltration was less than 60 mm, the occurrence of preferential flow could increase the connectivity of soil pores and enhance the water retention capacity of the soil. However, when the infiltration was greater than 60 mm, macropore flow might be formed, which would transport water directly to the deeper soil layers to reduce the water utilisation rate, increase the risk of groundwater contamination and possibly reduce the stability of the soil body to induce geological disasters such as landslides.

3.2. Effect of Preferential Flow on Solute Transport

The occurrence and development of preferential flow have an important effect on water and solute transport [26]. Br^- and NO_3^- were mixed with KI separately and infiltrated at different stages to investigate the effect of different infiltration amounts on solute transport. The detailed process of this experiment was performed in accordance with reference methods. The concentration distributions of applied Br^- and NO_3^- in P20, P40 and P60 are shown in Figures 3 and 4. The infiltration depths of the Br^- and NO_3^- solutes were almost identical with those of preferential flow and increased with the increase in infiltration water volume. These results were consistent with the findings of earlier studies showing that soil-water movement depends strongly on rain intensity [27]. However, the different distributions of Br^- and NO_3^- concentrations after the experiment suggested that different types of preferential flows had different effects on solute transport (Figures 3 and 4).

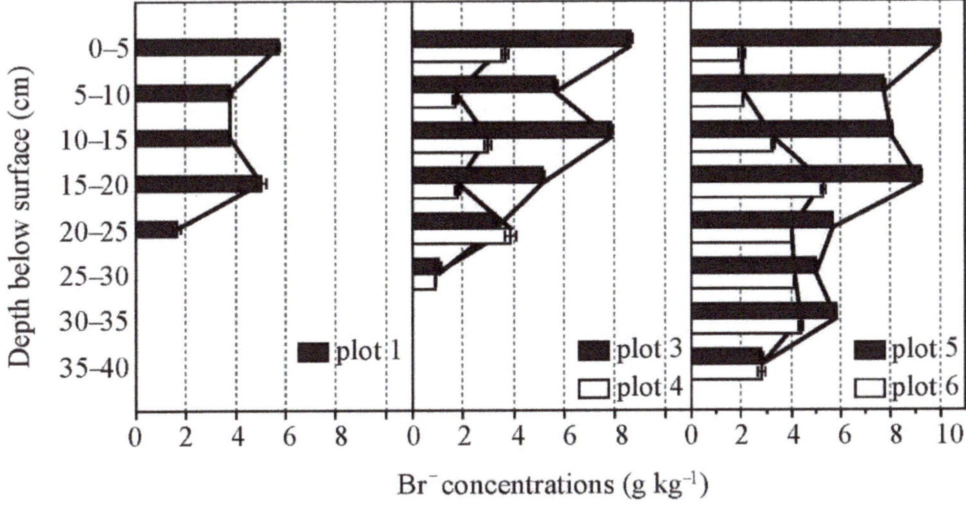

Figure 3. Changes in Br^- concentration with soil depth.

As shown in Figure 3, the concentration of Br^- at 0–40 cm soil depth under different infiltration intensities followed the order of plot 5 > plot 3 > plot 1, suggesting that increases in infiltration could increase in concentration. Additionally, in these plots, Br^- concentration distributions displayed small, W-shaped serrated patterns with the increase in soil depth and were highest in the topsoil layer. In other words, the Br^- concentration distributions in these plots showed similar trends and decreased with soil depth but peaked in different soil layers. In plot 1, Br^- concentration peaked at the soil depth of 20 cm. In plot 3, Br^- concentration peaked at the soil depth of 15 cm and then again at 25 cm. In plot 5, the first peak of the Br^- concentration was observed at the soil depth of 20 cm and the second at the soil depth of 35 cm. These results indicated that with the increase in infiltration amount, the peak of Br^- concentration appeared in deep soil layers and at near the end of the preferential flow. The last peak of Br^- concentration observed in each plot was most likely attributed to the solution reaching the end of the preferential flow paths where accumulation occurred and the wetting fronts continued to extend to the surrounding area [28]. Consequently, high Br^- concentration was found at about 5 cm before the maximum infiltration soil depth. The other peaks might be attributed to the fact that when the upper infiltration water volume was increased to 40 mm and 60 mm, infiltration exceeded the capacity for the vertical transport of preferential flow, resulting in the accumulation of water and the development of lateral infiltration. Dye coverage in the layer where the peak was higher than that in other layers also supported this interpretation (Figure 2), suggesting that the distribution pattern of preferential flow was the key to influencing solute transport.

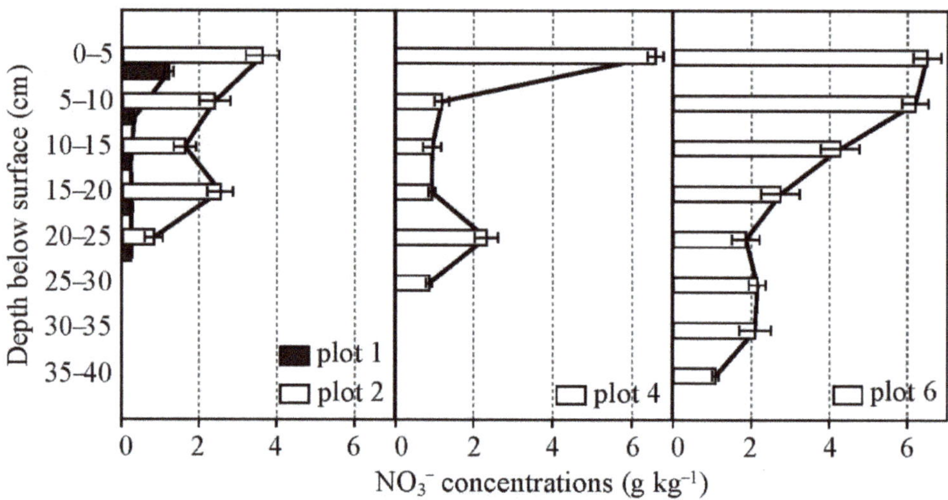

Figure 4. Changes in NO_3^- concentration with soil depth.

In comparison to plot 3 and plot 5, plot 4 and plot 6 used the solutions of KI and KNO_3 instead of KI and KBr in the final infiltration phase. The main purpose of this measure was to analyse the effect on the Br^- concentration in the soil for infiltration volumes of 20–40 mm and 40–60 mm, respectively. As shown in Figure 3, the concentration of Br^- in plot 4 was similar to that in plot 3, indicating that the infiltration volume of 20–40 mm did not change the distribution of Br^- concentration. The distribution of Br^- concentration in plot 6 was significantly different from that in the preferential flow dyeing area. In plot 6, unexpected increases in Br^- concentration were detected as the soil depth increased from 0 cm to 20, demonstrating that a large amount of Br^- in the surface layer was transported to the deeper soil by the preferential flow when the infiltration volume was at 40–60 mm. These results indicated that the solution from the second stage of infiltration mixed with the solution from the first stage of infiltration in the soil and transported downward. The third stage of infiltration pushed the pre-infiltrated solution downward with little interaction (as an approximately piston flow) at depths of 0–10 cm soil layers, probably because the first two stages of infiltration made the soil water saturated.

The NO_3^- concentrations in plot 2, plot 4 and plot 6 represented the distribution of NO_3^- at 0–20 mm, 20–40 mm and 40–60 mm of infiltration water, respectively (Figure 4). The NO_3^- concentration shown in plot 1 was the original soil NO_3^- content when the NO_3^--containing tracer was not added. The original NO_3^- concentration in the soil was low and therefore had a slight influence on the distribution of NO_3^- concentration in the solute transport experiments. The pattern of NO_3^- concentration in plot 2 was consistent with that of Br^- concentration in plot 1. This result indicated the absence of a significant difference in the concentration distribution of different solutes transported with preferential flow under a 20 mm infiltration volume. This is consistent with other studies that have shown rapid downward transport of water through the native preferential flow paths at low infiltration volume [21,22]. The significantly lower concentration of NO_3^- than that of Br^- at an infiltration of 20 mm may be attributed to the fact that NO_3^- is more readily available to plants and microorganisms than Br^- and converted into other forms, such as N_2 and NH_4^+. In plot 4, the highest NO_3^- concentration was detected in the topsoil, and NO_3^- concentration in other soil layer was significantly lower than that in plot 2, suggesting that the second stage solution was mainly concentrated on the soil surface and did not transport to deeper soil layers with the preferential flow. In plot 6, NO_3^- concentration decreased with the increase in soil layers without significant peaks. This result was consistent with previous studies indicating that increasing infiltration reduced the capillary effect of the soil layer and promoted the vertical connectivity of preferential flow paths, resulting in rapid solute transport to deep layers [1,29,30]. However, differently from previous studies, there were no peaks observed in this study and the NO_3^- concentration in plot 6 was significantly higher than that in plot 4, suggesting that the solute infiltration process in the third stage was accumulated and rapidly transported. This result may be due to the fact that when the amounts of infiltration volume reached a certain threshold, the formation of a penetrating macropore flow promoted the downward transport of solutes and reduced lateral infiltration.

These observations suggested that when the amount of water infiltration exceeded 40 mm, preferential flow infiltration depth and development increased, and solutes could infiltrate into deep soil layers along with macropore preferential flow during rainfall events. Therefore, in the future management of protective forest cultivation, fertiliser application and pesticide spraying before rain storms should be avoided as much as possible in order to reduce the pollution of shallow groundwater by pesticides and nutrients, etc.

4. Conclusions

Our results indicated that infiltration amounts significantly affected dye flow patterns and infiltration depth. When the infiltration volume was less than 40 mm, further increasing infiltration amounts increased the lateral infiltration of preferential flow and increased solute concentration. When the infiltration volume exceeded 40 mm, increasing infiltration amounts resulted in the development of highly continuous and effective preferential flow paths and increased the amount of solutes transported to deep soil layers through macropores. The results shown in this study could provide some suggestions for protected forest management and preferential flow hazard reduction. Nevertheless, the scale of this study was small, and the results might be influenced by soil heterogeneity. Further research should combine solute transport models with field tracing experiments to explore large-scale preferential flow and solute transport characteristics during natural rainfall events.

Author Contributions: Conceptualisation, M.L. and J.Y; methodology, M.L., R.Y. and J.Y; software, M.L.; validation, M.L. and J.Y; formal analysis, M.L., R.Y. and J.C.; investigation, M.L. and J.Y; resources, M.L., R.Y. and J.C.; data curation, M.L.; writing—original draft preparation, M.L., J.C., R.Y. and J.Y.; writing—review and editing, M.L. and J.Y.; visualisation, M.L. and J.Y.; supervision, M.L. and J.Y.; project administration, M.L.; funding acquisition, M.L., R.Y. and J.C. All authors have read and agreed to the published version of the manuscript.

Funding: This research was funded by the National Science and Technology Major Project (2018 ZX07101005) and the National Natural Science Foundation of China (32071839).

Institutional Review Board Statement: No studies involving humans or experimental animals were conducted in this research.

Informed Consent Statement: No studies involving humans or experimental animals were conducted in this research.

Data Availability Statement: The data presented in this study are available on request from the corresponding author. The data are not publicly available due to privacy or ethical restrictions.

Acknowledgments: Many thanks to the Forest Management Station of Simian Mountain for offering accommodations and supporting field experiments. We also gratefully acknowledge the editor and reviewers.

Conflicts of Interest: The authors declare no conflict of interest.

References

1. Wang, K.; Zhang, R. Heterogeneous soil water flow and macropores described with combined tracers of dye and iodine. *J. Hydrol.* **2011**, *397*, 105–117. [CrossRef]
2. Nimmo, J.R. Preferential flow occurs in unsaturated conditions. *Hydrol. Process.* **2012**, *26*, 786–789. [CrossRef]
3. Sheng, F.; Liu, H.; Kang, W.; Zhang, R.; Tang, Z. Investigation into preferential flow in natural unsaturated soils with field multiple-tracer infiltration experiments and the active region model. *J. Hydrol.* **2014**, *508*, 137–146. [CrossRef]
4. Legout, A.; Legout, C.; Nys, C.; Dambrine, E. Preferential flow and slow convective chloride transport through the soil of a forested landscape (Fougères, France). *Geoderma* **2009**, *151*, 179–190. [CrossRef]
5. Heijdena, G.d.; Legouta, A.; Polliera, B.; Bréchetb, C.; Rangera, J.; Dambrinec, E. Tracing and modeling preferential flow in a forest soil—Potential impact on nutrient leaching. *Geoderma* **2013**, *195–196*, 12–22. [CrossRef]
6. Jarvis, N.J. A review of non-equilibrium water flow and solute transport in soil macropores: Principles, controlling factors and consequences for water quality. *Eur. J. Soil Sci.* **2007**, *58*, 523–546. [CrossRef]
7. Forsmann, D.M.; Kjaergaard, C. Phosphorus release from anaerobic peat soils during convective discharge—Effect of soil Fe:P molar ratio and preferential flow. *Geoderma* **2014**, *223–225*, 21–32. [CrossRef]
8. Alaoui, A. Modelling susceptibility of grassland soil to macropore flow. *J. Hydrol.* **2015**, *525*, 536–546. [CrossRef]
9. Yi, J.; Yang, Y.; Liu, M.; Hu, W.; Zhang, D. Characterising macropores and preferential flow of mountainous forest soils with contrasting human disturbances. *Soil Res.* **2019**, *57*, 601–614. [CrossRef]
10. Vidon, P.; Cuadra, P.E. Impact of precipitation characteristics on soil hydrology in tile-drained landscapes. *Hydrol. Process.* **2010**, *24*, 1821–1833. [CrossRef]
11. Sheng, F.; Kang, W.; Zhang, R.; Liu, H. Characterizing soil preferential flow using iodine–starch staining experiments and the active region model. *J. Hydrol.* **2009**, *367*, 115–124. [CrossRef]
12. Chen, X.; Cheng, J. Application of Landscape Pattern Analysis to Quantitatively Evaluate the Spatial Structure Characteristics of PreferentialFlow Paths in Farmland. *Appl. Eng. Agric.* **2016**, *32*, 203–215. [CrossRef]
13. Grant, K.N.; Macrae, M.L.; Ali, G.A. Differences in preferential flow with antecedent moisture conditions and soil texture: Implications for subsurface P transport. *Hydrol. Process.* **2019**, *33*, 2068–2079. [CrossRef]
14. Murielle, G.; Sidle, R.C.; Alexia, S. The Influence of Plant Root Systems on Subsurface Flow: Implications for Slope Stability. *BioScience* **2011**, *61*, 869–879. [CrossRef]
15. Yao, J.; Cheng, J.; Sun, L.; Zhang, X.; Zhang, H. Effect of Antecedent Soil Water on Preferential Flow in Four Soybean Plots in Southwestern China. *Soil Sci.* **2017**, *182*, 1. [CrossRef]
16. Mooney, S.J.; Morris, C. A morphological approach to understanding preferential flow using image analysis with dye tracers and X-ray Computed Tomography. *Catena* **2008**, *73*, 204–211. [CrossRef]
17. Wang, F.; Chen, H.; Lian, J.; Fu, Z.; Nie, Y. Preferential Flow in Different Soil Architectures of a Small Karst Catchment. *Vadose Zone J.* **2018**, *17*, 2–10. [CrossRef]
18. Bargués Tobella, A.; Reese, H.; Almaw, A.; Bayala, J.; Malmer, A.; Laudon, H.; Ilstedt, U. The effect of trees on preferential flow and soil infiltrability in an agroforestry parkland in semiarid Burkina Faso. *Water Resour. Res.* **2014**, *50*, 3342. [CrossRef] [PubMed]
19. Kan, X.; Cheng, J.; Hou, F. Response of Preferential Soil Flow to Different Infiltration Rates and Vegetation Types in the Karst Region of Southwest China. *Water* **2020**, *12*, 1778. [CrossRef]
20. Hardie, M.; Doyle, R.; Cotching, W.; Holz, G.; Lisson, S. Hydropedology and Preferential Flow in the Tasmanian Texture-Contrast Soils. *Vadose Zone J.* **2013**, *12*, 108–112. [CrossRef]
21. Markus, W.; Felix, N. An experimental tracer study of the role of macropores in infiltration in grassland soils. *Hydrol. Process.* **2003**, *17*, 477–493. [CrossRef]
22. Yan, J.; Zhao, W. Characteristics of preferential flow during simulated rainfall events in an arid region of China. *Environ. Earth Sci.* **2016**, *75*, 566. [CrossRef]
23. Luo, Z.; Niu, J.; Xie, B.; Zhang, L.; Zhu, S. Influence of Root Distribution on Preferential Flow in Deciduous and Coniferous Forest Soils. *Forests* **2019**, *10*, 986. [CrossRef]
24. Pan, W.; Xu, Y.; Liu, Y.; Gao, L.; Yao, X. Quantitative determination of preferential flow characteristics of loess based on nonuniformity and fractional dimension. *Trans. CSAE* **2017**, *33*, 140–147. [CrossRef]
25. Schaik, N.L.M.B.V. Spatial variability of infiltration patterns related to site characteristics in a semi-arid watershed. *Catena* **2009**, *78*, 36–47. [CrossRef]
26. Makowski, V.; Julich, S.; Feger, K.H.; Breuer, L.; Julich, D. Leaching of dissolved and particulate phosphorus via preferential flow pathways in a forest soil: An approach using zero-tension lysimeters. *J. Plant. Nutr. Soil Sci.* **2020**, *183*, 238–247. [CrossRef]
27. Wu, Q.; Zhang, J.; Lin, W.; Wang, G. Appling dyeing tracer to investigate patterns of soil water flow and quantify preferential flow in soil columns. *Trans. CSAE* **2014**, *30*, 82–90. [CrossRef]

28. Ghafoor, A.; Koestel, J.; Larsbo, M.; Moeys, J.; Jarvis, N. Soil properties and susceptibility to preferential solute transport in tilled topsoil at the catchment scale. *J. Hydrol.* **2013**, *492*, 190–199. [CrossRef]
29. Wang, W.; Zhang, H.; Li, M.; Cheng, J.; Lu, W. Infiltration characteristics of water in forest soils in the Simian mountains, Chongqing City, southwestern China. *Front. For. China* **2009**, *4*, 338–343. [CrossRef]
30. Kramers, G.; Richards, K.G.; Holden, N.M. Assessing the potential for the occurrence and character of preferential flow in three Irish grassland soils using image analysis. *Geoderma* **2009**, *153*, 362–371. [CrossRef]

Article

Slope Erosion and Hydraulics during Thawing of the Sand-Covered Loess Plateau

Yuanyi Su [1,2], Peng Li [1,2,*], Zongping Ren [1,2], Lie Xiao [1,2], Tian Wang [1,2] and Yi Zhang [1,2]

[1] State Key Laboratory of Eco-hydraulics in Northwest Arid Region of China, Xi'an University of Technology, Xi'an 710048, China; suyuanyi666@163.com (Y.S.); renzongping@163.com (Z.R.); xiaosha525@163.com (L.X.); wlhuanjing@163.com (T.W.); 18202915856@163.com (Y.Z.)

[2] Key Laboratory of National Forestry Administration on Ecological Hydrology and Disaster Prevention in Arid Regions, Xi'an 710048, China

* Correspondence: nic@xaut.edu.cn; Tel./Fax: +86-29-8231-2658

Received: 20 July 2020; Accepted: 31 August 2020; Published: 1 September 2020

Abstract: Seasonal freeze-thaw processes have led to severe soil erosion globally. Slopes are particularly susceptible to changes in runoff, it can be useful to study soil erosion mechanisms. We conducted meltwater flow laboratory experiments to quantify the temporal and spatial distribution of hydraulic parameters on sandy slopes in relation to runoff and sediment yield under constant flow, different soil conditions (unfrozen slope: US; frozen slope: FS), and variable sand thickness. The results showed that sand can prolong initial runoff time, and US and FS have significantly different initial runoff times. There was a significant linear relationship between the cumulative runoff and the cumulative sediment yield. Additionally, hydrodynamic parameters of US and FS varied with time and spatially, as the distance between US and FS is linearly related to the top of the slope. We found that the main runoff flow pattern was composed of laminar flow and supercritical flow. There was a significant linear relationship between flow velocity and hydraulic parameters. The flow velocity is the best hydraulic parameter to simulate the trend of slope erosion process. This study can provide a scientific basis for a model of slope erosion during thawing for the Loess Plateau.

Keywords: loess; soil erosion; meltwater flow; runoff and sediment yield; hydraulic parameter

---

1. Introduction

The hydrodynamic properties of slopes have a decisive effect on runoff and sediment yield. Their study can help in understanding the process and mechanism of slope soil erosion and understanding the parameters of slope water dynamics, which are helpful in the construction of predictive models of slope soil erosion [1–3].

The Loess Plateau is a sand-covered landform that experiences substantial wind and water erosion [4–6]. Due to the differences in physical characteristics, infiltration, hydraulic conductivity, and water holding capacity between the surface sand layer and loess layer, a distinct sand-soil interface is formed between the sand layer and loess layer, and then the typical sand-soil dual structure is formed [7–10]. Zhang et al. [11] found in field rainfall experiments that the runoff and sediment yield processes on the sand-covered slopes are significantly different from those on loess slopes. Under light rain, the sand-covered slopes store rainfall, minimizing runoff, and the sediment content in any runoff that does occur is very large. Wu et al. [12] conducted a qualitative description of the sand-soil interface flow on the sand-covered slope through field surveys. Many scholars have studied the relationship between runoff and sediment yield, erosion processes, and the influence of sand layer size composition on runoff and sediment yield process through laboratory simulated rainfall experiments [6–8]. Tang et al. [13,14] quantitatively studied the spatiotemporal distribution of hydraulic parameters under different rainfall intensities and different sand thickness and their relationship with runoff and sediment yield. This study showed that the Reynolds Number can characterize the process of runoff and sediment yield on sand-covered slopes well [15]. However, studies on hydraulic characteristics of this particular landform at the slope scale during soil erosion is relatively limited.

The area of wind-water erosion on the Loess Plateau is in the middle latitude of the temperate zone, with an annual average precipitation of 300–600 mm. About 1/3 of days are below 0 °C each year, and it is windy and sandy in winter and spring with heavy rain in summer [16–19]. The hilly-gullied loess region is significantly affected by freezing and thawing, and snowmelt runoff erosion is a major manifestation of freeze-thaw erosion during thawing [20,21]. This erosion process is also found elsewhere throughout the world. In inland northeastern Oregon, 86% of soil erosion events are caused by freeze-thaw processes and snowmelt runoff [22]. In the northwest coastal region of the United States along the Pacific, over 90% of the total annual snow erosion is caused by melting snow [20]. When the frozen soil thaws, its shear strength decreases, and its erodibility increases, thus making the soil in the thawing period more susceptible to erosion [23–26]. The results of rainfall experiments by Sharratt et al. [27] showed that the impermeable frozen layer of soil prevents water infiltration during thawing, resulting in increased soil surface moisture content, surface runoff, with a high sediment content and soil erosion. The freeze-thaw process changes the structure of the topsoil and thereby influences the water erosion process as well [28,29]. At the same time, part of the hilly-gullied loess region subject to freeze-thaw erosion overlaps with the flakes of sand. Although this area of overlap is not large, it is widely distributed. Due to the overlapping of several types of erosion, "wind erosion, water erosion and freeze-thaw erosion", this superposition effect causes very serious soil erosion [30]. At present, scholars have done a lot of research on single-force erosion and wind-water composite erosion on the Loess Plateau. However, little research has been done on the problems of soil and water loss caused by multiple erosion types in the above areas.

Therefore, we studied sand-covered loess slopes using laboratory scouring experiments. We analyzed soil erosion characteristics and the spatiotemporal variation in hydraulic parameters of the sand-covered slopes when frozen or not, and described the relationship between hydraulic parameters and runoff and sediment yield. This study can provide a scientific basis for the construction of erosion forecast models for sand-covered loess slopes during thawing periods.

2. Materials and Methods

2.1. Material and Device

Two soil types were gathered from the Wangmaogou watershed (37°34′13″–37°36′03″ N, 110°20′46″–110°22′46″ E) of the Loess Plateau in Suide county, Shaanxi Province, China (Figure 1). Soil particle size was measured using a Mastersizer 2000 sediment particle size analyzer. The loess was comprised of 0.20% clay, 72.01% silt, and 27.79% sand. The sandy soil was comprised of 0.72% clay, 14.38% silt, and 84.9% sand. The soil was identified as a silt loam according to the soil classification standard of the United States Department of Agriculture. The dry bulk density of the soil was 1.3 g/cm^3, its organic matter content was 2.0 g/kg, and its saturated water content was 46.41%.

To measure soil erosion characteristics, we used a two-part experimental device that consisted of the frozen soil system and the scour experiment (Figure 2). The frozen soil system adopts a freeze-thaw test system implemented by the Xi'an University of Technology. The internal dimensions of the freeze-thaw were 4.5 m (length) × 2.5 m (width) × 2.5 m (height). Its internal temperature varied from −30–40 ± 1 °C. The temperature error was less than 2.0 °C, and there was a refrigeration and heating system, to maintain the experimental conditions. The system contained a runoff collection unit, soil box, sink, steady flow flume, and a water tank. The size of the soil box is 2 m long, 0.2 m wide, and 0.2 m deep with marks at 0.5 m increments on the side of the box to measure section velocities. From the top of the slope to the slope are S1, S2, S3, and S4. A sink of 2 m length, 0.2 m width, and 0.05 m depth was joined on the top of the soil box to make a stable concentrated flow.

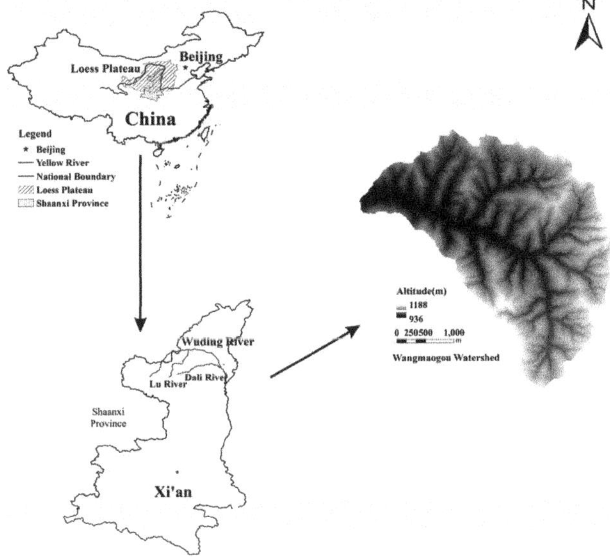

Figure 1. The geographical location of tested soil.

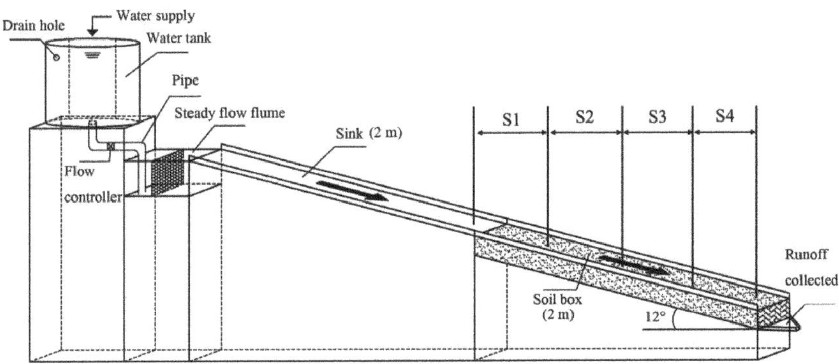

Figure 2. The scour experimental system.

2.2. Experimental Design

The scour experiments were conducted at the State Key Laboratory of Eco-hydraulics in the northwest arid region of China (Xi'an University of Technology) in Xi'an in April 2017. We used a local standard runoff plot and calculated the experimental flow rate from the mean precipitation during thawing as 1 L/min after correction. The scour experiment included unfrozen (US) and frozen slopes (FS) treatments and four sand thicknesses (0, 1, 2, and 3 cm). To ensure that the initial conditions of the scour experiments were consistent, the slope grade was set to 12°. The rainfall temperature was maintained at 10 °C and remained constant. Each experiment was replicated three times and we present the average of the three replicates. Each scour experiment lasted 15 min after the flow started (Table 1). Before the experiments, the soil samples were air-dried and passed through a 10 × 10 mm sieve to remove impurities such as plant roots. The soil samples were then moistened to a water content of 15% and covered with plastic film for 24 h to evenly distribute the soil moisture. A layer of gauze was laid on the bottom of the soil tank before the tank was filled with a 5 cm layer of sand. To mimic the soil's dry bulk density, the box was filled with soil in 5 cm layer intervals, and layers were mixed. To avoid confounds of being introduced by the box itself, the slope was designed to ensure erosion would flow through the middle of the box by lowering the slope and raising the sides. Once the box was filled, sand of different thicknesses (0, 1, 2, 3 cm) was layered above the soil in the box, along with water to increase the sand's water content. The soil tank was then frozen at −20 °C for 24 h and then placed on a scouring device bracket for testing. Since the room's temperature was higher than that of the freeze-thaw experiment system, the scour experiment also thawed.

Table 1. Design table of scour experimental.

State of Slope	Treatment	Depth of Sand (cm)	Flow Rate (L/min)	Slope (°)	Initial Soil Moisture Content (%)	Time (min)
Unfrozen slopes (US)	U0	0	1	12	15	15
	U1	1	1	12	15	15
	U2	2	1	12	15	15
	U3	3	1	12	15	15
Frozen slopes (FS)	F0	0	1	12	15	15
	F1	1	1	12	15	15
	F2	2	1	12	15	15
	F3	3	1	12	15	15

The flow rate was determined before the experiment began to ensure the difference between the actual flow and the intended treatment flow was less than 5% for three consecutive trials. After error testing, a formal test was performed by recording the initial flow time and collecting runoff and sediment samples every minute. Dye tracing (KMnO$_4$) was used to measure the flow velocity of different sections of the box by dividing the travel distance by the mean traveling time multiplied by an adjustment coefficient of 0.65. Samples were collected and the sediment was separated and then dried at 105 °C for 24 h and subsequently weighed.

2.3. Hydraulics Parameter Calculation and Methods

2.3.1. Calculation of Hydraulics Parameters

In this study, flow velocity (V), the Reynolds number (Re), Froude number (Fr), and the Darcy–Weisbach roughness coefficient (f) were selected as the research objects [31], then were calculated by the expression:

$$V = V_{\mathrm{m}} \times 0.65 \tag{1}$$

where V is the mean flow velocity (m/s), V_{m} is the observed velocity (m/s), and the flow travel distance is divided by the mean travelling time:

$$Re = \frac{VR}{v} \tag{2}$$

where v is the kinematic viscosity (m^2/s), and R is the hydraulic radius (m), which can be replaced by the value of average flow depth h:

$$h = \frac{Q}{VbT} \tag{3}$$

where Q is the total runoff in time T (m^3/s) and b is the width of water surface (m):

v (m^2/s) is the kinematic viscosity, it was calculated as follows:

$$v = \frac{0.01775}{1 + 0.0337t + 0.000221t^2} \tag{4}$$

where t is the water temperature (°C):

$$Fr = \frac{V}{\sqrt{gh}} \tag{5}$$

where g is the acceleration due to gravity (m/s^2), being 9.8 m/s^2:

$$f = \frac{8gh\sin\alpha}{V^2} \tag{6}$$

where α is the slope (°).

The coefficient of variation (*CV*) indicates the degree of data dispersion, the formula is as follows:

$$CV = (SD/Mean) \times 100\% \tag{7}$$

where *CV* is the coefficient of variation (%), *SD* is the standard deviation, and *Mean* is the average value.

2.3.2. Methods

The Photoshop (Adobe Photoshop CS4 Extended 11.0.1) was applied to design the experimental system. All statistical analyses were conducted using SPSS (IBM SPSS Statistics Version 21). Figures were generated in Origin 8.5.

3. Results

3.1. Erosion, Runoff, and Sediment Yield

3.1.1. Characteristics Values of Runoff and Sediment Yield

Table 2 showed the characteristic values of runoff and sediment yield under different treatments. Based on the initial runoff time of U0, the change in the initial runoff time under different treatments was calculated. It is calculated that the initial runoff time of U1, U2, and U3 is significantly longer than that of U0, and the initial runoff time of the slope surface under different sand thicknesses has been extended by 3.5 (U1), 4.73 (U2), and 6.36 (U3) times. The initial runoff time of F0 is 37.9% earlier than U0. The initial runoff time of F1, F2, and F3 did not change much compared with U0, but compared with U1, U2, and U3, the initial runoff time was much longer. Sand-covered slopes prolong initial runoff times and the effects become longer as the thickness of sand-cover increases. The initial runoff time of FS was significantly shorter than US. The total runoff under different treatments increased in the following order: U0 < U3 < U1 < U2 < F3 < F2 < F0 < F1. The *CV* of runoff under different treatments was between 2.48% and 22.14%, and the fluctuation range of the runoff was small, indicating that the impact of sand cover and soil freezing on the slope runoff process is small. The total sediment yield across different treatments declined in the order U0 < U1 < U2 < U3 < F0 < F1 < F2 < F3. Based on the total sediment yield of U0, the total sediment yield under different treatments are 3.37 (U1), 4.35 (U2), 4.96 (U3), 8.38 (F0), 8.88 (F1), 10.85 (F2), and 10.98 (F3) times. The *CV* of the sediment yield of US was between 27.8% and 63.3%, which indicates that the sediment yield process of US had a large degree of fluctuation. The increasing sand thickness increased the *CV* which indicates that the sediment yield of the slope varies drastically. The *CV* of FS was between 3.33% and 35.22%. Under the same conditions of sand cover thickness, the *CV* of FS was much smaller than that of US, which indicates that the sediment yield of FS was relatively stable.

Figure 2 showed the eroded topography under different treatments, with significant differences in surface morphology. On the unfrozen slopes and frozen slopes with different sand thicknesses, a rill appeared during the runoff process. However, under the same hydraulic conditions, the rill appeared in different shapes. For the US, the rill initially developed on the slope top and bottom, which extended to the slope middle at the same time. In U0, the rill showed a discontinuous distribution, and the development of the rill was primarily on the top (S1) and bottom (S4) of the slope. In U1, U2, and U3, the rills were continuous but shallow in depth. For the FS, the rill had the same characteristics. During the experimental processes, the rill initially only developed on the slope top and gradually extended to the slope bottom. The connected rill gradually appeared on the frozen slope (Figure 3).

Table 2. Initial runoff time, runoff and sediment yield under different treatments.

Treatment	Initial Runoff Time/(s)	Runoff		Sediment Yield	
		Total Runoff/L	CV/(%)	Total Sediment Yield/kg	CV/(%)
U0	39.38	9.64	22.14	0.93	27.80
U1	138.01	11.01	14.05	3.13	31.67
U2	186.36	11.3	15.51	4.03	51.16
U3	250.58	10.69	20.67	4.59	63.3
F0	24.45	13.82	2.48	7.76	3.33
F1	31.91	14.18	6.17	8.23	26.34
F2	34.02	13.46	13.86	10.04	26.22
F3	32.41	12.24	11.93	10.17	35.22

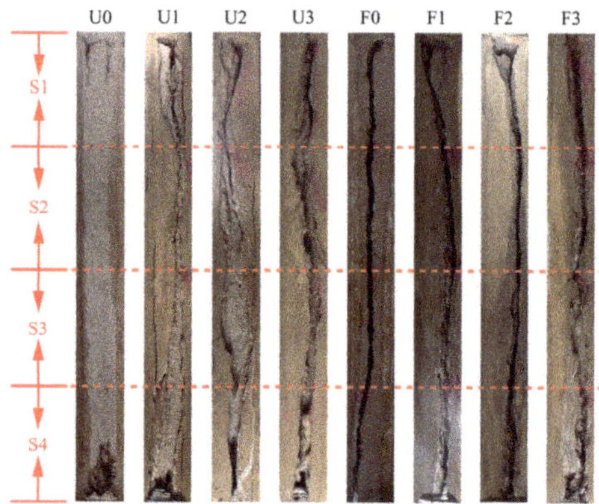

Figure 3. Eroded topography under different treatments. Note: S1 (2, 3, and 4), Section 1 (2, 3, and 4).

3.1.2. Correlation between Accumulative Runoff and Accumulative Sediment Yield

A function is fitted to the relationship between cumulative runoff and cumulative sediment yield for each experiment. The fitted equation is $M = CQ + D$, where M (kg) is the cumulative sediment yield, Q (L) is the cumulative runoff, and C and D are regression coefficients. All fitting equations were significant at $p < 0.001$ (Table 3). The coefficient C was defined as the sediment yield coefficient. MC was the mean of regression coefficient C. This coefficient obeys a certain change law. The MC of the US and FS were 0.31 and 0.67, respectively. The MC of the FS was 2.16 times than that of the US, indicating that the dependency of sediment yield on runoff was stronger for the FS than for the US.

Table 3. Cumulative runoff and cumulative sediment yield fitted equation.

Treatment	Fitted Equation		MC
U0	$M = 0.093\,Q + 0.029$	$R^2 = 0.997, p < 0.001$	
U1	$M = 0.283\,Q - 0.016$	$R^2 = 0.990, p < 0.001$	
U2	$M = 0.397\,Q - 0.035$	$R^2 = 0.972, p < 0.001$	0.31
U3	$M = 0.465\,Q + 0.263$	$R^2 = 0.930, p < 0.001$	
F0	$M = 0.565\,Q + 0.001$	$R^2 = 0.999, p < 0.001$	
F1	$M = 0.593\,Q - 0.361$	$R^2 = 0.987, p < 0.001$	
F2	$M = 0.740\,Q + 0.566$	$R^2 = 0.986, p < 0.001$	0.67
F3	$M = 0.778\,Q + 1.003$	$R^2 = 0.993, p < 0.001$	

3.2. Hydraulics of Slope Runoff

3.2.1. Spatiotemporal Variations of Flow Velocity

For the US, the flow velocity varies from 0.23 to 0.35 m/s during the tests, and its fluctuation range is small (Figure 4a). For the FS, the flow velocity varies from 0.18 to 0.35 m/s during the test, and its fluctuation range is relatively large (Figure 4). Under the condition of the same sand thickness, the mean values of flow velocity on the FS were 85.92% (F0/U0), 96.13% (F1/U1), 84.84% (F2/U2), and 88.47% (F3/U3) of the US, respectively. During the entire experiment, the flow velocity of the US and FS generally showed a downward trend (Figure 4). However, due to the conversion from erosion between inter-rill erosion to rill erosion during the erosion phase, the flow velocity changed due to the occurrence of rills. In the early stage of runoff, erosion was mainly between inter-rill erosion, the slope was relatively smooth, the runoff resistance was small, and the flow velocity was large. When the rill was generated on the slope, the flow velocity decreased significantly, and it occasionally rose with the backwater and reaches unconnected rills. In later stages of the experiment, the flow velocity tended to stabilize because the rill no longer developed. The variability of the flow velocity was mainly due to the increased resistance caused by the ground surface, the runoff energy consumption caused by the water flow down-cut, and the collapse of the soil on the side of the rill. Due to the abundance and looseness of sand, the fluctuation of U3 flow velocity is more severe.

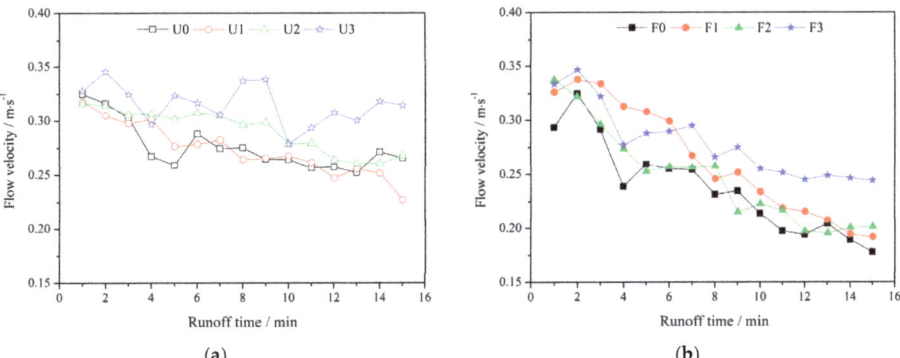

Figure 4. Variations in flow velocity under different treatments over time. (**a**) Unfrozen slope, (**b**) frozen slope.

In space, the V of the US and FS and the distance from the top of the slope can be represented by a linear function (US: $R^2 = 0.893$, $p > 0.05$; FS: $R^2 = 0.952$, $p < 0.05$). For the US, the V increases continuously as the distance from the top of the slope increases; for the FS, the V increases first to the maximum and then decreases as the distance from the top of the slope increases. At the same section, the V of the US decreases with the increase of the sand thickness, and then increases; the V of the FS increases with the increase in the sand thickness. Regression analysis showed that the V can be described by a linear function of the distance from the top of the slope (Table 4). The fitted equation showed that there is a certain range in which the V increases continuously as the slope length increases. Therefore, the slope length is very important. In some cases, precautions should be taken to mitigate the harm of downhill scour. Setting intercepting trenches and terraces on the slope can greatly reduce the slope length and thus reduce the erosion of runoff.

Table 4. Variations of flow velocity with the distance from the top of the slope (m/s).

Treatment	Distance from the Top of the Slope *E*/m				Fitted Equation
	0.5	**1**	**1.5**	**2**	
U0	0.26	0.29	0.29	0.32	
U1	0.21	0.27	0.33	0.29	$V_U = 0.071E + 0.212$
U2	0.22	0.31	0.31	0.35	$R^2 = 0.893, p > 0.05$
U3	0.23	0.33	0.37	0.39	
F0	0.19	0.20	0.26	0.33	
F1	0.18	0.21	0.31	0.37	$V_F = 0.091E + 0.146$
F2	0.19	0.22	0.29	0.28	$R^2 = 0.952, p < 0.05$
F3	0.24	0.25	0.33	0.32	

Note: The parameter "*E*" means the distance from the top of the slope, the same to below.

3.2.2. Spatiotemporal Variations of the Reynolds Number

From open channel hydraulics, the runoff flow is laminar when the *Re* < 500, the runoff flow is turbulent when the *Re* > 2000, and the runoff is transitional when the *Re* is between 500 and 2000. During the entire experiment, the *Re* ranged from 149.2 to 533.69 under US and FS, indicating that most of the runoff is laminar (Figure 5). For the US, the average value of the *Re* increased across conditions in this order: U0 < U1 < U2 < U3. There was little change in *Re* in the U0 treatment over time (range 155.63 to 204.82). The *Re* of U1, U2, and U3 increased over time, and the *Re* of U3 at the end of the experiment exceeded 500. For the FS, the *Re* in F1, F2, and F3 were approximately identical, all increasing slowly over time. The F0 increased rapidly from 0 to 5 min, and then showed a slow downward trend.

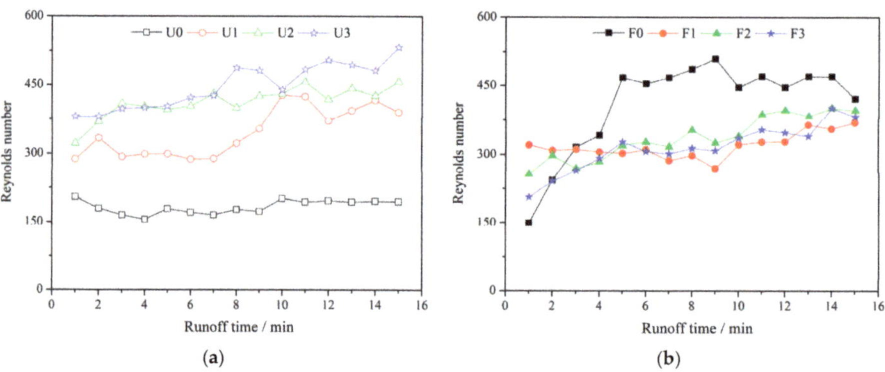

Figure 5. Variations in Reynolds number under different treatments over time. (a) Unfrozen slope, (b) frozen slope.

In space, the *Re* of the US and FS and the distance from the top of the slope can be represented by a linear function (US: $R^2 = 0.712, p > 0.05$; FS: $R^2 = 0.998, p < 0.01$). For the US, the *Re* was largest (*Re* > 500 for U1, U2, and U3) at 0.5 m from the top of the slope. As the distance from the top of the slope increased, the *Re* decreased. In the same section, the *Re* of the sand-covered slope was significantly larger than U0 (Table 5). For the FS, the *Re* was largest at 0.5 m from the top of the slope, and runoff was a transitional flow; only the *Re* of F0 is greater than 500. As the distance from the top of the slope increased, the *Re* decreased. As the distance from the top of the slope increased, the *Re* decreased. In the same section, the *Re* of F0 was greater than that of the sand-covered slope. At the same sand thickness, the *Re* of F0 was greater than U0. The *Re* of sand-covered slopes of US was larger than that of FS.

Table 5. Variations of Reynolds number with the distance from the top of the slope.

Treatment	Distance from the Top of the Slope E/m				Fitted Equation
	0.5	**1**	**1.5**	**2**	
U0	283.55	194.56	172.46	181.68	
U1	570.51	380.66	433.42	309.84	$Re_U = -135.01E + 484.59$
U2	636.61	375.33	371.24	392.56	$R^2 = 0.712, p > 0.05$
U3	587.71	463.96	388.73	410.48	
F0	518.21	499.82	444.29	430.38	
F1	433.68	410.35	340.95	350.17	$Re_F = -57.09E + 489.25$
F2	396.99	499.04	399.28	347.70	$R^2 = 0.998, p < 0.01$
F3	488.91	328.52	326.88	371.09	

3.2.3. Spatiotemporal Variations of the Froude Number

The critical value of subcritical flow and supercritical flow is 1. If the Fr is greater than 1, it is a supercritical transition of flow, otherwise, it is a subcritical flow. The Fr gradually decreased over time (Figure 6). The Fr in the US ranged from 3.88 to 5.24, and the Fr of the FS ranged from 2.93 to 4.85. This indicated that the runoff on the slope is supercritical during the experiment. For the US, the Fr of U0 was greater than the Fr of the sand-covered slope. The changes in the Fr of U1, U2, and U3 over time were roughly the same. For the FS, the Fr decreased faster with time as compared to the US, and showed a clear layering phenomenon.

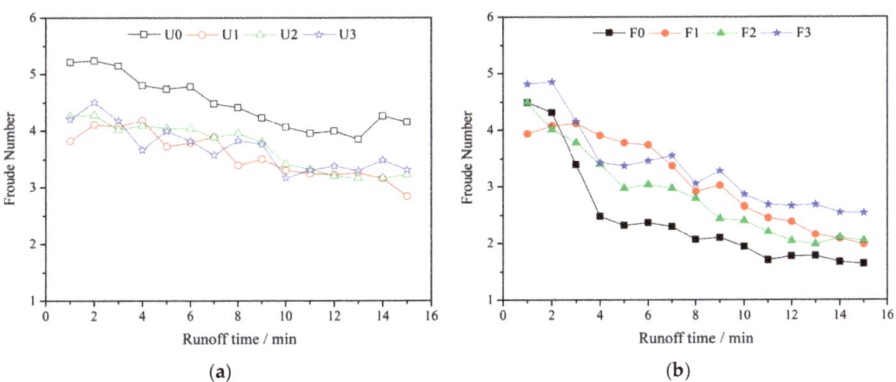

Figure 6. Variations in the Froude number under different treatments over time. (**a**) Unfrozen slope, (**b**) frozen slope.

In space, the Fr of the US and FS and the distance from the top of the slope can be represented by a linear function (US: $R^2 = 0.899, p > 0.05$; FS: $R^2 = 0.956, p < 0.05$). The Fr under different treatments increased with the distance from the top of the slope (except U2 and F3). The maximum Fr of U2 and F3 occurred at 1.5 m from the top of the slope. For the US, the Fr of U0 in different sections was greater than the Fr of sand-covered slopes. Under the same section, the Fr had no obvious change law with the increase of sand thickness (Table 6). Under the same sand thickness, the Fr of the US was larger than the Fr of the FS.

Table 6. Variations of Froude number with the distance from the top of the slope.

Treatment	Distance from the Top of the Slope *E*/m				Fitted Equation
	0.5	1	1.5	2	
U0	3.37	4.47	4.67	5.32	
U1	1.86	2.89	3.67	3.66	$Fr_U = 1.488E + 1.656$
U2	1.40	3.74	3.76	4.34	$R^2 = 0.899, p > 0.05$
U3	1.76	2.98	4.07	4.29	
F0	1.35	1.45	1.92	3.18	
F1	1.25	1.68	3.39	4.03	$Fr_F = 1.362E + 0.656$
F2	1.37	1.62	2.72	2.86	$R^2 = 0.956, p < 0.05$
F3	1.74	2.38	3.56	3.24	

3.2.4. Spatiotemporal Variations of the Darcy-Weisbach Roughness Coefficient

As shown in Figure 7, the *f* has volatility, but generally increases gradually with time. The range of the *f* for the US and the FS is 0.06~0.56 and 0.08~1.81, respectively. For the US, there was little change in *f* of U0, and the change of the *f* of sand-covered slope with time showed strong fluctuation. This phenomenon may be caused by sediment pick-up. This occurred due to the back water and increased resistance, and forced the *f* value to increase during the experiment. For the FS, the *f* of F0 increased sharply with time, and the change of the *f* of F3 with time was relatively gentle. In the case of frozen soil, the average value of the *f* decreased with the increasing sand thickness.

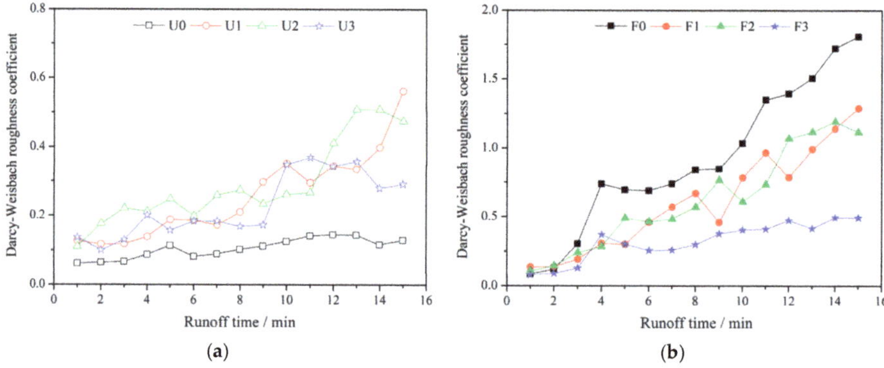

Figure 7. Variations in the Darcy-Weisbach roughness coefficient under different treatments over time. (**a**) Unfrozen slope, (**b**) frozen slope.

In space, the *f* of the US and FS and the distance from the top of the slope can be represented by a linear function (US: $R^2 = 0.719, p > 0.05$; FS: $R^2 = 0.972, p < 0.05$). Under different treatments, the *f* decreased as the slope length increased. Given the same thickness of sand cover and the same section, *f* was greater in FS than that in US (Table 7).

Table 7. Variations of Darcy-Weisbach roughness coefficient with the distance from the top of the slope.

Treatment	Distance from the Top of the Slope E/m				Fitted Equation
	0.5	1	1.5	2	
U0	0.21	0.09	0.08	0.06	
U1	0.72	0.25	0.13	0.13	$f_U = -0.348E + 0.694$
U2	1.04	0.14	0.13	0.09	$R^2 = 0.719, p > 0.05$
U3	0.65	0.24	0.11	0.09	
F0	1.71	1.57	0.85	0.26	
F1	1.57	0.95	0.33	0.16	$f_F = -0.754E + 1.695$
F2	1.18	1.09	0.45	0.36	$R^2 = 0.972, p < 0.05$
F3	0.78	0.43	0.17	0.19	

3.3. Quantification of Hydrodynamic Parameters of Slope Erosion under US and FS

3.3.1. Relationship between Flow Velocity and Hydraulic Parameters under US and FS

The flow velocity is one of the basic factors that affect the hydraulic parameters such as Reynolds number (Re), Froude number (Fr) and Darcy-Weisbach roughness coefficient (f). Figure 8 showed the relationship between flow velocity and hydraulic parameters. The Re decreased with increasing V in both the US ($R^2 = 0.797$) and the FS ($R^2 = 0.871$), while the Fr increased with increasing V in the US ($R^2 = 0.913$) and the FS ($R^2 = 0.977$). The f decreased with increasing V on the US ($R^2 = 0.857$) and the FS ($R^2 = 0.946$). Comparing the fitting coefficients of the various relations, it can be found that due to the influence of soil freezed, the flow velocity has less influence on the hydrodynamic parameters. All the determination coefficients (R^2) were high (from 0.797 to 0.977), and there was a significant linear relationship between flow velocity and hydraulic parameters ($p < 0.01$). This may be due to the fact that hydrodynamic parameters are also affected by flow depth and sediment concentration.

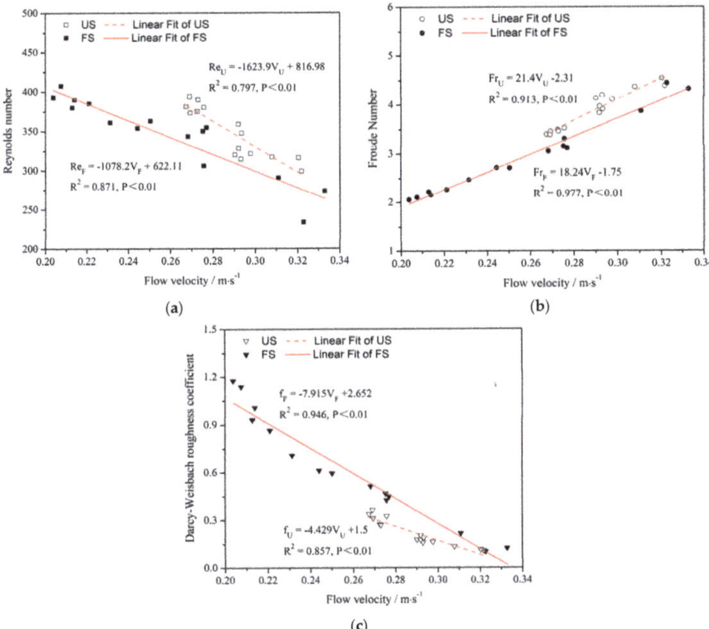

Figure 8. Relationship between flow velocity and hydraulic parameters. (**a**) *Re* with *V*, (**b**) *Fr* with *V*, (**c**) *f* with *V*.

3.3.2. Interrelations of Flow Velocity and Hydraulic Parameters with Runoff Rate Response under US and FS

Table 8 showed the correlation between runoff rate with flow velocity (V) and hydraulics parameters (Re, Fr, and f) under different treatments. It can be seen from the equation that the above flow velocity and hydraulic parameters can be used to describe the runoff process under experimental conditions to a certain extent (Table 8). In terms of fitting effect for the runoff process of the US, the test parameters can be arranged in the order of $V > Fr > Re > f$. For the runoff process on the FS, the test parameters can be arranged in the order of $Fr > Re > V > f$. The runoff rate, flow velocity, and hydraulics parameters have a significant linear relationship ($p < 0.01$), and R^2 is above 65%. By fitting the data of flow velocity, hydraulic parameters, and sediment yield rate, it is found that although there is a certain relationship between them, this relationship is not significant ($p > 0.05$).

Table 8. Correlation between runoff rate with flow velocity and hydraulic parameters.

Runoff/L	Hydraulic Parameters			
	V/(m/s)	Re	Fr	f
US	$R = -6.394V + 2.559$ $R^2 = 0.813, p < 0.01$	$R = 0.0033Re - 0.443$ $R^2 = 0.725, p < 0.01$	$R = -0.278Fr + 1.788$ $R^2 = 0.767, p < 0.01$	$R = 1.207f + 0.445$ $R^2 = 0.655, p < 0.01$
FS	$R = -1.474V + 1.273$ $R^2 = 0.787, p < 0.01$	$R = 0.0013Re + 0.437$ $R^2 = 0.85, p < 0.01$	$R = -0.084Fr + 1.141$ $R^2 = 0.866, p < 0.01$	$R = 0.167f + 0.791$ $R^2 = 0.668, p < 0.01$

Note: R was the runoff (L).

4. Discussion

4.1. Effects of Slope, Sand Cover, and Soil Freezing on Soil Erosion

For the US, the initial runoff time of the sand-covered slope increased, and the effect was clearer with the increasing sand thickness (Table 2). This is consistent with the research results of Zhang et al. and Tang et al. [8,13]. This relationship occurred due to the high porosity of aeolian sand soil [6]. The greater the sand thickness, the greater the water storage effect, and ultimately the initial runoff time increases greatly. For the FS, the initial runoff time was significantly reduced under different sand thicknesses because the water present in the soil surface layer and the water in the soil pores condense to form an "ice cap". In early stages, the "ice cap" hindered the inflow and infiltration, resulting in a significant reduction in the initial runoff time [32,33].

The total runoff and total sediment yield on the slope are related to the degree of erosion on the slope during soil erosion [7]. We found the total runoff under different treatments was 1.02 to 1.28 times than that of U0, and the total sediment yield under different treatments was 1.97 to 10.94 times than that of U0 (Table 2). Frozen soil and sand cover on the slope both lead to changes in the total runoff and total sediment yield. The reasons are as follows: (1) When the temperature drops below 0 °C, the water stored in the sand layer and the water in the soil pores freeze into ice, and the volume expands, which reduces soil stability. The bottom layer of the water-tight layer forms an impervious layer, the runoff flows along the contact surface, the friction between the water flow and the slope surface is reduced, and a small ditch is quickly formed, which increases the amount of erosion and eventually leads to increased soil erosion [34–38]; (2) frozen soil greatly shortens the initial runoff time and the appearance time of rill, which makes it easier to generate runoff on the slope surface, and also more likely to generate a fine ditch, leading to increased erosion [39–42]. The upper-most sand layer can prolong the initial runoff time and store more water. When the slope begins to produce water, the stored water will drain along with the runoff and carry more sediment, increasing sediment yield [4–6]. For the US and FS, the sediment yield of sand covered slope is 3.37~4.96 and 1.06~1.13 times of U0 and F0, respectively. In this study, the sediment yield under different treatments increased with the increasing runoff. There was a linear relationship between the cumulative runoff and cumulative sediment yield under different treatments (Table 3). This study further supported the previously reported relationships [43].

4.2. Effects of Slope Sand Covered and Soil Freezed on Hydraulic Parameters

Runoff on the slope is the driving force of soil erosion [44]. The movement of sediment particles will be affected by the runoff [45]. Therefore, there is a close relationship between slope runoff, sediment movement, and hydraulic parameters. In this study, the underlying condition is the main factor affecting hydraulic properties. The hydraulic characteristics of the slope are mainly affected by various factors such as sand cover and soil frozen [15,46]. Sand cover on the slope changed the infiltration capacity of the soil, which in turn changed the runoff of the slope. Therefore, the topography of the sand cover slope changed greatly during the experiment (Figure 2). Compared with the rill formed by U0, the sand-covered slope in the US formed wide and deep due to runoff erosion, and the runoff depth is increased, resulting in changes in hydraulic characteristics. The research results of Tang et al. [15] showed that the flow pattern of water greatly influences the erosion of sand-covered slopes. This study showed that flow velocity and hydraulics parameters can describe the runoff process under different treatments. Among them, R^2 of V, Re, and Fr all reached more than 70% (Table 8). Since the data of flow velocity can be obtained directly by the experiment, the flow velocity can be used to better describe the runoff process of the US and FS. However, some seemingly effective hydraulic parameters cannot explain the process of sediment yield on the FS. Frozen soil condenses the water in the soil surface layer and pores into an "ice cap" that hinders runoff infiltration. The infiltration capacity of the slope is reduced, the runoff on the slope is larger, and the erosion is greater [47–49]. Therefore, during the experiment, narrow and deep ditches were quickly formed on the FS, resulting in changes in hydraulic characteristics. In addition, the limited observation technology during the erosion process leads to poor fitting of hydraulic parameters and sediment yield on the FS. Despite these shortcomings, the results of this study can still provide a reference for the establishment of a model of erosion on sand-covered loess slopes during thawing.

4.3. Implications for the Relationship between Hydraulic Parameters and Slope Erosion

Global climate change will cause the local permafrost area to melt in advance in the seasonal freeze-thaw area, thus changing the erosion situation in this area. Therefore, in the past few decades, soil erosion resulting from climate warming has captured strong attention in cold regions [49–51]. During the thawing period of winter and spring, the soil erosion in the wind water erosion crisscross zone of the Loess Plateau is usually the result of the combined action of water erosion, wind erosion, and freeze-thaw erosion. However, the problem of soil erosion caused by the combined action is far more than the harm of single action erosion itself. The superposition of different types of erosion has led to huge changes in soil erosion [30,52–55]. In future research, the analysis and quantitative description of composite erosion and single erosion should be encouraged. While it is necessary to measure the impact of each type of erosion on total erosion, it is also necessary to analyze the relationship between hydraulic parameters, runoff, and sediment yield. According to the different effectiveness of hydraulic parameters, selecting the appropriate hydraulic parameters to establish an evaluation model is key [56,57]. This information will help us clarify the feedback relationship between the soil erosion process and the hydrodynamic process from a scientific perspective. At the same time, erosion changes in local areas can also be predicted more comprehensively.

5. Conclusions

Flume tests were performed to study the mechanism of hydrodynamics erosion on the steep sand-covered Loess slopes during the thawing period to improve our understanding of the mechanisms of runoff and sediment yield and to establish a model of soil erosion during thawing. The results showed that the initial runoff time increases with the increase of sand thickness. Under the same sand thickness, the initial runoff time of FS is significantly shorter than the US. For the US, the total sediment yield of different sand thicknesses was significantly higher than that of U0. The cumulative runoff and sediment yield of different treatments can be expressed as a function of $M = CQ + D$. During the entire experiment, the flow velocity in the US and FS treatments generally showed a downward trend. The distance between the hydraulic parameters of US and FS and the top of the slope can be expressed as a linear function. The main flow pattern of runoff was composed of laminar flow and supercritical flow. Linear equations can be used to describe the relationship between flow velocity and the main hydraulic parameters including Reynolds number, Froude number, and Darcy-Weisbach roughness coefficient. Different hydrodynamic parameters show varying degrees of effectiveness in describing slope erosion processes. Flow velocity is the best hydraulic parameter to simulate the trend of slope erosion process.

Author Contributions: Y.S. and P.L. conceived the main idea of the paper. Z.R., L.X., T.W., and Y.Z. designed and performed the experiment. Y.S. wrote the manuscript and all authors contributed in improving the paper. All authors have read and agreed to the published version of the manuscript.

Funding: This research was funded and supported by the National Natural Science Foundation of China (grant no. 51779204), the National Key Research and Development Program of China (grant no. 2017YFC0504501), and the Shaanxi Province Innovation Talent Promotion Plan Project Technology Innovation Team (grant no. 2018TD-037).

Acknowledgments: We thank the reviewers for their useful comments and suggestions. All authors have read and agreed to the published version of the manuscript.

Conflicts of Interest: The authors declare no conflict of interest.

Abbreviations

The following abbreviations are used in this manuscript:

US	unfrozen slope
U0 (1,2,3)	unfrozen slope, the thickness of sand covering is 0 (1,2,3) cm
FS	frozen slope
F0 (1,2,3)	frozen slope, the thickness of sand covering is 0 (1,2,3) cm

References

1. Cao, L.X.; Zhang, K.L.; Dai, H.L.; Liang, Y. Modeling interrill erosion on unpaved roads in the loess plateau of China. *Land Degrad. Dev.* **2015**, *26*, 825–832. [CrossRef]
2. Wang, G.Y.; Joho, I.; Yang, Y.S.; Chen, S.M.; Judi, K.; Xie, J.S.; Lin, W.L. Extent of soil erosion and surface runoff associated with large-scale infrastructure development in Fujian Province, China. *Catena* **2012**, *89*, 22–30. [CrossRef]
3. Cheng, B.; Lv, Y.; Zhan, Y.; Su, D. Constructing China's roads as works of art: A case study of "Esthetic Greenway" construction in the Shennongjia Region of China. *Land Degrad. Dev.* **2013**, *26*, 324–330. [CrossRef]
4. Xu, J.X. The wind-water two-phase erosion and sediment-producing processes in the middle Yellow River basin, China. *Sci. China* **2000**, *43*, 176–186. [CrossRef]
5. Xu, J.X.; Yang, J.S.; Yan, Y.X. Erosion and sediment yields as influenced by coupled eolian and fluvial processes: The Yellow River, China. *Geomorphology* **2006**, *73*, 1–15. [CrossRef]
6. Xie, L.Y.; Bai, Y.J.; Zhang, F.B.; Yang, M.Y.; Li, Z.B. Effects of thickness and particle size composition of overlying sand laye on runoff and sediment yield on sand-covered loess slopes. *Acta Pedol. Sin.* **2017**, *54*, 60–72.
7. Xu, G.C.; Tang, S.S.; Lu, K.X.; Li, P.; Li, Z.B.; Gao, H.D.; Zhao, B.H. Runoff and sediment yield under simulated rainfall on sand-covered slopes in a region subject to wind-water erosion. *Environ. Earth Sci.* **2015**, *74*, 2523–2530. [CrossRef]
8. Zhang, F.B.; Bai, Y.J.; Xie, L.Y.; Yang, M.Y.; Li, Z.B.; Wu, X.R. Runoff and soil loss characteristics on loess slopes covered with aeolian sand layers of different thicknesses under simulated rainfall. *J. Hydrol.* **2017**, *549*, 244–251. [CrossRef]
9. Zhang, F.B.; Yang, M.Y.; Li, B.B.; Li, Z.B.; Shi, W.Y. Effects of slope gradient on hydro-erosional processes on an aeolian sand-covered loess slope under simulated rainfall. *J. Hydrol.* **2017**, *553*, 447–456. [CrossRef]
10. Zhang, X.; Li, Z.B.; Li, P.; Tang, S.S.; Wang, T.; Zhang, H. Influences of sand cover on erosion processes of loess slopes based on rainfall simulation experiments. *J. Arid. Land* **2018**, *10*, 39–52. [CrossRef]
11. Zhang, L.P.; Tang, K.L.; Zhang, C.P. Research on soil wind erosion laws in loess hilly-gully region covered by Sheet Sand. *J. Soil Water Conserv.* **1999**, *5*, 40–45.

12. Wu, X.R.; Zhang, F.B.; Wang, Z.L. Variation of sand and loess properties of binary structure profile in Hilly Region covered by sand of the Loess Plateau. *J. Soil Water Conserv.* **2014**, *28*, 190–193.

13. Tang, S.S.; Li, Z.B.; Ren, Z.P.; Yao, J.W.; Tang, H. Experimental study on the process of runoff and sediment yield on sand-covered slope. *J. Soil Water Conserv.* **2015**, *29*, 25–28.

14. Tang, S.S.; Li, Z.B.; Li, C.; Zhao, B.H. Runoff and Sediment yield process on sand covered slope under simulated rainfall. *J. Northwest A F Univ. Nat. Sci. Ed.* **2016**, *44*, 139–146.

15. Tang, S.S.; Li, Z.B.; Lu, K.X.; Liu, Y.; Su, Y.Y.; Ma, Y.Y. Relationship between hydrodynamic parameters and runoff and sediment yield on sand-covered slope in rainfall simulation study. *Trans. Chin. Soc. Agric. Eng.* **2017**, *33*, 136–143.

16. Zhao, B.H.; Li, Z.B.; Li, P.; Xu, G.C.; Gao, H.D.; Cheng, Y.T.; Chang, E.H.; Yuan, L.; Zhang, Y.; Feng, Z.H. Spatial distribution of soil organic carbon and its influencing factors under the condition of ecological construction in a hilly-gully watershed of the Loess Plateau, China. *Geoderma* **2017**, *296*, 10–17. [CrossRef]

17. Shi, P.; Qin, Y.; Liu, Q.; Zhu, T.T.; Li, Z.B.; Li, P.; Ren, Z.P.; Liu, Y.; Wang, F.C. Soil respiration and response of carbon source changes to vegetation restoration in the Loess Plateau, China. *Sci. Total Environ.* **2019**, *707*, 135507. [CrossRef]

18. Shi, P.; Feng, Z.H.; Gao, H.D.; Li, P.; Xiao, L. Has "Grain for Green" threaten food security on the Loess Plateau of China? *Ecosyst. Health Sustain.* **2020**, *6*, 1709560. [CrossRef]

19. Zhang, Y.; Li, P.; Liu, X.J.; Xiao, L.; Shi, P.; Zhao, B.H. Effects of farmland conversion on the stoichiometry of carbon, nitrogen, and phosphorus in soil aggregates on the Loess Plateau of China. *Geoderma* **2019**, *351*, 188–196. [CrossRef]

20. Wang, T.; Li, P.; Li, Z.B.; Hou, J.M.; Xiao, L.; Ren, Z.P.; Xu, G.C.; Yu, K.X.; Su, Y.Y. The effects of freeze–thaw process on soil water migration in dam and slope farmland on the Loess Plateau, China. *Sci. Total Environ.* **2019**, *666*, 721–730. [CrossRef]

21. Wang, T.; Li, P.; Liu, Y.; Hou, J.M.; Li, Z.B.; Ren, Z.P.; Cheng, S.D.; Zhao, J.H.; Hinkelmann, R. Experimental investigation of freeze-thaw meltwater compound erosion and runoff energy consumption on loessal slopes. *Catena* **2020**, *185*, 104310. [CrossRef]

22. Kirkby, M.J. Modeling Water Erosion Processes. In *Soil Erosion*; Kirkby, M.J., Morgan, R.P.C., Eds.; Wiley: Chichester, UK, 1980; pp. 183–196.

23. Zuzel, J.; Pikul, J. Effects of straw mulch on runoff and erosion from small agricultural plots in northeastern oregon. *Soil Sci.* **1993**, *156*, 111–117. [CrossRef]

24. Edwards, L.M.; Burney, J.R. The effect of antecedent freeze-thaw frequency on runoff and soil loss from frozen soil with and without subsoil compaction and ground cover. *Can. J. Soil Sci.* **1989**, *69*, 799–811. [CrossRef]

25. Fan, H.; Liu, Y.; Xu, X.; Wu, M.; Zhou, L. Simulation of rill erosion in black soil and albic soil during the snowmelt period. *Acta Agric. Scand. Sect. B Soil Plant. Sci.* **2017**, *67*, 510–517. [CrossRef]

26. Ting, J.M.; Torrence, M.R.; Ladd, C.C. Mechanisms of strength for frozen sand. *J. Geotech. Eng.* **1983**, *109*, 1286–1302. [CrossRef]

27. Sharratt, B.S.; Lindstrom, M.J.; Benoit, G.R.; Young, R.A.; Wilts, A. Runoff and soil erosion during spring thaw in the northern US Corn Belt. *J. Soil Water Conserv.* **2000**, *55*, 487–494.

28. Henry, H.A.L. Soil freeze—Thaw cycle experiments: Trends, methodological weaknesses and suggested improvements. *Soil Biol. Biochem.* **2007**, *39*, 977–986. [CrossRef]

29. Pawluk, S. Freeze-thaw effects on granular structure reorganization for soil materials of varying texture and moisture content. *Can. J. Soil Sci.* **1988**, *68*, 485–494. [CrossRef]

30. Shi, Z.H.; Fang, N.F.; Wu, F.Z. Soil erosion processes and sediment sorting associated with transport mechanisms on steep slopes. *J. Hydrol. Amst.* **2012**, *454*, 123–130. [CrossRef]

31. Zhang, L.T.; Gao, Z.L.; Yang, S.W.; Yang, S.W.; Li, Y.H.; Tian, H.W. Dynamic processes of soil erosion by runoff on engineered landforms derived from expressway construction: A case study of typical steep spoil heap. *Catena* **2015**, *128*, 108–121. [CrossRef]

32. Sharratt, B.S.; Lindstrom, M.J. Laboratory simulation of erosion from a partially frozen soil. In *Soil Erosion*; American Society of Agricultural and Biological Engineers: Honolulu, HI, USA, 2001; pp. 159–162.

33. Pikul, J.L.; Aase, J.K. Fall contour ripping increases water infiltration into frozen soil. *Soil Sci. Soc. Am. J.* **1998**, *62*, 1017. [CrossRef]

34. Ma, Q.; Zhang, K.; Jabro, J.D.; Ren, L.; Liu, H. Freeze–thaw cycles effects on soil physical properties under different degraded conditions in Northeast China. *Environ. Earth Sci.* **2019**, *78*, 321. [CrossRef]

35. Kimaro, D.N.; Poesen, J.; Msanya, B.M.; Deckers, J.A. Magnitude of soil erosion on the northern slope of the Uluguru Mountains, Tanzania: Interrill and rill erosion. *Catena* **2008**, *75*, 38–44. [CrossRef]
36. Wischmeier, W.H.; Smith, D.D. Predicting rainfall erosion losses: A guide to conservation planning. In *Agriculture Handbook (USA)*; Department of Agriculture, Science and Education Administration: Washington, DC, USA, 1978; Volume 537, pp. 5–8.
37. Xiao, L.; Zhang, Y.; Li, P.; Xu, G.C.; Shi, P.; Zhang, Y. Effects of freeze-thaw cycles on aggregate-associated organic carbon and glomalin-related soil protein in natural-succession grassland and Chinese pine forest on the Loess Plateau. *Geoderma* **2019**, *334*, 1–8. [CrossRef]
38. Xiao, L.; Yao, K.H.; Li, P.; Liu, Y.; Zhang, Y. Effects of freeze-thaw cycles and initial soil moisture content on soil aggregate stability in natural grassland and Chinese pine forest on the Loess Plateau of China. *J. Soils Sediments* **2019**, *20*, 1222–1230. [CrossRef]
39. Øygarden, L. Rill and gully development during extreme winter runoff event in Norway. *Catena* **2003**, *50*, 217–242. [CrossRef]
40. Saxton, K.E.; Mccool, D.K.; Papendick, R.I. Slot mulch for runoff and erosion control. *J. Soil Water Conserv.* **1981**, *36*, 44–47.
41. Sun, B.Y.; Xiao, J.B.; Li, Z.B.; Ma, B.; Zhang, L.T.; Huang, Y.L.; Bai, L.F. An analysis of soil detachment capacity under freeze-thaw conditions using the Taguchi method. *Catena* **2018**, *162*, 100–107. [CrossRef]
42. Ferrick, M.G.; Gatto, L.W. Quantifying the effect of a freeze–thaw cycle on soil erosion: Laboratory experiments. *Earth Surf. Process. Landf.* **2005**, *30*, 1305–1326. [CrossRef]
43. Wang, T.; Li, P.; Ren, Z.P.; Xu, G.C.; Li, Z.B.; Yang, Y.Y.; Yao, J.W. Effects of freeze-thaw on soil erosion processes and sediment selectivity under simulated rainfall. *J. Arid Land* **2017**, *9*, 234–243. [CrossRef]
44. Asadi, H.; Ghadiri, H.; Rose, C.W. An investigation of flow-driven soil erosion processes at low streampowers. *J. Hydrol.* **2007**, *342*, 134–142. [CrossRef]
45. Asadi, H.; Moussavi, A.; Ghadiri, H. Flow-driven soil erosion processes and the size selectivity of sediment. *J. Hydrol. Amst.* **2011**, *406*, 73–81. [CrossRef]
46. Bao, Y.X.; Wang, X.; Zhou, L.L.; Chen, Z.Q.; Zhang, K. A study on the temporal and spatial evolutionary processes of the dynamic responses of hydrodynamic erosion parameters on freeze-thaw slopes. *J. Soil Water Conserv.* **2017**, *31*, 103–110.
47. Layton, J.B.; Skidmore, E.L.; Thompson, C.A. Winter-associated changes in dry-soil aggregation as influenced by management. *Soil Sci. Soc. Am. J.* **1993**, *57*, 1568. [CrossRef]
48. Bullock, M.S.; Nelson, S.D.; Kemper, W.D. Soil Cohesion as affected by freezing, water content, time and tillage. *Soil Sci. Soc. Am. J.* **1988**, *52*, 770. [CrossRef]
49. Bochove, E.V.; Danielle, P.; Pelletier, F. Effects of freeze-thaw and soil structure on nitrous oxide produced in a clay Soil. *Soil Sci. Soc. Am. J.* **2000**, *64*, 1638–1643. [CrossRef]
50. Dagesse, D.F. Freezing-induced bulk soil volume changes. *Can. J. Soil Sci.* **2010**, *90*, 389–401. [CrossRef]
51. Sahin, U.; Anapali, O. Short communication: The effect of freeze-thaw cycles on soil aggregate stability in different salinity and sodicity conditions. *Span. J. Agric. Res.* **2007**, *5*, 431–434. [CrossRef]
52. Cheng, Y.T.; Li, P.; Xu, G.C. The effect of soil water content and erodibility on losses of available nitrogen and phosphorus in simulated freeze-thaw conditions. *Catena* **2018**, *166*, 21–33. [CrossRef]
53. Wang, S.J. Characteristics of freeze and thaw weathering and its contribution to sediment yield in middle Yellow River basin. *Bull. Soil Water Conserv.* **2004**, *24*, 1–5.
54. Zhang, R.F.; Wang, X.; Fan, H.M.; Zhou, L.L.; Wu, M.; Liu, Y.H. Study on the regionalization of freeze-thaw zones in China and the erosion characteristics. *Sci. Soil Water Conserv.* **2009**, *7*, 24–28.
55. Wang, F.; Fan, H.M.; Guo, C.J.; Zhou, L.L.; Wu, M.; Liu, Y.H.; Chen, Y.B. Comparison and analysis of climate and environment variation in two main freeze-thaw erosion regions in China. *Ecol. Environ.* **2008**, *1*, 173–177.
56. Bryan, R.B. Soil erodibility and processes of water erosion on hillslope. *Geomorphology* **2000**, *32*, 385–415. [CrossRef]
57. Knapen, A.; Poesen, J.; Govers, G.; Gyssels, G.; Nachtergaele, J. Resistance of soils to concentrated flow erosion: A review. *Earth Sci. Rev.* **2007**, *80*, 75–109. [CrossRef]

Article

Effect of Freeze-Thaw Cycles on Soil Detachment Capacities of Three Loamy Soils on the Loess Plateau of China

Jian Lu [1], Baoyang Sun [2,3,*], Feipeng Ren [2,3], Hao Li [2,3] and Xiyun Jiao [1,*]

[1] College of Agricultural Engineering, Hohai University, Nanjing 210098, China; lujian@mwr.gov.cn
[2] Changjiang River Scientific Research Institute, Changjiang Water Resources Commission, Wuhan 430010, China; feipengren2006@mail.bnu.edu.cn (F.R.); haol@whu.edu.cn (H.L.)
[3] Engineering Technology Research Center of Mountain Flood Geological Disaster Prevention and Control, Ministry of Water Resources, Wuhan 430010, China
* Correspondence: sunbx@mail.crsri.cn (B.S.); xyjiao@hhu.edu.cn (X.J.); Tel.: +86-189-9559-9642 (B.S.)

Abstract: Soil detachment is the initial phase of soil erosion and is of great significance to study in seasonal freeze-thaw regions. In order to elucidate the effects mechanism of freeze-thaw cycles on soil detachment capacity of different soils, a sandy loam, a silt loam, and a clay loam were subjected to 0, 1, 5, 10, 15, and 20 freeze-thaw cycles before they were scoured. The results revealed that with increased freeze-thaw cycles, soil bulk density and water-stable aggregates content decreased after the first few times and then kept nearly stable after about 10 cycles, especially for sandy loam. The shear strength of all soils gradually decreased as freeze-thaw cycles increased, except the values of clay loam increased subsequent to the 5th and 15th cycles. After the 20th cycle, the degree of decline of silt loam was the greatest (77.72%), followed by sandy loam (63.18%) and clay loam (39.77%). The soil organic matter of clay loam was much greater than silt loam and sandy loam and all significantly increased after freeze-thaw. Soil detachment capacity of silt loam and sandy loam was positively correlated with freeze-thaw cycle, which was contrary to findings for clay loam. The values of clay loam increased at first and then decreased during the cycles, reaching minimum values at about the 15–20th cycle. After the 20th cycle, the values of sandy loam and silt loam significantly increased 1.62 and 4.74 times over unfrozen, respectively, which was greater than clay loam (0.53 times). A nonlinear regression analysis indicated that the soil detachment capacity of silt loam could be estimated well by soil properties ($R^2 = 0.87$, $p < 0.05$). This study can provide references for the study of the soil erosion mechanism in seasonal freeze-thaw regions.

Keywords: freeze-thaw cycles; loamy soil; soil property; soil detachment capacity; Loess Plateau

Citation: Lu, J.; Sun, B.; Ren, F.; Li, H.; Jiao, X. Effect of Freeze-Thaw Cycles on Soil Detachment Capacities of Three Loamy Soils on the Loess Plateau of China. *Water* **2021**, *13*, 342. https://doi.org/10.3390/w13030342

Academic Editor: Csaba Centeri

Received: 11 January 2021
Accepted: 26 January 2021
Published: 29 January 2021

1. Introduction

Soil erosion has become one of the most critical environmental problems influencing sustainable development and agricultural productive capacity [1,2]. It comprises a series of complex physical processes including detachment, entrainment, transport, and the deposition of soil particles as the result of one or more natural or anthropogenic erosive forces [3]. During the initial stages of soil erosion, soil detachment is defined as the process by which constituent particles are separated from the matrix at a particular location on the surface by erosive agents [4].

Soil detachment capacity is affected by slope, overland flow hydraulics parameters, land use, biocrusts, and soil properties [5–9]. Among them, soil properties exert a profound influence on soil erosion, to an even greater extent than flow discharge and slope under some circumstances [10]. The process of soil detachment involves the interaction between flow and soil particles. Water shear stress increased as result of the adhesion between soil particles when soil particles were detached [6,7]. Additionally, the adhesion between soil particles was positively proportional to clay content, soil bulk density, shear strength, water stable aggregates content, and soil organic matter. However, the increase in sand

189

content, soil porosity, and water content causes a decrease in adhesion [9–12]. These soil properties were important indices for evaluating soil erodibility [13]. Previous research has demonstrated a 2 to 3 times higher rate of soil erodibility during the winter-spring thaw period than the rest of the year [14], while other studies have shown that temporal variation in this variable might result from freeze-thaw action [15,16].

In recent decades, as global climate has tended to be warmer, the effects of freeze-thaw in areas of high latitude and high elevation have been intensified [17,18]. Freeze-thaw erosion does not always occur, but freeze-thaw action can provide effective material sources for other erosion forces by affecting soil properties, and its distribution range is larger than that of freeze-thaw erosion [19]. Generally, frozen soil melts from the surface downward, and an impermeable layer forms at the boundary, with the underlying soil remaining frozen [20]. Due to decreased friction at this border, water can easily flow [21]. Via laboratory simulation experiments, Ferrick and Gatto [22] found that average groove depth, width, and degree of powdery soil erosion following freeze-thaw were significantly greater than those of soils without freeze-thaw. Barnes et al. [23] used an erosion needle method to monitor the impact of freeze-thaw cycles on gully erosion of clay soil in the field over prolonged periods of time, finding freeze-thaw significantly increased erosion of gullies, especially the lateral walls.

During a period of soil thawing, the frozen soil starts thawing with the process of water and heat transfer, and the water changes and migrates in solid, liquid, and gas phases [20,24]. Due to the different densities of water and ice, the constant phase change of soil water causes ice crystal growth and water migration, the frequent frost heaving and thawing of soil leading to changes in soil structures and properties [25]. However, with the increase of freeze-thaw cycles, research results on the changes of soil properties have not been completely consistent as a result of soil texture. For example, soil aggregates—as important components of soil structure, its composition, and stability—influence soil erodibility [26]. Since the 1950s, many studies have investigated the effects of freeze-thaw action on the stability of soil aggregates. It has been shown that freeze-thaw significantly decreased water stable aggregates [27–29]. However, the treatment of freeze-thaw cycles led to disaggregation of micro aggregates and thus enhanced the formation of surface sealing that reduced splash erosion [30]. Hence, the effects of freeze-thaw on the stability of soil aggregates have been studied extensively over several decades; results have proved contradictory because of differences in soil texture, structure, chemical properties, and freeze-thaw cycles.

On the Loess Plateau of China from March to May every year are the periods of thawing; the plateau belongs to a seasonal freeze-thaw area, so freeze-thaw, snowmelt, and rainfall agents occur interactively or simultaneously. After the freeze-thaw cycles in early spring, soil properties—such as bulk density, water stable aggregates, and shear strength change—that lead to erosion easily occur when rain intensity or snowmelt is not too great [31]. However, limited research has been carried out to date to understand how freeze-thaw affects soil properties, and thus soil detachment capacity. The main aim of this study was therefore to research the effects of freeze-thaw cycles on soil properties and soil detachment capacity, and then to quantify the relationship between soil detachment capacity and properties of sandy loam, silt loam, and clay loam under conditions of freeze-thaw. This study can provide a scientific basis for the study of the mechanism of complex erosion of freeze-thaw and water during thawing periods.

2. Materials and Methods

2.1. Soil Samples

The three soils used in this study (loessal soil, aeolian sandy soil, anthropogenic-alluvial soil) were collected from the central and north of Loess Plateau and had the widest distribution or were most affected by human activity (Figure 1). The minimum and maximum daily temperatures can reach about −26 °C in February and 37 °C in July, and mean annual precipitation is about 360 mm. During the freeze-thaw periods, the minimum and maximum recorded daily temperatures in this region can reach about $-10\,^{\circ}\mathrm{C}$ and $22\,^{\circ}\mathrm{C}$ (March and April), and maximum daily precipitation can be as high as 198 mm.

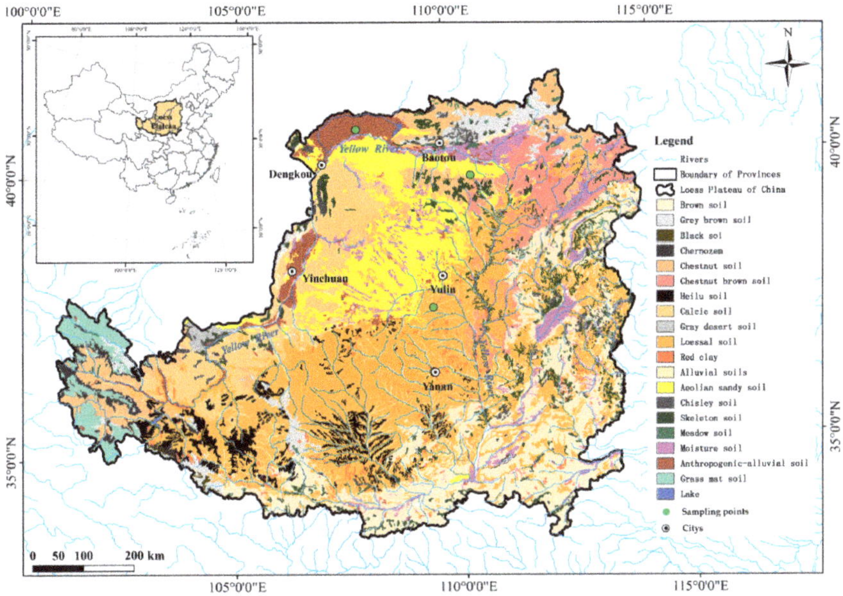

Figure 1. Location of sampling points and soil types of Loess Plateau in China.

All the above three soils were taken from 0–20 cm of the surface of abandoned land. Based on United States textural classification standards, the aeolian sandy soil, loessal soil, and anthropogenic-alluvial soil used in this study were sandy loam, silt loam, and clay loam, respectively (Table 1).

Table 1. Information on three soil samples.

Soil Samples	Geographic Coordinates		Soil Texture %		
			Clay (<0.002)	Silt (0.002–0.05)	Sand (>0.05)
Aeolian sandy soil	110°31′17″ E	39°58′12″ N	11.14	14.84	74.02
Loessal soil	109°15′46″ E	36°46′28″ N	20.17	61.04	18.79
Anthropogenic-alluvial soil	107°40′47″ E	40°54′21″ N	32.18	23.68	44.14

2.2. Determination of Soil Properties

The soil samples were collected by a ring knife with a volume of 100 cm^3. During the sampling process, the handle was placed on the ring knife. The edge of the ring knife was pressed down vertically into the soil with consistent force until the ring knife was filled with soil samples. Then the ring knife filled with soil was taken out and the excess soil around the ring knife was carefully cut off. Finally, the soil sample was put into a dried cylindrical aluminum box in the oven. The soil moisture content and bulk density were measured by oven-drying. Soil particle size distribution was determined by using laser diffraction (Malvern Mastersizer 2000, Malvern, UK) [32,33]. Water stable aggregates (>0.25 mm) were measured by the wet-sieving method. The content of larger than 0.25 mm aggregate was computed from the size distribution of aggregates [29]. Soil organic matter was measured by potassium bichromate by external heating. Shear strength was measured by miniature adhesion instrument (Pocket shear 15.10, Royal Eijkelkamp Company, Giesbeek, The Netherlands). Soil unconfined compressive strength (CS) before and after each freeze-thaw cycle was measured by soil firmness meter. The tests of all soil properties were repeated three times.

2.3. Design of Freeze-Thaw and Scour Simulation Experiments

To remove stones, grass, and other debris, soil samples were stored in polyvinyl chloride (PVC) cylindrical boxes (10 cm diameter, 5 cm depth) based on their bulk densities in the field. There were nine holes at the bottom of each box and gauze covered box bottoms before being filled with the soil samples. According to the measured average value of the samples taken in the field, all the test soil samples were configured with an initial mass water content of about 10%. Then they were frozen at $-10\,°C$ for 12 h and thawed at room temperature between 5 and 10 °C for 12 h to simulate the natural phenomenon of night freezing and day thawing. The slope gradient of experiments was controlled at $15°$, flow discharge constant was held at 6 L min^{-1}, and the experiments used soils with three textures (i.e., sandy loam, silt loam, and clay loam). The samples were subjected to six distinct freeze-thaw cycles (0, 1, 5, 10, 15, and 20 times) and utilized a full-factorial design that required 54 tests, each comprising three replications. All tests were carried out between November and March in order to ensure that air and water temperatures were similar to those recorded in the field during thawing.

Soil detachment capacities were obtained by performing flow scouring experiments. To do this, a scouring device comprised of a water supply tank and flume (400 cm in length, 15 cm in width, and 5 cm in depth) made of PVC material (Figure 2) was used. Fine sand was adhered to this flume to simulate field surface roughness, and a flowmeter was used to control the scouring flow from the water supply tank. Each sample was placed in the test section (10 cm in diameter, Figure 2) of the flume bed, located at a distance of 0.3 m from the flume outlet, and the slope gradient and flow discharge were adjusted to 15° and 6 L min^{-1} prior to each experiment. Tests were timed as soon as they began and ended when the depth of the eroded soil in the soil sample box reached 2 cm [9–11]. The wet soil was then oven-dried at 105 °C for 24 h and then weighed.

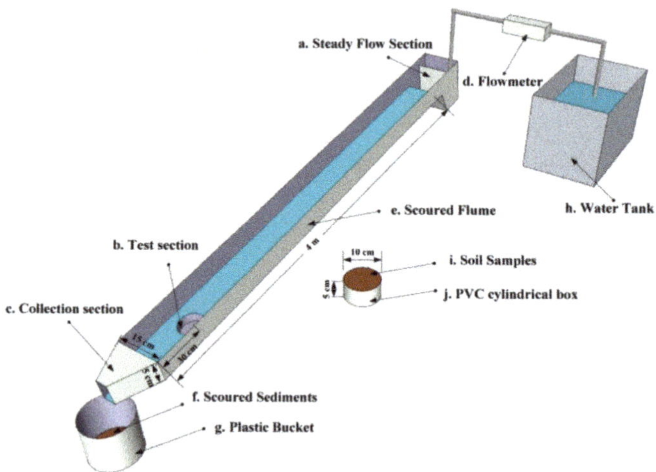

Figure 2. Schematic diagram of experimental setup.

Soil detachment capacities was then calculated as follows:

$$SDC = \frac{w_w - w_d}{A \cdot T} \tag{1}$$

where *SDC* is soil detachment capacity (g m^{-2} s^{-1}), w_w is the dry weight of soil before testing (g), w_d is the dry weight after testing (g), *T* is the test's duration(s), and *A* is the sample cross-sectional area (m^2).

2.4. Data Analyses

All data were analyzed by using SPSS 22.0 (International Business Machine Company, Chicago, IL, USA), and detected significant differences in mean physical properties and soil detachment capacity between soil types and number of freeze-thaw cycles were analyzed via a one-way analysis of variance (ANOVA) followed by least significant difference tests (LSD) ($p < 0.05$) and two-way ANOVA. A nonlinear regression method was used to estimate the relationships between soil detachment capacity and the physical properties of soils. Determination coefficients (R^2) and Mann–Whitney U tests were used to evaluate the effectiveness of the models. The figure plotting was conducted by Origin v. 2020 (OriginLab Corp., Northampton, MA, USA).

3. Results

3.1. The Effects of Freeze-Thaw Cycles on Soil Properties

On average, for bulk density, the values of sandy loam were the greatest (1.39 g cm^{-3}), followed by silt loam (1.29 g cm^{-3}) and clay loam (1.17 g cm^{-3}). With the increase of freeze-thaw cycles, bulk density of the three soils decreased significantly and gradually tended to be stable after the 10th freeze-thaw cycle, especially for sandy loam (Figure 3). It could be seen that bulk density of the three soils all had no significant difference between the 1st and 5th cycles, which was similar to the values between 15th and 20th cycles. After a 20th freeze-thaw cycle, the bulk density of sandy loam, silt loam, and clay loam decreased 5.87%, 8.99%, and 9.07% under unfrozen, respectively.

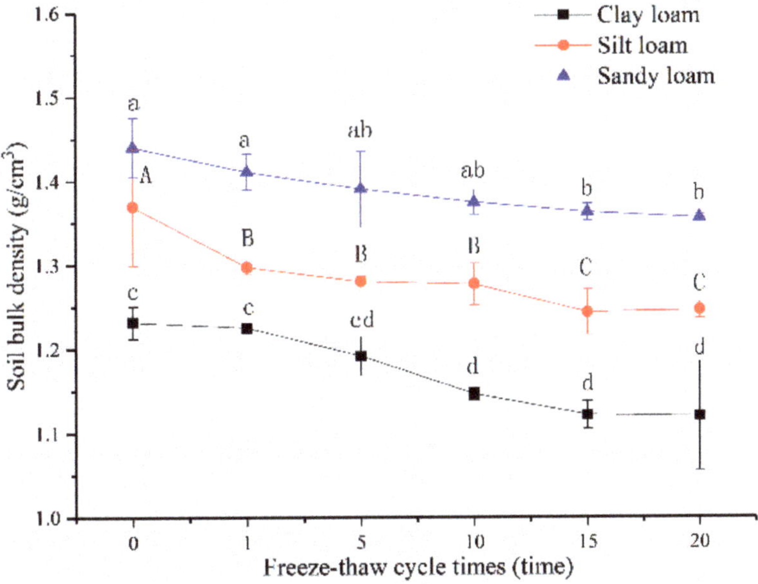

Figure 3. Relationships between soil bulk density and freeze-thaw cycles of three soils. Note: The letters indicated whether the differences in the test results were significant or not. "a, b" stands for sandy loam, "A, B" stands for silt loam, "c, d" stands for clay loam. And different letters indicate a significant difference of test results among the different freeze-thaw cycles and soils at the 0.05 level.

Contrasted with bulk density, the water stable aggregates content (>0.25 mm) of clay loam was the greatest with a mean of 50.58%, followed by silt loam (32.70%) and sandy loam (11.73%) before freeze-thaw. Water stable aggregates content of clay loam ranged from 45.72% to 36.19% with a mean of 41.26% after the 1st to 20th freeze-thaw cycle, from 25.04% to 16.83% with a mean of 20.70% for silt loam, and from 15.13% to 8.08% with a mean of 11.05% for sandy loam. The values of clay loam and silt loam decreased significantly after a 1st freeze-thaw cycle, then increased initially and decreased afterward, especially for silt loam; there were significant differences between each of their freeze-thaw cycles (Figure 4). However, the values of sandy loam increased after the 1st freeze-thaw cycle, and there was basically no significant change after the 10th freeze-thaw cycle. After a 20th freeze-thaw cycle, the water stable aggregates of silt loam, clay loam, and sandy loam decreased 48.54%, 28.44%, and 7.04% under unfrozen, respectively.

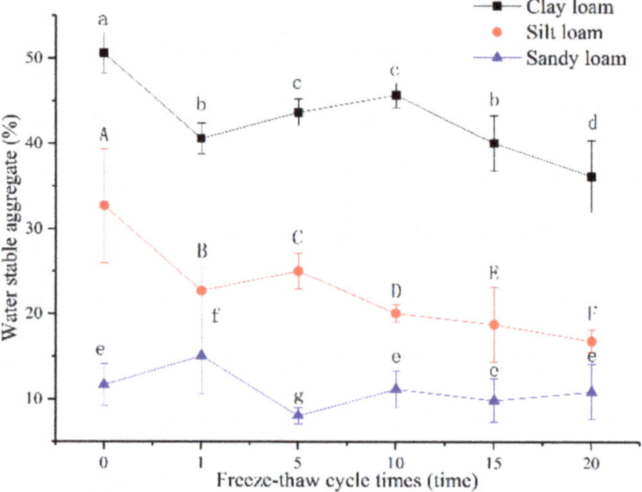

Figure 4. Relationships between soil water stable aggregates and freeze-thaw cycles of three soils. Note: The letters indicated whether the differences in the test results were significant or not. "a, b, c, d" stands for clay loam, "A, B, C, D, E, F" stands for silt loam, "e, f, g" stands for sandy loam. Different letters indicate a significant difference of test results among the different freeze-thaw cycles and soils at the 0.05 level.

The soil organic matter of clay loam was greater than silt loam and sandy loam, no matter before or after freeze-thaw, and the means of the three soils all significantly increased with the increase in freeze-thaw cycle ($p < 0.05$, Figure 5). Significant changes in the means of soil organic matter were no longer observed for all three soils after a 15th freeze-thaw cycle. After a 20th freeze-thaw cycle, the means of sandy loam, silt loam, and clay loam increased 1.27, 1.49 and 1.28 times over unfrozen, respectively.

The shear strength of silt loam was the largest (13.49 ± 1.27 kPa) before freeze-thaw, followed by clay loam (11.67 ± 0.84 kPa), and last was sandy loam (6.38 ± 0.83 kPa). The shear strength of all soils gradually decreased as freeze-thaw cycles increased, especially after the 1st time, except the values for clay loam increased subsequent to the 5th and 15th cycles (Figure 6). After a 20th cycle, the degree of decline of silt loam was the greatest (77.72%), followed by sandy loam (63.18%), and clay loam (39.77%).

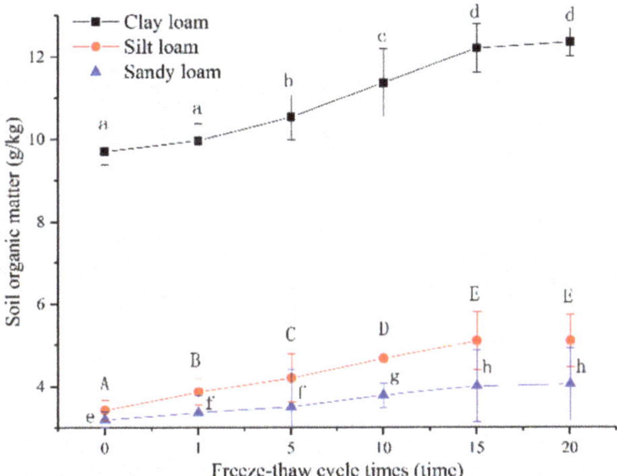

Figure 5. Relationships between soil organic matters and freeze-thaw cycles of three soils. Note: The letters indicated whether the differences in the test results were significant or not. "a, b, c, d" stands for clay loam, "A, B, C, D, E" stands for silt loam, "e, f, g, h" stands for sandy loam. Different letters indicate a significant difference of test results among the different freeze-thaw cycles and soils at the 0.05 level.

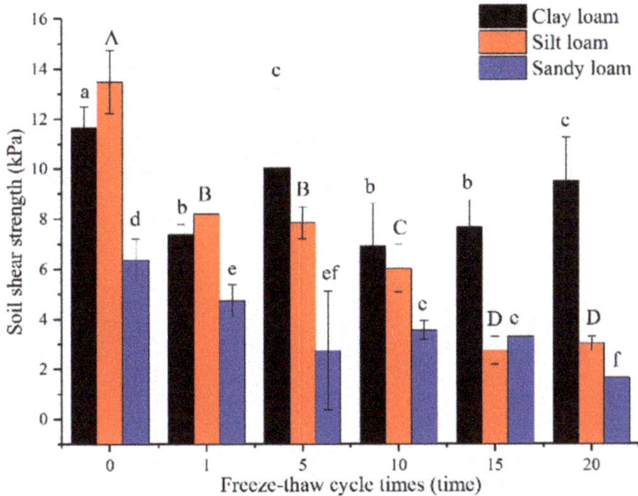

Figure 6. Relationships between soil shear strength and freeze-thaw cycles of three soils. Note: The letters indicated whether the differences in the test results were significant or not. "a, b, c" stands for clay loam, "A, B, C, D" stands for silt loam, "d, e, f" stands for sandy loam. Different letters indicate a significant difference of test results among the different freeze-thaw cycles and soils at the 0.05 level.

The mean soil compressive strength before the freeze-thaw of clay loam (1.68 ± 0.14 KPa) and silt loam (1.65 ± 0.05 KPa) were significantly higher than sandy loam (1.14 ± 0.19 KPa, $p < 0.05$). The values of clay loam decreased after the 1st freeze-thaw cycle, then increased after the 5th freeze-thaw cycle, but no significant differences were found between before and after freeze-thaw cycles ($p > 0.05$). However, this variable for silt loam initially increased and then decreased after the 1st cycle, and no significant change in sandy loam was recorded before the 10th freeze-thaw cycle. Different freeze-thaw cycles exerted no significant influence on the CS of sandy loam ($p > 0.05$, Figure 7).

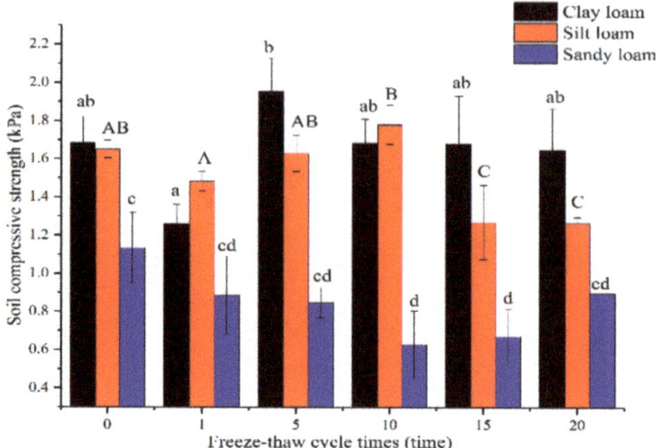

Figure 7. Relationship between soil compressive strength and freeze-thaw cycles of three soils. Note: The letters indicated whether the differences in the test results were significant or not. "a, b" stands for clay loam, "A, B, C" stands for silt loam, "c, d" stands for sandy loam. Different letters indicate a significant difference of test results among the different freeze-thaw cycles and soils at the 0.05 level.

3.2. Freeze-Thaw Cycles Impacts on Soil Detachment Capacity

The soil detachment capacities of three soils were calculated by using Equation (1). The mean value of sandy loam (370.09 ± 53.61 g m^{-2} s^{-1}) was greater than that of clay loam (251.30 ± 39.87 g m^{-2} s^{-1}) and silt loam (144.90 ± 10.28 g m^{-2} s^{-1}) before freeze-thaw cycles. The mean values of sandy loam and silt loam from the 1st to the 20th freeze-thaw cycle significantly increased 1.38 and 3.56 times over unfrozen ($p < 0.05$). However, the values decreased 1.29 times under unfrozen for clay loam. The means of sandy loam and silt loam had no significant difference and were all 2.6 times over clay loam after a 20th freeze-thaw cycle (Figure 8).

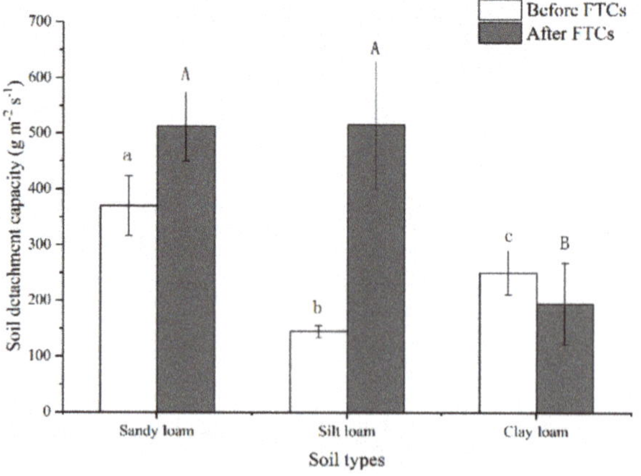

Figure 8. Soil detachment capacities before and after freeze-thaw cycles of three soils. Note: The letters indicated whether the differences in the test results were significant or not. "a, b, c" stands for before freeze-thaw cycles (FTCs), "A, B" stands for after freeze-thaw cycles. Different letters indicate a significant difference of test results among the different freeze-thaw cycles and soils at the 0.05 level.

The variation of three soils soil detachment capacities with freeze-thaw cycles increased and is shown in Figure 9. The soil detachment capacity of clay loam increased after the 1st freeze-thaw cycle, but then decreased gradually with the increase of freeze-thaw cycles. The values after the 10th cycle were significantly less than the early stages of freeze-thaw cycles (0–5 times, $p < 0.05$). After the 20th cycle, soil detachment capacity of clay loam decreased 47.47% under unfrozen. The soil detachment capacity of sandy loam and silt loam significantly increased 1.39 and 2.81 times over unfrozen after the 1st freeze-thaw cycle, while as the number of cycles increased, there was no longer significant change until the 15th cycle. After the 15th cycle, soil detachment capacity of silt loam became larger than sandy loam, and after a 20th cycle the values of sandy loam and silt loam significantly increased 1.62 and 4.74 times over unfrozen, respectively. After the 20th cycle, the soil detachment capacity of silt loam was the greatest (709.65 ± 44.14 g m^{-2} s^{-1}), then sandy loam (601.33 ± 56.60 g m^{-2} s^{-1}), and clay loam was last (132.00 ± 26.25 g m^{-2} s^{-1}).

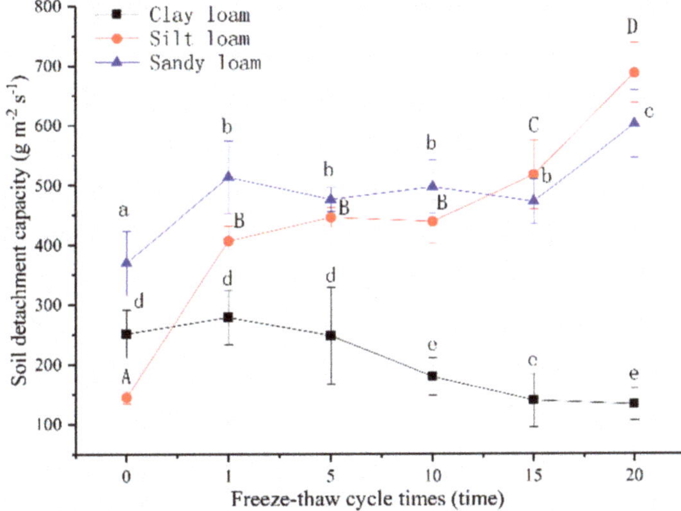

Figure 9. Variation of soil detachment capacity with freeze-thaw cycles. Note: The letters indicated whether the differences in the test results were significant or not. "a, b, c" stands for sandy loam, "A, B, C, D" stands for sandy loam, "d, e" stands for clay loam. Different letters indicate a significant difference of test results among the different freeze-thaw cycles and soils at the 0.05 level.

3.3. Relationship between Soil Detachment Capacity and Soil Properties

Under the condition of freeze-thaw, there were significant correlations between soil detachment capacities and different soil properties based on correlation analysis ($p < 0.05$). The relationship between soil detachment capacity of sandy loam and soil properties could only be well fitted by a linear function of shear strength ($p < 0.05$, Table 2). However, soil detachment capacity of silt loam had a significant relationship with soil organic matter, bulk density, water stable aggregates, and shear strength ($p < 0.05$), and it was negatively correlated with bulk density and shear strength and positively correlated with soil organic matter and water stable aggregates. Soil detachment capacity of clay loam also could be fitted by a linear and exponential function of bulk density and soil organic matter for clay loam ($p < 0.05$), but it was negatively correlated with soil organic matter, which was contrary to findings for bulk density.

Table 2. Regression analysis of soil detachment capacities of three soils and soil properties under the condition of freeze-thaw.

Soil Properties	Sandy Loam		Silt Loam		Clay Loam	
	Regression Equation	R^2	Regression Equation	R^2	Regression Equation	R^2
Soil organic matter	-	-	$y = 20.19e^{0.69x}$	0.74 *	$y = 3617.3e^{-0.27x}$	0.63 **
Bulk density	-	-	$y = -2165.1x + 3238.5$	0.54 *	$y = 915.48x - 877.94$	0.49 *
Water stable aggregates	-	-	$y = 3255.1e^{-0.09x}$	0.85 **	-	-
Shear strength	$y = -38.88x + 634.25$	0.49 *	$y = -44.28x + 762.83$	0.80 **	-	-

Note: The number of samples for each soil property was 18; y is soil detachment capacity, x is soil properties. * Significant at $p < 0.05$, ** Significant at $p < 0.01$.

Generally, soil detachment capacities of different soils are difficult to obtain because measured soil erosion processes under different soils, slopes, and flow discharges are needed. Thus, it is helpful to develop regression equations based on easily measured parameters of soils. A nonlinear regression analysis indicated that soil detachment capacity (*SDC*) of silt loam could be estimated well by soil organic matter (*SOM*), bulk density (*BD*), water stable aggregates (*WSA*), and shear strength (*SS*) (Equation (2), Figure 10). Additionally, linear regression analysis indicated that soil detachment capacity of clay loam could be estimated well by soil organic matter and bulk density (Equation (3), Figure 11).

$$SDC_{silt} = SOM^{0.68} \times BD^{-2.63} \times WSA^{-2.05} \times SS^{0.23} \times 8.85 \times 10^4 \tag{2}$$

$$SDC_{clay} = -51.29 \times SOM\text{-}20.96 \times BD + 785.57 \tag{3}$$

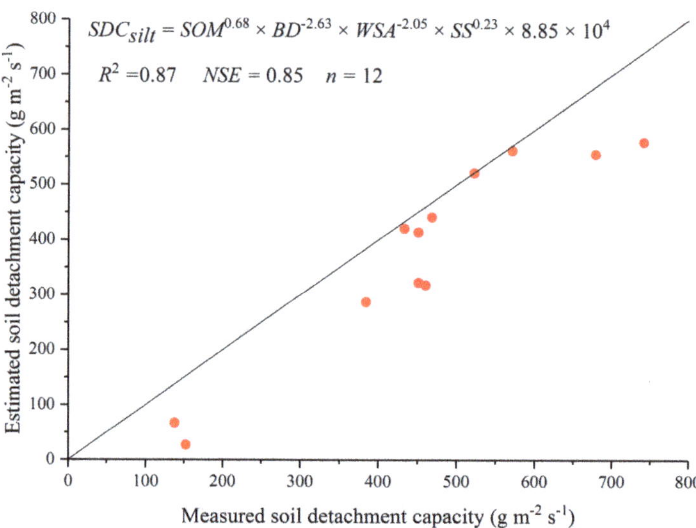

Figure 10. Measured soil detachment capacity vs. estimated ones using Equation (2) of silt loam.

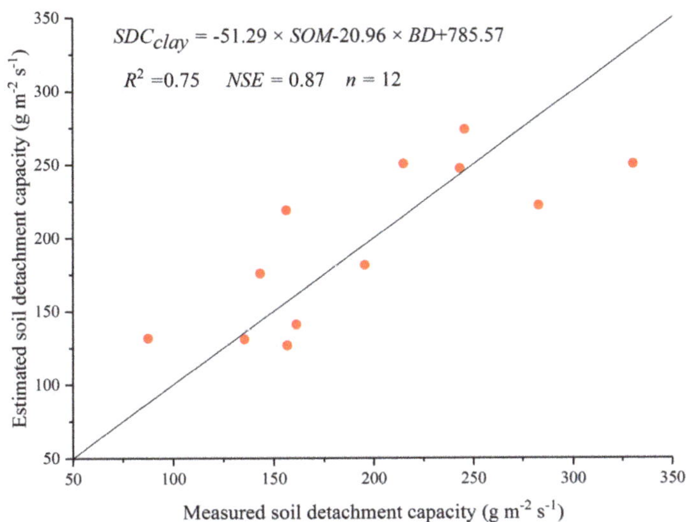

Figure 11. Measured soil detachment capacity vs. estimated ones using Equation (3) of clay loam.

4. Discussion

4.1. Effect of Freeze-Thaw on Soil Properties

With increasing subsequent freeze-thaw cycles (10–15 times), soil bulk density decreased to stable values (Figure 3) which indicates that freeze-thaw cycles had limited effect on soil bulk density. Frequent phase changes of soil water slowly altered the porosity ratio of soil, causing modifications in soil bulk density. However, there exists a critical freeze-thaw cycle related to the soil remodeling properties and the limited degree of freeze-thaw [34,35]. On the other hand, development of variation tendency and the stable state of soil bulk density depend on soil texture and initial state. Bulk density of saline–sodic soils would increase when the initial value was lower, while soils with larger initial values would become looser in structure and decrease in bulk density [36]. In this study, the bulk density of all three soils decreased with the increase of freeze-thaw cycles, which may be due to the initial moisture content and that there were no impurities in the soils. Starkloff et al. [37] found that looser sandy soil was more affected by freeze-thaw than silt soil of Nordic countries by X-ray scanner, which contrasts with this study.

Generally, soil moisture transport and phase transition during freeze-thaw process can lead to the fragmentation of aggregates. However, the influence of freeze-thaw cycles on aggregates was not only related to soil texture but also to the size of aggregates. Smaller aggregates were reformed under the action of soil deformation compression and soil moisture adsorption after the large aggregates were broken. There were studies that have shown that it can reach a peaking after 2–3 freeze-thaw cycles [38]. Additionally, the negative effects of freeze-thaw on water stable aggregates are more pronounced for aggregates larger than 0.25 mm, in comparison with micro aggregates [39]. For sandy loam in this study, during the freeze-thaw process, fragmentation and recombination occurred frequently, and the aggregate content reached the peak value before the 5th freeze-thaw cycle (Figure 4). After 10th freeze-thaw cycle, it had no significant change, maybe because of the lower content of water stable aggregates [40]. For silt loam and clay loam in this study, aggregates contents were relatively high, and the crushing effect was greater than recombination at the beginning of freeze-thaw. Nonetheless, aggregates contents peaked after the 10th and 5th freeze-thaw cycle for clay loam and silt loam, respectively (Figure 4). Overall, freeze-thaw decreased the aggregates stability of all soils, but the effect was more severe on the silt soil (reduced by 48.54%), which was similar to existing research [27].

Freeze-thaw breaks soil aggregates, exposes carbohydrates, fatty acids and sterols, and increases their contact with and utilization by microorganisms. Extractable nutrients have been observed to increase by 2 to 3 times [41]. Moreover, the increase of fine particulate or clay, which has large surface areas, has strong adsorption capacity for organic matter, resulting in redistribution or dissolution of soil organic matter [42]. Thus, the phase changes of water led to contraction of organic matter, destruction of bonds with soil particles, and led to increased release of soil organic matter in this study (Figure 5). Especially for silt loam, the increase of organic matter content (1.49 times) was related to the decrease of aggregates content (48.54%) and was the largest among the three soils. However, as freeze-thaw cycles increased, soil organic matter would not rise significantly at all after the 10th cycles due to its relatively small proportion of total soil carbon and the limited effect of freeze-thaw cycles.

After freeze-thaw cycles, an important index to gauge soil resistance to erosion, the shear strength of three soils significantly decreased, being supportive to most previous studies showing that soil structure becomes looser, and the mechanical properties and microstructure of soils change significantly [43–47]. Soil cohesion was determined by the tensile forces of menisci at the particle contact surfaces as well as the contact relationships between solid particles and liquid films [15]. In the process of freeze-thaw, the frequent phase changes of the water between soil particles led to tensile forces being destroyed and the extent of damage varied with freeze-thaw cycles and soil texture. The shear strength of sandy loam and silt loam decreased after the first few freeze-thaw cycles and then kept nearly stable after about the 5th to 15th cycles, which was similar to the aforementioned other soil properties (Figure 6). However, shear strength of clay loam still increased after the 15th freeze-thaw cycle, mainly because of the indirect effect of its higher soil aggregates content after the 10th cycle and the increased soil organic matter [48].

4.2. Effect Mechanism of Freeze-Thaw on Soil Detachment Capacity

Soil detachment is the initial stage of soil erosion, and the most affected by freeze-thaw. Soil erosion amounts during periods of thawing were higher than those of other seasons, and they could reach 2–3 times [21–23]. Similarly, this study showed that soil detachment capacity of sandy loam and silt loam under freeze-thaw was larger (by approximately 1.38 and 3.56 times) than that under unfrozen, which contrasts with clay loam (Figure 8). Obviously, soil properties in this experiment proved to be the most important factor that significantly influenced soil detachment capacity, determined by soil types and the number of freeze-thaw cycles. The process of soil detachment in the Qinghai-Tibet plateau was significantly lower than that of other regions of China, which was related to the degree of freeze-thaw and soil properties, but the specific mechanism of the effect was unclear [49].

As the most important factors, shear strength and water stable aggregates were strongly negatively correlated with soil erodibility [50–52]. Soil detachment capacity significantly increased as a result of reductions in soil shear strength, and water stable aggregates were destroyed by freeze-thaw [13]. In the case of sandy loam, for example, soil detachment capacity was basically larger than silt loam and clay loam without freeze-thaw cycles in this study. These discrepancies were mainly due to the fact that shear strength, water stable aggregates, and soil organic matter of sandy loam were less, while the tensile forces of coarse particles and the contact relationships between them and liquid films were weaker in some cases than others [50,53]. Additionally, the soil properties had not changed significantly, as freeze-thaw cycles increased after it had been destroyed by the first freeze-thaw (Figures 3–7). This means that soil detachment capacity of sandy loam did not significantly increase with the increase of freeze-thaw cycles and only had a significant negative correlation with shear strength (Table 2).

Likewise, soil detachment capacity of silt loam was the greatest after the 20th freeze-thaw cycle in this study (Figure 9). Higher powder content has been related to greater formation of capillaries, leading to more rapid moisture migration and greater damage due to freeze-thaw [50]. Therefore, silt loam was most prone to erosion when exposed to freeze-thaw. The process of soil detachment was complex, and in our experiments, it was significantly influenced and well estimated by bulk density, soil organic matter, water stable aggregates, and shear strength (Table 2, Figure 10). Compared to sandy loam and silt loam, the decreased soil detachment capacity of clay loam after freeze-thaw mainly depended on soil organic matter and soil bulk density mainly because increased freeze-thaw times, the variation trend of water stable aggregates, and the relative complexity of shear strength cannot represent the change of soil detachment capacity. Additionally, soil detachment capacity cannot be determined well by a single factor, and the comprehensive effect of various factors should be considered [9–11].

5. Conclusions

The effects of freeze-thaw cycles on the properties and soil detachment capacity of three loamy soils were examined by using the artificial freeze-thaw and scour experiments. Soil bulk density, water stable aggregates, and shear strength of three soils were negatively correlated with freeze-thaw cycles. However, the change of soil organic matter was the opposite. After 10 to 15 freeze-thaw cycles, soil properties were basically stable. Soil detachment capacity of silt loam and sandy loam were positively correlated with freeze-thaw cycles, which was contrary to findings for clay loam. The mean soil detachment capacity of sandy loam was the greatest before and after freeze-thaw. After the 15th freeze-thaw cycle, the soil detachment capacity of sandy loam was exceeded by silt loam. A nonlinear regression function could be used to describe the relationship between the soil detachment capacity of silt loam and soil organic matter, bulk density, water stable aggregates, and shear strength. Soil organic matter and bulk density were the best hydraulic parameters to simulate soil detachment capacity of clay loam.

Author Contributions: Conceptualization, J.L. and X.J.; methodology, J.L.; software, H.L.; validation, J.L., B.S., F.R., H.L., and X.J.; formal analysis, J.L.; investigation, B.S., F.R., and H.L.; resources, X.J.; data curation, H.L.; writing—original draft preparation, J.L.; writing—review and editing, B.S.; project administration, B.S.; funding acquisition, B.S. All authors have read and agreed to the published version of the manuscript.

Funding: This research was funded by the Program of National Key Research and Development of China (2018YFC0407601-03), Central Public-interest Scientific Institution Basal Research Fund of China (No. CKSF2019292/SH+TB and CKSF2019179/TB), and the National Natural Science Foundation of China (41807066 and 41877082).

Institutional Review Board Statement: Not applicable.

Informed Consent Statement: Not applicable.

Data Availability Statement: The data presented in this study are available on request from the corresponding author. The data are not publicly available due to the funded project is not finished yet.

Acknowledgments: All authors thank editors and reviewers for insightful comments on the original manuscript.

Conflicts of Interest: The authors declare no conflict of interest.

References

1. Luca, M. Govern our soils. *Nature* **2015**, *528*, 32–33.
2. Mohammed, S.; Al-Ebraheem, A.; Holb, I.J.; Alsafadi, K.; Dikkeh, M.; Pham, Q.B.; Linh, N.T.T.; Szabo, S. Soil management effects on soil water erosion and runoff in central Syria—A comparative evaluation of general linear model and random forest regression. *Water* **2020**, *12*, 25–29. [CrossRef]
3. Nearing, M.A.; Bradford, J.M.; Parker, S.C. Soil Detachment by shallow flow at low slopes. *Soil Sci. Soc. Am. J.* **1991**, *55*, 351–357. [CrossRef]
4. Zhang, G.H.; Liu, B.Y.; Liu, G.B.; Nearing, M.A. Detachment of undisturbed soil by shallow flow. *Soil Sci. Soc. Am. J.* **2003**, *67*, 713–719. [CrossRef]
5. Chang, E.; Li, P.; Li, Z.; Su, Y.; Zhang, Y.; Zhang, J.; Liu, Z.; Li, Z. The impact of vegetation successional status on slope runoff erosion in the Loess Plateau of China. *Water* **2019**, *11*, 2614. [CrossRef]

6. Knapen, A.; Poesen, J.; Govers, G.; Nachtergaele, J. Resistance of soils to concentrated flow erosion: A review. *Earth-Sci. Rev.* **2007**, *80*, 75–109. [CrossRef]

7. Wang, D.D.; Wang, Z.L.; Shen, N.; Chen, H. Modeling soil detachment capacity by rill flow using hydraulic parameters. *J. Hydrol.* **2016**, *535*, 473–479. [CrossRef]

8. Zhang, G.H.; Liu, G.B.; Tang, K.M.; Zhang, X.C. Flow Detachment of Soils under Different Land Uses in the Loess Plateau of China. *Trans. ASABE* **2008**, *51*, 883–890.

9. Zhang, G.H.; Ding, W.F.; Pu, J.; Li, J.M.; Qian, F.; Sun, B.Y. Effects of moss-dominated biocrusts on soil detachment by overland flow in the Three Gorges Reservoir Area of China. *J. Mt. Sci.* **2020**, *17*, 2418–2431. [CrossRef]

10. Yi, W.; Cao, L.X.; Fan, J.B.; Lu, H.; Liang, Y. Modelling Soil Detachment of Different Management Practices in the Red Soil Region of China. *Land Degrad. Dev.* **2017**, *28*, 1496–1505.

11. Wang, B.; Zhang, G.H.; Shi, Y.Y.; Zhang, X.C.; Ren, Z.P.; Zhu, L.J. Effect of natural restoration time of abandoned farmland on soil detachment by overland flow in the Loess Plateau of China. *Earth Surf. Process. Landf.* **2013**, *38*, 1725–1734. [CrossRef]

12. Chen, Z.X.; Guo, M.M.; Wang, W.L. Variations in Soil Erosion Resistance of Gully Head along a 25-Year Revegetation Age on the Loess Plateau. *Water* **2020**, *12*, 3301. [CrossRef]

13. Nciizah, A.D.; Wakindiki, I.I.C. Physical indicators of soil erosion, aggregate stability and erodibility. *Arch. Agron. Soil Sci.* **2015**, *61*, 827–842. [CrossRef]

14. Chow, T.L.; Rees, H.W.; Monteith, J. Seasonal distribution of runoff and soil loss under four tillage treatments in the upper St. John River valley New Brunswick, Canada. *Can. J. Soil Sci.* **2000**, *80*, 649–660. [CrossRef]

15. Kok, H.; Mccool, D.K. Quantifying freeze/thaw-induced variability of soil strength. *T ASAE* **1990**, *33*, 501–506. [CrossRef]

16. Bajracharya, R.M.; Lal, R.; Hall, G.F. Temporal variation in properties of an uncropped, ploughed Miamian soil in relation to seasonal erodibility. *Hydrol. Process.* **1998**, *12*, 1021–1030. [CrossRef]

17. Wang, R.; Zhu, Q.K.; Ma, H. Spatial-temporal variations in near-surface soil freeze-thaw cycles in the source region of the Yellow River during the period 2002–2011 based on the Advanced Microwave Scanning Radiometer for the Earth Observing System (AMSR-E) data. *J. Arid Land* **2017**, *9*, 850–864. [CrossRef]

18. Teng, H.F.; Liang, Z.Z.; Chen, S.C.; Liu, Y.; Rossel, R.A.; Chappell, A.; Yu, W.; Shi, Z. Current and future assessments of soil erosion by water on the Tibetan Plateau based on RUSLE and CMIP5 climate models. *Sci. Total Environ.* **2018**, *635*, 673–686. [CrossRef]

19. Fan, H.M.; Cai, Q.G. Review of research progress in freeze-thaw erosion. *Sci. Soil Water Conserv.* **2003**, *1*, 50–55.

20. Chen, J.; Zheng, X.; Zang, H.; Ping, L.; Ming, S. Numerical Simulation of Moisture and Heat Coupled Migration in Seasonal Freeze-thaw Soil Media. *J. Pure Appl. Microbiol.* **2013**, *7*, 151–156.

21. Gatto, L.W. Soil freeze–thaw-induced changes to a simulated rill: Potential impacts on soil erosion. *Geomorphology* **2000**, *32*, 147–160. [CrossRef]

22. Ferrick, M.G.; Gatto, L.W. Quantifying the effect of a freeze-thaw cycle on soil erosion: Laboratory experiments. *Earth Surf. Process. Landf.* **2005**, *30*, 1305–1326. [CrossRef]

23. Barnes, N.; Luffman, I.; Nandi, A. Gully Erosion and Freeze-Thaw Processes in Clay-Rich Soils, Northeast Tennessee, USA. *Georesj* **2016**, *9*, 67–76. [CrossRef]

24. Fu, Q.; Hou, R.J.; Li, T.X. Soil Moisture-heat transfer and its action mechanism of freezing and thawing soil. *Trans. Chin. Soc. Agric. Mach.* **2016**, *47*, 99–110.

25. Sun, B.Y.; Li, Z.B.; Xiao, J.B.; Zhang, L.T.; Ma, B.; Li, J.M.; Cheng, D.B. Research progress of the effect of freeze-thaw on soil physical and chemical properties and wind and water erosion. *Chin. J. Appl. Ecol.* **2019**, *30*, 337–347.

26. Barthes, B.; Roose, E. Aggregate stability as an indicator of soil susceptibility to runoff and erosion, validation at several levels. *Catena* **2002**, *47*, 133–149. [CrossRef]

27. Kværno, S.H.; Oygarden, L. The influence of freeze–thaw cycles and soil moisture on aggregate stability of three soils in Norway. *Catena* **2006**, *67*, 175–182. [CrossRef]

28. Oztas, T.; Fayetorbay, F. Effect of freezing and thawing processes on soil aggregate stability. *Catena* **2003**, *52*, 1–8. [CrossRef]

29. Li, G.Y.; Fan, H.M. Effect of Freeze-Thaw on Water Stability of Aggregates in a Black Soil of Northeast China. *Pedosphere* **2014**, *24*, 285–290. [CrossRef]

30. Sadeghi, S.H.; Raeisi, M.B.; Hazbavi, Z. Influence of freeze-only and freezing-thawing cycles on splash erosion. *Int. Soil Water Conserv. Res.* **2018**, *6*, 275–279. [CrossRef]

31. Wang, T.; Li, P.; Liu, Y.; Li, Z.B. Experimental investigation of freeze-thaw meltwater compound erosion and runoff energy consumption on loessal slopes. *Catena* **2020**, *185*, 104310. [CrossRef]

32. Koiter, A.J.; Owens, P.N.; Petticrew, E.L. The role of soil surface properties on the particle size and carbon selectivity of interrill erosion in agricultural landscapes. *Catena* **2017**, *153*, 194–206. [CrossRef]

33. Wang, L.; Shi, Z.H. Size Selectivity of eroded sediment associated with soil texture on steep slopes. *Soil Sci. Soc. Am. J.* **2015**, *79*, 917. [CrossRef]

34. Dagesse, D.F. Freezing-induced bulk soil volume changes. *Can. J. Soil Sci.* **2010**, *90*, 389–401. [CrossRef]

35. Jie, Z.; Tang., Y. Experimental inference on dual-porosity aggravation of soft clay after freeze-thaw by fractal and probability analysis. *Cold Reg. Sci. Technol.* **2018**, *153*, 181–196.

36. Sahin, U.; Angin, I.; Kiziloglu, F.M. Effect of freezing and thawing processes on some physical properties of saline–sodic soils mixed with sewage sludge or fly ash. *Soil Tillage Res.* **2008**, *99*, 254–260.

37. Starkloff, T.; Larsbo, M.; Stolte, J. Quantifying the impact of a succession of freezing-thawing cycles on the pore network of a silty clay loam and a loamy sand topsoil using X-ray tomography. *Catena* **2017**, *156*, 365–374. [CrossRef]
38. Lehrsch, G.A. Freeze-Thaw Cycles Increase Near-Surface Aggregate Stability. *Soil Sci.* **1998**, *163*, 63–70.
39. Wang, D.Y.; Ma, W.; Niu, Y.H. Effects of cyclic freezing and thawing on mechanical properties of Qinghai–Tibet clay. *Cold Reg. Sci. Technol.* **2007**, *48*, 34–43. [CrossRef]
40. Edwards, L.M. The effect of alternate freezing and thawing on aggregate stability and aggregate size distribution of some Prince Edward Island soils. *Eur. J. Soil Sci.* **2010**, *42*, 193–204. [CrossRef]
41. Song, Y.; Yu, X.F.; Zhou, Y.C. Progress of freeze-thaw effects on carbon, nitrogen and phosphorus cycling in soils. *Soils Crop.* **2016**, *5*, 78–90.
42. Mohanty, S.K.; Saiers, J.E.; Ryan, J.N. Colloid-facilitated mobilization of metals by freeze-thaw cycles. *Environ. Sci. Technol.* **2014**, *48*, 977–984. [PubMed]
43. Ni, W.K.; Shi, H.Q. Influence of freezing-thawing cycles on micro-structure and shear strength of loess. *J. Glaciol. Geocryol.* **2014**, *36*, 922–927.
44. Zhang, W.B.; Ma, J.Z.; Tang, L. Experimental study on shear strength characteristics of sulfate saline soil in Ningxia region under long-term freeze-thaw cycles. *Cold Reg. Sci. Technol.* **2019**, *160*, 48–57.
45. Nguyen, T.H.; Cui, Y.J.; Valery, F. Effect of freeze-thaw cycles on mechanical strength of lime-treated fine-grained soils. *Transp. Geotech.* **2019**, *21*, 100281.
46. Dong, X.H.; Zhang, A.J.; Lian, J.B.; Guo, M.X. Study of shear strength deterioration of loess under repeated freezing-thawing cycles. *J. Glaciol. Geocryol.* **2010**, *32*, 767–772.
47. Liu, J.; Chang, D.; Yu, Q. Influence of freeze-thaw cycles on mechanical properties of a silty sand. *Eng. Geol.* **2016**, *210*, 23–32. [CrossRef]
48. Schjonning, P.; Lamande, M.; Keller, T. Subsoil shear strength–Measurements and prediction models based on readily available soil properties. *Soil Tillage Res.* **2020**, *200*, 104638. [CrossRef]
49. Li, M.Y.; Xiao, H.; Huan, H.; Shao, Y.; Xia, Z. Modelling soil detachment by overland flow for the soil in the Tibet Plateau of China. *Sci. Rep.* **2019**, *9*, 8063–8073. [CrossRef]
50. Formanek, G.E.; Mccool, D.K.; Papendick, R.I. Freeze-thaw and consolidation effects on strength of a wet silt loam. *Trans. ASAE-Am. Soc. Agric. Eng.* **1984**, *27*, 1749–1752. [CrossRef]
51. Sun, L.; Zhang, G.H.; Luan, L.L.; Liu, F.L. Temporal variation in soil resistance to flowing water erosion for soil incorporated with plant litters in the Loess Plateau of China. *Catena* **2016**, *145*, 239–245. [CrossRef]
52. Boswell, E.P.; Balster, N.J.; Bajcz, A.W.; Thompson, A.M. Soil aggregation returns to a set point despite seasonal response to snow manipulation. *Geoderma* **2020**, *357*, 113954. [CrossRef]
53. Flerchinger, G.N.; Lehrsch, G.A.; McCool, D.K. Freezing and thawing. In *Processes Encyclopedia of Soils in the Environment*; Elsevier Inc.: Amsterdam, The Netherlands, 2005; pp. 104–110.

Article

Estimation of Potential Soil Erosion and Sediment Yield: A Case Study of the Transboundary Chenab River Catchment

Muhammad Gufran Ali [1], Sikandar Ali [1,*], Rao Husnain Arshad [1], Aftab Nazeer [2,3,*], Muhammad Mohsin Waqas [4], Muhammad Waseem [5], Rana Ammar Aslam [6], Muhammad Jehanzeb Masud Cheema [7], Megersa Kebede Leta [8] and Imran Shauket [6]

[1] Department of Irrigation and Drainage, Faculty of Agricultural Engineering and Technology, University of Agriculture, Faisalabad 38000, Pakistan; mgufranali008@gmail.com (M.G.A.); husnain.arshad@uaf.edu.pk (R.H.A.)

[2] Department of Water Management, Delft University of Technology, 2600 GA Delft, The Netherlands

[3] Department of Agricultural Engineering, Bahauddin Zakariya University, Multan 60800, Pakistan

[4] Department of Agricultural Engineering, Khwaja Fareed University of Engineering and Information Technology, Rahim Yar Khan 64200, Pakistan; mohsin.waqas@kfueit.edu.pk

[5] Department of Civil Engineering, Ghulam Ishaq Khan Institute of Engineering Sciences and Technology, Topi 23460, Pakistan; muhammad.waseem@giki.edu.pk

[6] Department of Structures and Environmental Engineering, Faculty of Agricultural Engineering and Technology, University of Agriculture, Faisalabad 38000, Pakistan; ammar.aslam@uaf.edu.pk (R.A.A.); imranshaukat@uaf.edu.pk (I.S.)

[7] Department of Land and Water Conservation Engineering, Faculty of Agricultural Engineering and Technology, PMAS Arid Agriculture University, Rawalpindi 46000, Pakistan; mjm.cheema@uaar.edu.pk

[8] Faculty of Agriculture and Environmental Sciences, University of Rostock, 18059 Rostock, Germany; megersa.kebede@uni-rostock.de

* Correspondence: sikandar_ali@uaf.edu.pk (S.A.); a.nazeer@tudelft.nl (A.N.); Tel.: +92-32-1765-9581 (S.A.); +31-64-906-5653 (A.N.)

Citation: Ali, M.G.; Ali, S.; Arshad, R.H.; Nazeer, A.; Waqas, M.M.; Waseem, M.; Aslam, R.A.; Cheema, M.J.M.; Leta, M.K.; Shauket, I. Estimation of Potential Soil Erosion and Sediment Yield: A Case Study of the Transboundary Chenab River Catchment. *Water* **2021**, *13*, 3647. https://doi.org/10.3390/w13243647

Academic Editor: Csaba Centeri

Received: 25 October 2021
Accepted: 15 December 2021
Published: 18 December 2021

Publisher's Note: MDPI stays neutral with regard to jurisdictional claims in published maps and institutional affiliations.

Abstract: Near real-time estimation of soil loss from river catchments is crucial for minimizing environmental degradation of complex river basins. The Chenab river is one of the most complex river basins of the world and is facing severe soil loss due to extreme hydrometeorological conditions, unpredictable hydrologic response, and complex orography. Resultantly, huge soil erosion and sediment yield (SY) not only cause irreversible environmental degradation in the Chenab river catchment but also deteriorate the downstream water resources. In this study, potential soil erosion (PSE) is estimated from the transboundary Chenab river catchment using the Revised Universal Soil Loss Equation (RUSLE), coupled with remote sensing (RS) and geographic information system (GIS). Land Use of the European Space Agency (ESA), Climate Hazards Group InfraRed Precipitation with Station (CHIRPS) data, and world soil map of Food and Agriculture Organization (FAO)/The United Nations Educational, Scientific and Cultural Organization were incorporated into the study. The SY was estimated on monthly, quarterly, seasonal, and annual time-scales using sediment delivery ratio (SDR) estimated through the area, slope, and curve number (CN)-based approaches. The 30-year average PSE from the Chenab river catchment was estimated as 177.8, 61.5, 310.3, 39.5, 26.9, 47.1, and 99.1 tons/ha for annual, rabi, kharif, fall, winter, spring, and summer time scales, respectively. The 30-year average annual SY from the Chenab river catchment was estimated as 4.086, 6.163, and 7.502 million tons based on area, slope, and CN approaches. The time series trends analysis of SY indicated an increase of 0.0895, 0.1387, and 0.1698 million tons per year for area, slope, and CN-based approaches, respectively. It is recommended that the areas, except for slight erosion intensity, should be focused on framing strategies for control and mitigation of soil erosion in the Chenab river catchment.

Keywords: RUSLE; soil erosion; sediment yield; Chenab river; remote sensing; GIS

1. Introduction

Soil is a precious natural resource [1], plays a key role in the functioning ecosystem [2,3], and provides valuable goods and services [4,5] essential for human security [6,7]. Soil erosion is a natural geomorphic process and environmental problem [8,9] arising from anthropogenic activities [10] agricultural intensification, deforestation, land degradation, and global climate change [11,12]. Soil erosion is also considered as one of the significant threats to the ecosystem [13–15], as it not only causes soil erosion from upper catchments and deposition in rivers and lakes through the geologic ages worldwide [16,17], but also carries nutrients, pesticides, chemicals, etc., and cause groundwater contamination [18–20]. It has been estimated that about 56% of global soil is being degraded by light to severe forms of soil erosion caused has by water [21]. Accelerated forms of soil erosion by water become a global problem [22,23], that not only cause rivers' catchment problems [24,25] but also act as barriers to achieving the United Nations Sustainable Development Goals [26]. Therefore, estimation of soil erosion by water from river catchment is in dire need [27], so that proper soil erosion mitigation options can be focused [28].

Estimation of soil erosion (PSE and SY) from large and complex rivers' catchments has always been a big challenge to researchers worldwide [6,29–31]. Initially, soil erosion research was conducted more than seven decades ago using north American datasets [32–34]. Several mathematical models, conceptual, empirical, process oriented, and physically based, have been applied for soil erosion modeling/ estimation at different spatiotemporal scales [35–42]. Scientists are also working on process-oriented soil erosion models such as the Water Erosion Prediction Project [43,44], European Soil Erosion Model [45], Limburg Soil Erosion Model [46], and Pan European Soil Erosion Risk Assessment [47]. The research community is also improving the empirical model known as Universal Soil Loss Equation (USLE) [48–52] which is not only practically sound [53,54] but can also be applied over complex and large river basins [11,55–57]. The USLE and RUSLE are being applied successfully for estimation of PSE and SY from rivers' catchments throughout the world under changing spatiotemporal conditions [6,58–63]. Large-scale soil erosion modeling has been performed by using the RUSLE model in Europe [64], Canada [65], Australia [66], and China [67].

The RUSLE was developed to estimate soil erosion by water in temperate climates, and is an empirical model founded on the USLE [68]. The RUSLE model estimates the average annual rate of soil erosion from complex river basins for multiple scenarios including management practices, cropping systems, and erosion control practices [69]. The RUSLE can also be used for estimation of average annual soil erosion rate from ungauged river catchments using local hydrometeorological information and catchment characteristics [70]. The RUSLE model considers the effect of many factors such as rainfall erosivity, soil erodibility, slope length and slope steepness, cover management, and conservation practices [71]. The soil loss occurs in three steps. Soil erosion starts with the detachment of soil particles, followed by transport and sequent deposition [72]. The RUSLE model neither estimates gully/channel erosion nor discusses the sediment deposition, so there is a need to introduce SDR for estimation of sediment delivered to the outlet from the catchment [73]. The SDR is estimated by researchers based on area, slope, and CN approaches in rivers' catchments globally [74–79]. The RUSLE model is applied along with the SDR for estimation of sediment yield from rivers' catchments [80–84].

The use of RUSLE for soil erosion needs detailed spatiotemporal in situ data [85] which is not possible, as the transboundary Chenab river catchment is divided among Pakistan (402 km^2), Jammu and Kashmir (20,139.7 km^2), and India (7939.27 km^2) [86]. This clearly reflects that 27.87%, 70.71%, and 1.42% area of the catchment lies in India, Jammu and Kashmir (Indian Control), and Pakistan, respectively. In such a complex transboundary river catchment, the global gridded and RS-based datasets can be appropriate alternatives for research purposes [87–89]. Recently, researchers have employed RS and GIS technologies for evaluation of PSE and SY [90]. RS and GIS technologies provide detailed information with a spatiotemporal resolution appropriate for quantifying soil erosion at a regional/local scale [91,92]. Moreover, use of RS and GIS can reasonably account for the spatiotemporal variability of parameters and catchment heterogeneity [65]. For estimation of water-based erosion, RUSLE coupled with RS and GIS is the most commonly adopted and feasible technique to quantify the magnitude and spatial distribution of soil erosion/loss from rivers' catchments [62,93–97].

A 17 to 27-year study [98] of 9 sediment stations within the Chenab river catchment revealed that there are very high erosion rates. A soil erosion study using the USLE model in similar nearby area also revealed high, very high, and severe soil erosion [99]. Researchers applied the sediment yield index model to the catchment of the Marusudar tributary of the Chenab river, which revealed a high rate of soil erosion [100]. At present, no soil erosion study has been conducted on the complete transboundary Chenab river catchment using the RUSLE model. Keeping in view all the soil erosion issues of the transboundary river catchments, the present study aims to estimate the spatially distributed PSE of the transboundary Chenab river catchment using the RUSLE model integrated with RS and GIS. The spatial distribution of PSE is one of the main targets of this study, so 55 sub-basins were created from the Chenab river catchment. The study also aims to estimate SDR using area, slope, and CN-based approaches, for estimation of sediment yield from the Chenab river catchment. Both PSE and SY were estimated on annual, seasonal, quarterly, and monthly time scales from 1991 to 2020.

2. Material and Methods

2.1. Study Area

The Chenab river catchment is shown in Figure 1. The geographic extent of the Chenab river catchment lies 74°–77.85° E and 32°–34.3° N, while elevation ranges from 240 to 7085 m, and average slope of the river in the catchment is about 25 m/km. It originates in the Kangra and Kulu district of the Himachal Pradesh, India. The Bhaga and Chandra streams emerge from mega snowfields on opposing sides of the Baralcha pass and join the Tandi, Jammu and Kashmir. The climate in the catchment typically comprises two seasons in a year. The Rabi season spans from November to April, the Kharif from May to October. Another climate classification also exists consisting of four seasons, Fall (September, October, November), Winter (December, January, February), Spring (March, April, May), and Summer (June, July, August). The snow-dominant Chenab river catchment [88] receives 65% of precipitation in the monsoon (June, July, August) or pre-monsoon (March, April, May), while 26% precipitation is received in the winter season [101]. The higher altitude of the upper and middle parts of the Chenab river catchment are snow-dominant regions. The mean annual rainfall varies from 278.5 to 2214.9 mm as shown in Figure 2.

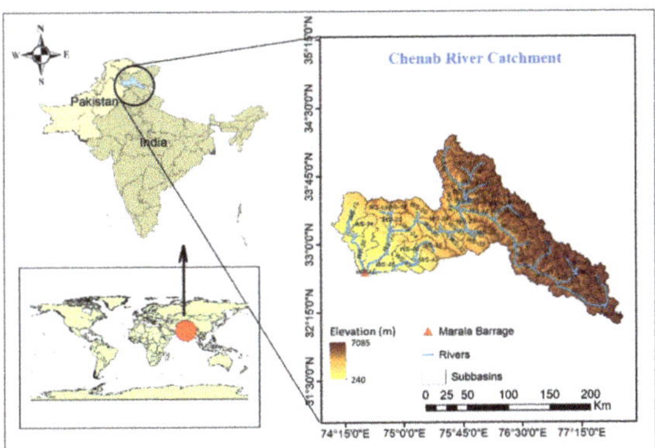

Figure 1. Topographic map of the Chenab River catchment representing river, 55 sub-basins and outlet.

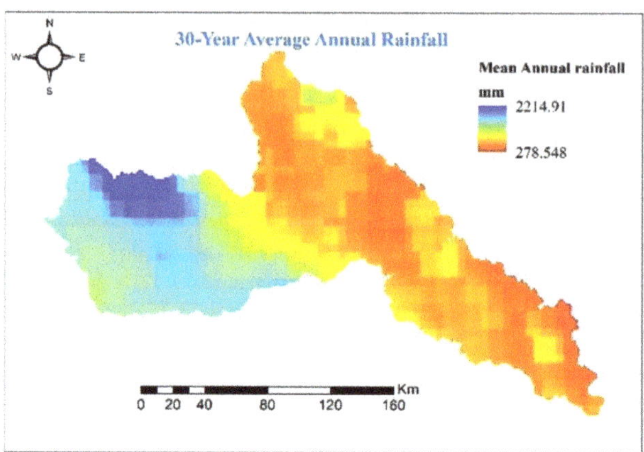

Figure 2. The 30-year average annual rainfall of the Chenab river catchment.

2.2. Methodology

The overall methodology of the research for estimation of PSE and SY is presented in the flowchart (Figure 3). The topographic information, soil data, land use information, and precipitation data were used to estimate all the factors to be used in the RUSLE model for estimation of PSE. The SDRs were estimated based on area, slope and CN approaches, and the SDRs were further used along with PSE to estimate the SY.

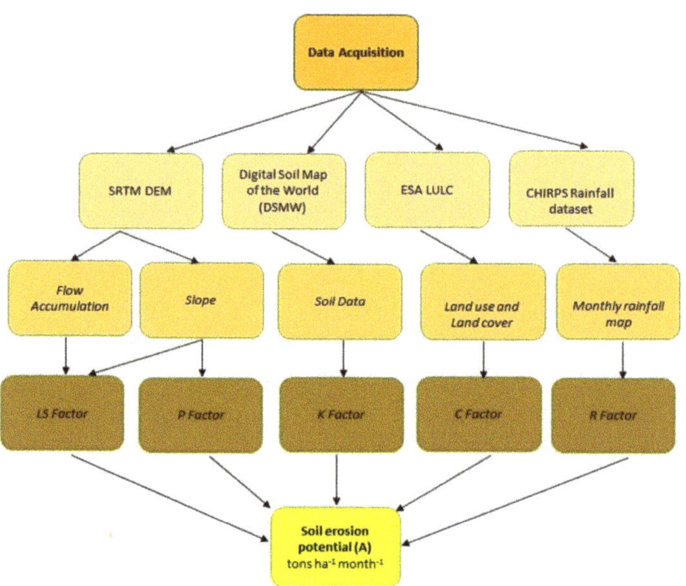

Figure 3. Flow chart for estimation of potential soil erosion and sediment yield.

2.2.1. Datasets Used in the Study

We used the Shuttle Radar Topography Mission Digital Elevation Model (90 m) for watershed delineation, calculation of length and slope factors, and support practice factor based on slope-contour approach. The land use change detection of the river catchment is a major challenge [102] and helps to understand hydrological processes and associated systems of the river basins [103–106]. Due to lack of in situ data, the global land cover map was used at a spatial resolution of 300 m, produced by the ESA Climate Change Initiative [107]. We estimated land cover management factors using the land use map produced by ESA at a spatial resolution of 300 m. The soil types and texture are also important, along with land use, to understand the hydrological response of the river catchment [108,109]. The Digital Soil Map of the world (DSMW) was used in this study, which is produced by the FAO/ UNESCO. The erosivity factor for 30 years (1991–2020) was estimated using the CHIRPS precipitation data at a spatial resolution of 0.25°.

2.2.2. RUSLE Model for Estimation of Potential Soil Erosion

Wischmeier and Smith developed the USLE for estimation of soil erosion [68]. The new equation (RUSLE) replaces USLE's distinctive rainfall or runoff factor as the rainfall erosivity factor [71]. We used RUSLE to estimate soil erosion at monthly, quarterly, seasonally, and yearly time scales on a surface slope based on the runoff model, soil type, farming practices, topography (slope), and supervision techniques [110]. The RUSLE is an empirical equation that estimates PSE in tons per hectare (Equation (1)).

$$A = R \times K \times L \times S \times C \times P \tag{1}$$

where A is estimated monthly soil erosion (ton ha^{-1} month^{-1}), R is rainfall erosivity factor (MJ mm ha^{-1} h^{-1} month^{-1}), K is soil erodibility factor (t MJ^{-1} ha^{-1} mm^{-1}), L is slope length factor, S is slope steepness factor, C is cover management factor, and P is supporting practices. All these factors of RUSLE were mapped in GIS raster format at quarterly, seasonally, and annual time scales.

Rainfall Runoff Erosivity Factor (R)

Without soil surface protection, the rainfall erosivity factor (R) triggers sheet and rill erosion. Soil loss significantly depends on rainfall because it detaches soil particles from the ground surface and transports them to the river channel [111]. Heavy rainfall events having large droplets size can quickly detach soil particles, compared to droplets of smaller size. The bulk of sheet or rill erosion occurs due to the high runoff generated by a heavy rainfall storm. Rainfall has a significant effect on soil erosion due to the kinetic energy that each raindrop contains, which causes soil particles to detach from the ground surface. We used monthly CHIRPS data for 30 consecutive years (1991–2020) over the Chenab river catchment. In order to estimate the rainfall-runoff erosivity factor of the Chenab river catchment, we calculated R-factor by using Equation (2) developed by Jung et al. [112].

$$R = 0.0378 \times X^{1.4190} \tag{2}$$

where R is rainfall runoff erosivity factor (MJ mm ha^{-1} h^{-1} month^{-1}), and X is monthly rainfall amount (mm).

Soil Erodibility Factor (K)

The soil erodibility factor represents the soil's vulnerability to degradation as evaluated under normal unit plot conditions. Soil erodibility is the quantity of soil loss per unit of rainfall erosive energy. We estimated the K factor by using Equation (3) developed by [113]. This depends on soil contents of organic carbon, silt, sand, and clay, obtained from FAO soil data.

$$K = 0.1317 \cdot f_{csand} \cdot f_{cl-si} \cdot f_{orge} \cdot f_{hisand} \tag{3}$$

$$f_{csand} = \left(0.2 + 0.3\, exp\left[-0.256 \times ms \times \left(1 - \frac{msilt}{100} \right), \right] \right) \tag{4}$$

$$f_{cl-si} = \left(\frac{msilt}{mc + msilt} \right)^{0.3} \tag{5}$$

$$f_{orge} = \left(1 - \frac{0.0256 \times orgC}{orgC + exp[3.72 - 2.95 \times orgC]} \right) \tag{6}$$

$$f_{hisand}\left(1 - \frac{0.7 \times \left(1 - \frac{ms}{100}\right)}{\left(1 - \frac{ms}{100}\right) + exp\left[-5.51 + 22.9\left(1 - \frac{ms}{100}\right)\right]}\right) \tag{7}$$

where ms is % sand content in topsoil, $msilt$ is % silt content in topsoil, mc is % clay content in top soil, and $orgC$ is % organic carbon content in top soil. We estimated the K factor by Equation (3). A maximum of four soil types were identified in the soil map using the FAO soil data. The soil erodibility factor was estimated based on the sand, silt, and clay percentages.

Slope-Length and Slope-Steepness Factor (LS)

The LS factor represents the effects of topography on soil erosion by including slope-length factor (L) and slope-steepness factor (S), both of which influence overland flow velocity [114]. The Topographic slope-length (L) and slope-steepness (S) reflect a ratio of soil erosion under defined conditions compared to soil loss at a site with a "normal" slope steepness of 9% and a slope length of 22.6 m [62]. Because the LS-factor causes high runoff velocity and hence more runoff volume, the highest slope has the greatest danger of soil erosion. The equation of LS is given as;

$$LS\left(\frac{\lambda}{22.13}\right)^{m} \times \left(\frac{\sin\beta}{0.0896}\right)^{n} \tag{8}$$

$$\beta = \frac{\sin\theta/0.0896}{3 \times \sin\theta^{0.8} + 0.56} \tag{9}$$

where

θ is Slope of watershed (degree), n is 1.3
λ = slope-length (m) = flow acc. $\times$ cell size
m = variable slope length exponent = $\frac{\beta}{1+\beta}$

Cover Management Factor (C)

The dimensionless cover and management factor plays its role in reducing soil erosion, and depends on the land use patterns of the area [115]. The C factor is the ratio between soil loss from areas with protective cover and management to soil loss from continuously clean tilled fallow land [116]. The C factor varies from 0 to 1 depending on land use characteristics, excluding water bodies [117].

Supporting Conservation Practice Factor (P)

The P-factor is the ratio of soil loss induced by each type of conservation technique to the comparable erosion generated by uphill and downhill sloped cropping [110,118]. It modifies the volume and water discharge, hence affecting the magnitude of soil erosion. The support conservation techniques used in the catchment, such as contouring, terracing, strip cropping, etc., are referred to as the support practice factor. The P factor has a value between 0 and 1, with 0 representing very good preservation practice, and one indicating no preservation technique [45].

2.2.3. Estimation of Sediment Yield

$$SY = SDR \times A \tag{10}$$

where SY is in tons/month, SDR (fraction), and A is PSE (tons/month).

Estimation of Sediment Delivery Ratio

The SDR is the ratio of sediments delivered at outlet to the gross erosion of the catchment upstream of the measurement location [119], and SDR represents several processes which are involved in estimation of SY [120]. Although the United States Department of Agriculture has published a handbook [121] in which the SDR is linked to drainage areas, there is no perfect process for estimating SDR. A variety of factors can influence SDR, including sediment load, texture, proximity to the mainstream, channel density, basin area, slope, length, land use, rainfall, and runoff. The Soil Conservation Service (SCS) curve is the established relationship between SDR and drainage area. For example, a watershed with a higher channel density has a higher SDR than a catchment with a lower channel density. The SDR of a watershed with steep slopes is greater than that of a watershed with flat and large valleys. The size of the area of interest should also be defined in order to predict SDR. The higher the area size, the smaller the fraction of SDR since large areas have more chance of trapping soil particles. The SDR equations have been derived by several researchers in different river basins of the world [122–125]. The researchers correlated the SDR with area [121] and developed their equation as follows:

$$SDR = 0.5656 \times A^{-0.11} \tag{11}$$

where A is the area of watershed in sq. miles (mi^2).

The researchers [126] also correlated the SDR with the slope and developed the equation as follows:

$$SDR = 0.627 \times S^{0.403} \tag{12}$$

where S is the slope of watershed in degree.

The relief-length ratio is prepared by getting the maximum and minimum value of the elevation of the catchment, then by taking the difference of maximum and minimum elevation and finally dividing it by the length of the river. SDR against CN is prepared according to the empirical equation. According to the research [74], the SDR is related to watershed area, relief-length ratio, and SCS CN, and the following equation was derived in a study conducted on 15 Texas basins:

$$SDR = 1.366 \times 10^{-11} \times A^{-0.0998} \times ZL^{0.3629} \times CN^{5.444} \tag{13}$$

where A is the area in km^2 and ZL is the relief-length ratio in m/km.

The hydrologic soil group and ground cover are used to calculate the CN. In general, the most reliable findings are achieved when each sub-catchment is homogeneous, with as few CNs as possible. When a large number of CNs are averaged into a single sub-catchment, the results are not necessarily the same as when multiple sub-catchments are produced and combined together. A single sub-catchment, for example, will only have one peak; however, merging many sub-catchments might result in a multi-peak hydrograph.

3. Results

3.1. Factors of the RUSLE Model

The 30-year average rainfall erosivity factor maps at annual, seasonal (rabi, kharif), and quarterly time scales (fall, winter, spring, and summer) are presented in Figure 4.

The peak rainfall erosivity factor was 826.8, 277.9, 558.5, 199, 115.3, 217.7, and 426.8 MJ mm^{-1} ha^{-1} h^{-1} for annual, rabi, kharif, fall, winter, spring, and summer, respectively. The spatial distribution of all the maps in Figure 4 reveals that the upper parts of the catchment have lower rainfall erosivity, and the lower parts of the catchment have very high erosivity because of high rainfall over the lower part as shown in Figure 2. Higher rainfall erosivity has been observed in kharif season (May to October), and in summer season. The maps of rabi season (November to April), fall, winter and spring seasons reveal that higher erosivity has been observed in higher latitudes in the lower parts of the catchments.

The estimated soil erodibility factor values using %sand, %silt, %clay, and % OC are given in Table 1. The four soil types were found in the river catchment, and K value ranges from 0.0174 to 0.023.

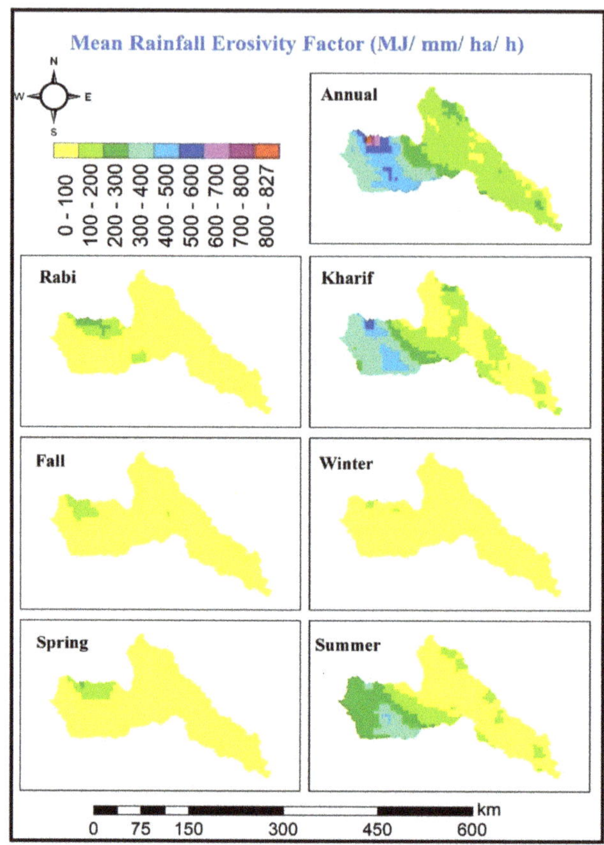

Figure 4. 30-year average annual, seasonal, and quarterly rainfall erosivity factors for the Chenab river catchment.

Table 1. Soil erodibility factor value calculated for soil type in the Chenab river catchment.

Soil Unit Symbol	% Sand Topsoil	% Silt Topsoil	% Clay Topsoil	% OC Topsoil	K Factor Value
I	58.9	16.2	24.9	0.97	0.0196588
Be	36.4	37.2	26.4	1.07	0.0223028
Lo	76	9.9	14.1	0.41	0.0174135
Jc	39.6	39.9	20.6	0.65	0.0232663

The 30-year average soil erodibility, slope-length and slope-steepness, cover management, and supporting conservation practice factors map of the Chenab river catchment is presented in Figure 5. The map of K factor reveals that most of the catchment is below 0.0197, while some portions of the catchment are in areas of high soil erodibility (greater than 0.0197).

The LS factor map of the Chenab river catchment is presented in Figure 5, which reveals that most of the catchment lies under two main classes, 0 to 5 in upper and middle parts of the catchment, and 13 to 18 in lower parts of the catchment. Therefore, the effect of LS on the PSE is not higher, as a higher LS factor area is not common in the catchment. The land use map of Chenab river catchment prepared using land use data of ESA is presented in Figure 6. The upper parts of the river catchment are covered with snow, consolidated bare land, grassland, and mosaic trees and shrubs. The middle to lower parts of the catchment are covered with tree cover and mosaic cropland, while the lowest parts of the catchment near to the outlet are covered with irrigated cropland. The cover management factor values were estimated (Table 2) using the land use information of ESA.

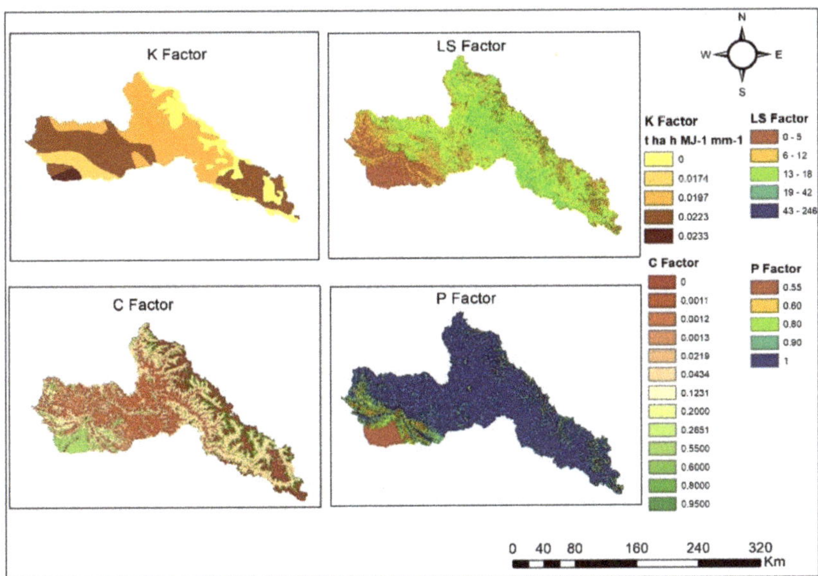

Figure 5. Soil erodibility, slope-length and slope-steepness, cover management, and supporting conservation practice factors in the Chenab river catchment.

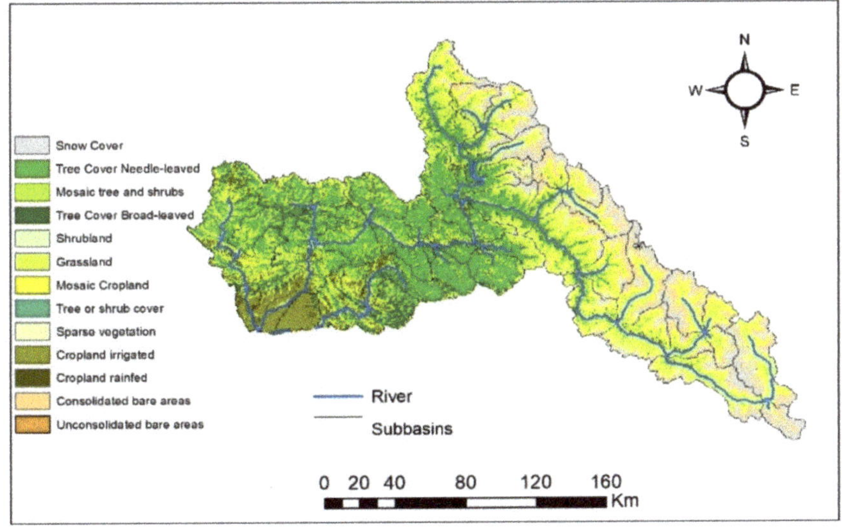

Figure 6. Land use of the Chenab river catchment developed using European Space Agency land use.

Table 2. Cover management factor values of the Chenab river catchment.

Land Use and Land Cover Class	C Factor Value	Source
Snow Cover	0	[127]
Tree Cover Needle-leaved	0.0011	[127]
Mosaic tree and shrubs	0.0012	[127]
Tree Cover Broad-leaved	0.0013	[127]
Shrub land	0.0219	[127]
Grassland	0.0434	[127]
Mosaic Cropland	0.1231	[127]
Tree or shrub cover	0.2000	[127]
Sparse vegetation	0.2651	[127]
Cropland irrigated	0.5500	[127]
Cropland rainfed	0.6000	[127]
Consolidated bare areas	0.8000	[127]
Unconsolidated bare areas	0.9000	[127]

The cover management factor map was developed based on the C value of various vegetative coverings, as illustrated in Figure 5. The snow-covered areas of the upper catchment are under the lowest C values, while the bare areas in of the upper catchment are under high C values, while most of the upper to middle parts of the catchment are under medium C value (0.0434 to 0.1232). The middle to lower parts of the catchment are under lower C values, and the extreme lower parts of the catchment are under slightly higher C values.

In this research, data from the literature [128] was obtained for estimation of conservation management factor against % slope as presented in Table 3. As in situ crop information cannot be obtained, the average of the conservation practices was used in order to estimate the P value to be used in RUSLE. The P factor map for the Chenab river catchment is presented in Figure 5. Most of the catchment is under a high value of P, while the catchment areas near to the outlet are under lower values of P.

Table 3. Conservation management factor values for contouring, strip cropping, and terracing against slope.

Slope (%)	Contouring	Strip Cropping	Terracing
0.0–7.0	0.550	0.270	0.100
7.0–11.3	0.600	0.300	0.120
11.3–17.6	0.800	0.400	0.160
17.6–26.8	0.900	0.450	0.180
26.8>	1.000	0.500	0.200

3.2. Potential Soil Erosion

The 30-year annual average PSE from the Chenab river catchment is presented in Figure 7. The 30-year average PSE from the Chenab river catchment was estimated as 177.8, 61.5, 310.3, 39.5, 26.9, 47.1, and 99.1 tons/ha for annual, rabi, kharif, fall, winter, spring, and summer time scales, respectively. The minimum values of PSE are on the snow cover areas, and lower values are seen also on tree cover and irrigated croplands. The bare areas near the rivers and the upper latitudes of the lower parts of the catchments are under medium PSE. On an annual time-scale, there is less area of catchment which has more than 10 PSE. The kharif season shows higher PSE values compared to the rabi season, mainly due to heavy precipitation in monsoon season and high magnitude of runoff due to rainfall-runoff and snowmelt. The PSE in summer season is higher in different areas of the catchment as compared to the fall, winter, and spring seasons.

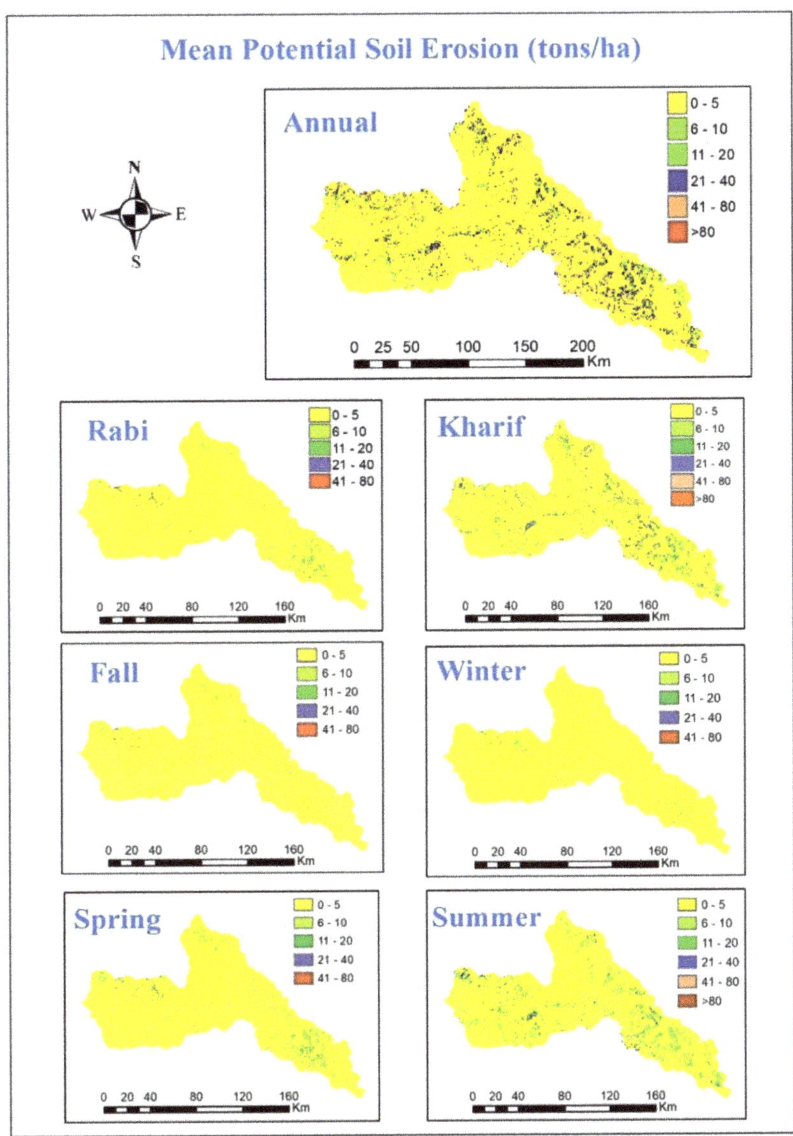

Figure 7. 30-year average potential soil erosion from the Chenab river catchment at different time-scales.

The PSE distribution for all the mentioned time-scales is grouped into six soil erosion intensity classes (slight, moderate, high, very high, severe, and very severe) as per the literature [129] as shown in Figure 8.

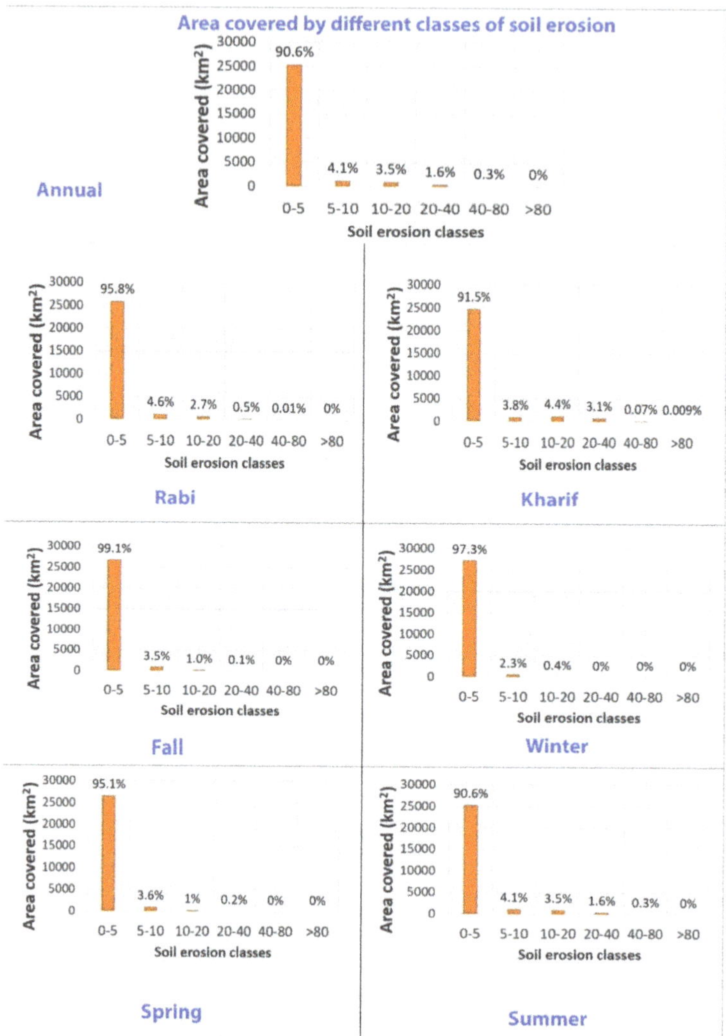

Figure 8. Catchment area percentage under different soil erosion classes based on PSE (ton/ha).

These six classes have value ranges from 0–5, 5–10, 10–20, 20–40, 40–80, and >80 tons/ha, respectively. On an annual time-scale, most PSE is under slight intensity, while areas are also under moderate, high, and very high, but the percentage of area in these three intensity classes is less. In the kharif season, there is more area in high and very high intensity classes as compared to the rabi season. In the summer season, there is area under very high intensity of erosion as compared to fall, winter, and spring season. It is obvious from Figure 8 that the percentage of area in 0–5 class is more than 90% for all the time periods.

3.3. Sediment Yield

The SY from the Chenab river catchment was estimated using area, slope, and CN-based approaches. The CN for the Chenab river catchment is presented in Figure 9. The CN values in the snow-covered areas and some lower parts of the Chenab river catchment are higher, the central parts of the catchment are under medium CN values, and some lower parts of the catchment are under low CN values. The sub-catchment-wise SDR based area, slope, and CN are presented in Figure 10. There is reliable agreement between the SDRs based on area and slope, while the slope-based SDR is higher than the other. The CN-based SDR was higher than the others, while the higher CN-based SDR values were observed in sub-catchment numbers 10, 11, 12, 23, and 47.

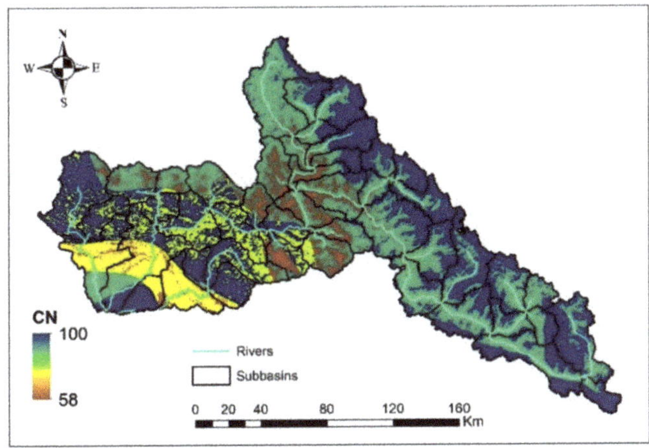

Figure 9. CN map of the Chenab river catchment.

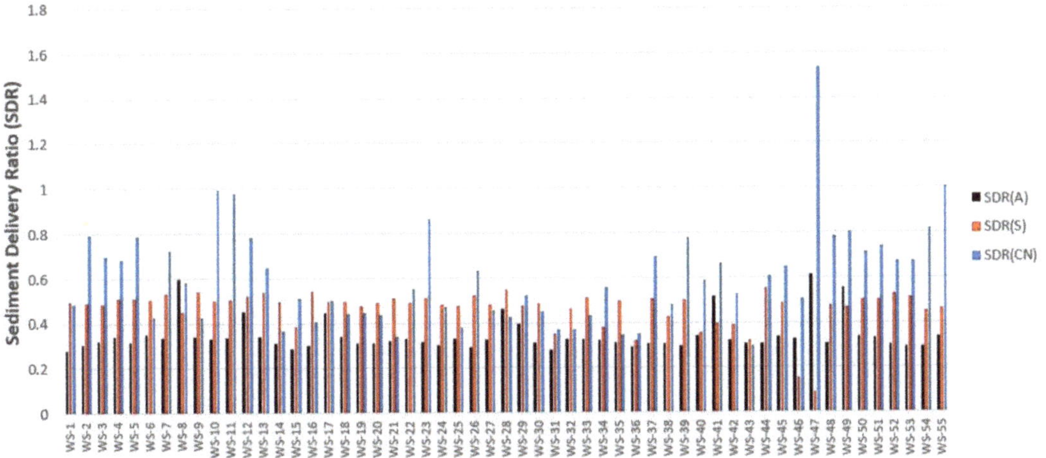

Figure 10. Mean Sediment delivery ratio in each micro-catchment.

In the area-based SDR, watershed (WS)-47 exhibited the highest SDR of 0.61 having a 0.5 sq. mile area, which contributes 0.0045% of the total area of the catchment, while the WS-1 had the minimum SDR of 0.281 consisting of 565.8 sq. miles area which contributes 5.2% of the total area of the catchment/watershed. In the slope-based SDR, the WS-44 exhibited the highest SDR of 0.551 having a 270.118 sq. miles area which contributes 2.48% of the total area of the watershed, while the WS-47 had the minimum SDR of 0.08 consisting of a 0.5 sq. mile area which contributes 0.0045% of the total area of the watershed. In the CN-based SDR, the overall ratio seemed high as compared to area and slope module, in which WS-47 contributed the highest SDR of 1.53 having a 0.5 sq. mile area which contributes 0.0045% of the total area of the watershed. On the other hand, the WS-43 had the minimum SDR consisting of 265.98 sq. miles area which contributes 2.44% of the total area of the watershed.

The annual sediment yield pattern over the last 30 years is presented in Figure 11. A similar pattern among area, slope, and CN-based annual SY has been observed. The highest SY was observed in 2014, and it showed that the sediment load was highest due to intense precipitation and surface runoff. In 2014, the area, slope and C-based SY was observed as 7,886,149, 12,032,723, and 14,550,254 tons, respectively. However, in 1998, the SY had the lowest value of 1,183,469, 1,733,782, and 2,067,188 tons based on area, slope and CN, respectively. The time series trends analysis of SY indicated an increase of 0.0895, 0.1387, and 0.1698 million tons per year for area, slope, and CN-based approaches, respectively.

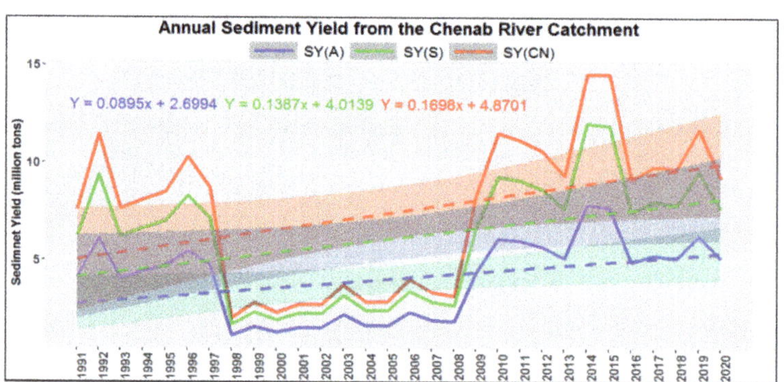

Figure 11. Annual sediment yield from the Chenab river catchment 1991 to 2020.

The SY over the last 30-year in Rabi and Kharif season is presented in Figure 12.

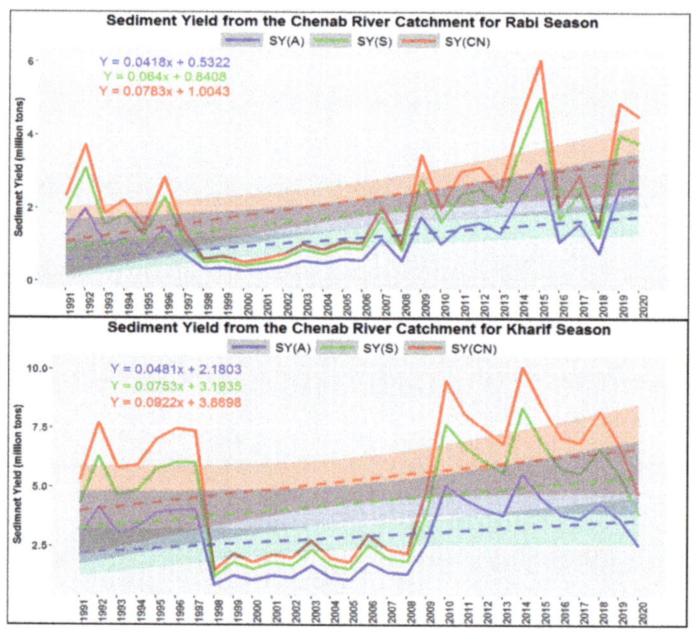

Figure 12. Sediment yield from the Chenab river catchment for rabi and kharif seasons from 1991–2020.

The overall contribution of the kharif season was higher than that of the rabi season. In 2014, the highest SY of the kharif season based on area, slope, and CN was 5,574,869, 8,405,752, and 10,124,635 tons, respectively. In 2015, the highest SY of the rabi season based on area, slope, and CN was 3,264,869, 5,015,752, and 8,104,635 tons, respectively. The SY time series trends analysis of the rabi season indicated an increase of 0.0418, 0.064, and 0.0783 million tons per year for area, slope, and CN-based approaches, respectively. The SY time series trends analysis of the kharif season indicated an increase of 0.0481, 0.0753, and 0.0922 million tons per year for area, slope, and CN-based approaches, respectively.

The SY for fall, winter, spring, and summer is presented in Figure 13.

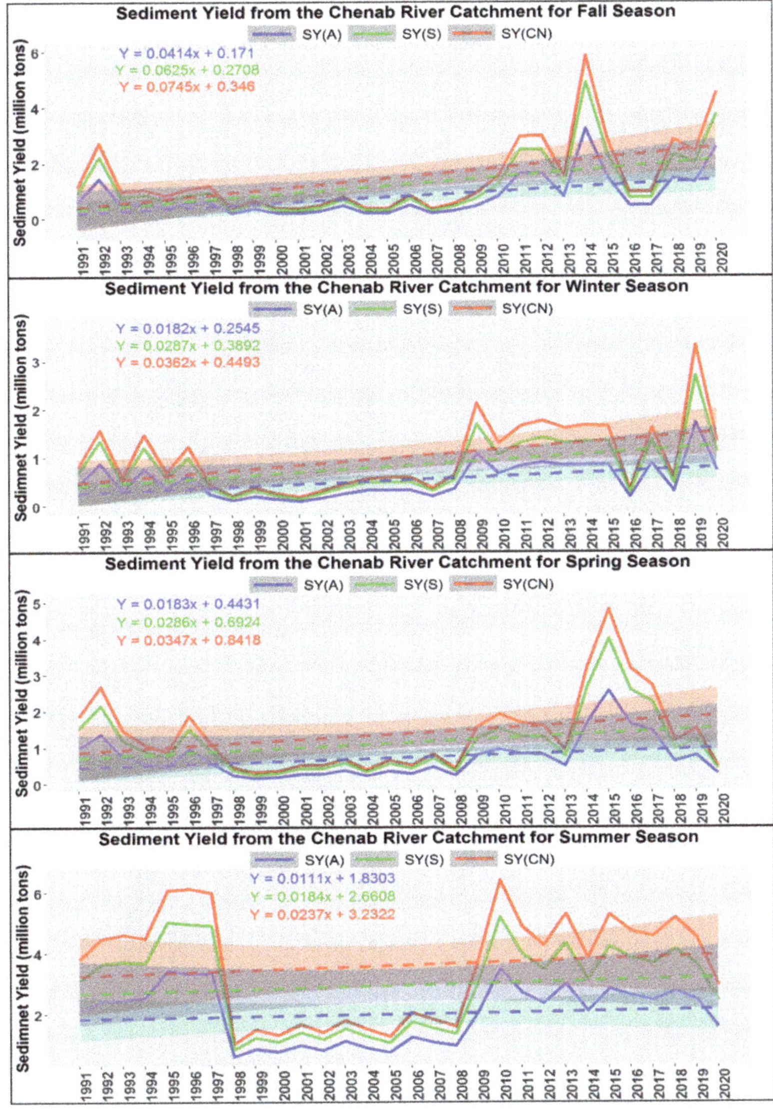

Figure 13. Sediment yield from the Chenab river catchment for fall, winter, spring, and summer seasons from 1991–2020.

The SY for fall period was higher in 1992, 2010, 2011, 2012, 2014, 2018, 2019, and 2020. The SY for winter period was higher in 1992, 1994, 1996, 2008–2014, 2017, 2019, and 2020. The SY for spring period was higher in 1991–1993, 1996, 2009–2012, 2014–2017. The SY for the summer season was higher in 1991–1997 and 2009–2020.

The monthly sediment yield for the last 30 years is presented in Figure 14. The highest SY values of 3,125,536 tons, 4,745,077 tons, and 5,643,741 tons for area, slope, and CN, respectively, were observed in September 2014. Similarly, the lowest values of 11,484 tons, 17,904, and 21,811 tons for area, slope, and CN, respectively, were observed in November 1998. The time series trends analysis of SY indicated an increase of 0.0006, 0.001, and 0.0012 million tons per month for area, slope, and CN-based approaches, respectively.

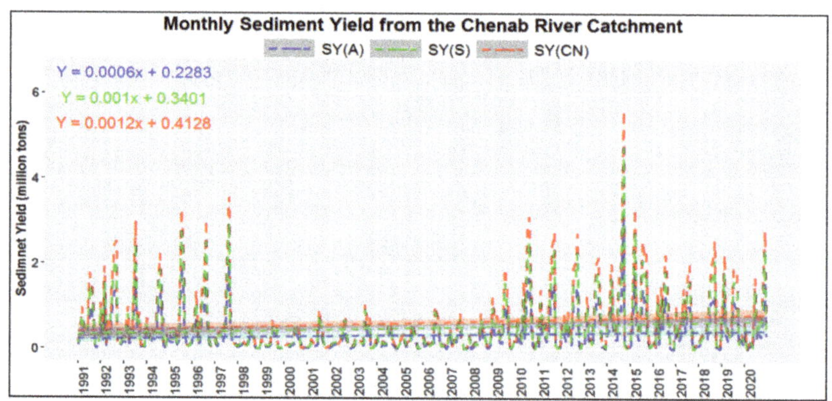

Figure 14. Monthly sediment yield from the Chenab river catchment from 1991 to 2020.

4. Discussion

Soil erosion is a severe issue, particularly in the Chenab river catchment, where various variables contribute to fast soil erosion and sedimentation. The rate of runoff and sedimentation is accelerated by factors such as the region's steep slope, temperature, velocity of flowing water, and environmental conditions [130]. The impact of a raindrop on the soil surface and the cutting force of running water causes soil particles to detach. Raindrop splash, despite having a minimal effect, triggers downslope transport of eroded soil particles [131]. High intensity rainfall makes the detachment of soil particles quicker and causes the mass movement of sediments along the runoff generated. Early in the rainy season, when the rainfall is intense but the vegetation has not developed enough to protect the soil, is the most favorable time for erosion. In general, the time between plowing and crop emergence is referred to as the farmer's interval [132]. The CHIRPS precipitation data was incorporated in this research as it has better performance over areas that usually receive high rainfall [133].

Slope also varies, decrease in slope causing velocity to decrease, and ultimately to sediment transport decreases [134], which further increases the rate of sediment deposition. The topography and high elevation are usually the main reason for high intensity precipitation [135]. The K factor of the study area varied from 0.019 to 0.023. The soil, having low moisture content and permeability, represents a low K factor value. Soil erosion is closely concerned with the state of land use and agriculture practices, and cover management, as most of the Chenab river catchment was under 0 to 5 tons ha^{-1}. The annual average PSE rates throughout the world are estimated as 12 to 15 tons ha^{-1}, while the river catchment areas with a PSE of lower than 3 tons ha^{-1} year^{-1} are generally below the estimated tolerable soil loss level and should be exempt from any mitigating activities.

Though different soil erosion studies have been carried out on the Chenab river catchment with various models and techniques, this is the first time that estimation of PSE and SY for the last 30 years has been performed using the RUSLE model and SDRs on annual, seasonal, quarterly and monthly time-scales The SDR is estimated using area, slope, and CN-based approaches, and slope-based estimations are being applied by researchers around the world [136–144]. It is also observed in the study that SDR decreased with increase in area or stream length, and this is also observed by other researchers [124,145]. Moreover, topography-based SDR (slope) is scale dependent, and the scale dependency of such SDR has also been observed by researchers [146–148]. The RUSLE and SDR (mainly slope-based) are being applied by the scientific community for estimation of PSE and SY [149–155]. The RUSLE is used at multiple spatial scales by dividing a pixel into sub-regions with similar features and linking them to a GIS data structure [110]. Such models are now commonly in use to create an environment-based information system that allows for the estimation and evaluation of various management scenarios [156]. This methodology, however, has certain limitations but provides reliable results by identifying the high PSE areas. The RUSLE model was selected and applied to nearby similar rivers' catchments [97,135,157].

The complete lower parts of the catchments are under high rainfall erosivity, therefore soil and water conservation measures and crop management practices are needed in order to reduce the rainfall erosivity, which will ultimately reduce the overall PSE and SY. The organic content of the soil reduces soil erodibility because increase in organic matter reduces the susceptibility of soil detachment, and increases infiltration, which further reduce runoff and thus soil erosion. The organic content should be increased in high K values areas by incorporating manure. With the novelty of providing soil loss estimates at finer spatial and temporal scales, the findings of this study can be useful for assessing soil erosion in other data-scarce areas, and can be helpful to resource conservation experts for making informed decisions.

5. Conclusions

The following conclusions have been derived from a 30-year study of soil erosion from the Chenab river catchment using RUSLE and SDR approaches;

- The analysis results depicted that the range of average annual PSE was from 0.0 to 177.8 tons/ha, while, in the Rabi and Kharif season, the range of average PSE was between 0.0 to 61.5 tons/ha and 0.0 to 310.2 tons/ha. Similarly, in Fall, Winter, Spring, and Summer timescales, the range of average PSE was from 0 to 39.5, 26.9, 47.1, and 99.1 tons/ha, respectively.
- The time series trends analysis of SY indicated an increase of 0.0895, 0.1387, and 0.1698 million tons per year for area, slope, and CN-based approaches, respectively.
- The annual SY estimated by area, slope and CN was highest in 2014 with 7,886,149, 12,032,723, and 14,550,254 tons, respectively. The average PSE of Kharif season was highest (70%), followed by Fall (41%), Rabi (30%), Summer (26%), Spring (22%), and Winter (11%) season.
- The annual SY estimated by area, slope and CN was minimum in 1998 with 1,183,469, 1,733,728, and 2,067,188, respectively. The average PSE of Kharif season was highest for soil loss (71%), followed by Summer (49.5%), Rabi (29%), Spring (22%), Fall (18%), and Winter (10.5%), respectively.
- Inter-comparison of SY estimated by SDR based on slope and CN showed the consistency pattern and thus proved the authenticity of empirical models, but the SY estimated by area-based SDR was less compared to the slope and CN-based SDR Approaches.

Author Contributions: Conceptualization, M.G.A., R.H.A. and S.A.; Methodology, M.G.A., A.N. and S.A.; Software, R.H.A. and M.M.W.; Formal Analysis, A.N., M.W. and R.A.A.; Investigation, M.M.W., R.A.A. and M.J.M.C.; Resources, M.G.A., M.J.M.C. and S.A.; Data Curation, A.N., M.W. and M.K.L.; Writing—Original Draft Preparation, M.G.A., A.N. and I.S.; Writing—Review & Editing, S.A., M.J.M.C. and I.S.; Visualization, M.G.A., I.S. and M.K.L.; Supervision, A.N. and R.H.A. All authors have read and agreed to the published version of the manuscript.

Funding: This research has not received any external funding.

Institutional Review Board Statement: Not applicable.

Data Availability Statement: Data belongs to the authors.

Conflicts of Interest: The authors declare no conflict of interest.

References

1. Bini, C. Soil: A precious natural resource. In *Conservation of Natural Resources*; Kudrow, N.J., Ed.; Nova Science Publishers: Hauppauge, NY, USA, 2009; pp. 1–48.
2. Pimentel, D.; Kounang, N. Ecology of Soil Erosion in Ecosystems. *Ecosystems* **1998**, *1*, 416–426. [CrossRef]
3. Zhao, G.; Mu, X.; Wen, Z.; Wang, F.; Gao, P. Soil erosion, conservation, and eco-environment changes in the Loess Plateau of China. *Land Degrad. Dev.* **2013**, *24*, 499–510. [CrossRef]
4. Costanza, R.; d'Arge, R.; De Groot, R.; Farber, S.; Grasso, M.; Hannon, B.; Limburg, K.; Naeem, S.; O'neill, R.V.; Paruelo, J. The value of the world's ecosystem services and natural capital. *Nature* **1997**, *387*, 253–260. [CrossRef]
5. Swinton, S.M.; Lupi, F.; Robertson, G.; Hamilton, S.K. Ecosystem services and agriculture: Cultivating agricultural ecosystems for diverse benefits. *Ecol. Econ.* **2007**, *64*, 245–252. [CrossRef]
6. Alewell, C.; Borrelli, P.; Meusburger, K.; Panagos, P. Using the USLE: Chances, challenges and limitations of soil erosion modelling. *Int. Soil Water Conserv. Res.* **2019**, *7*, 203–225. [CrossRef]
7. Amundson, R.; Berhe, A.A.; Hopmans, J.W.; Olson, C.; Sztein, A.E.; Sparks, D.L. Soil and human security in the 21st century. *Science* **2015**, *348*, 1261071. [CrossRef] [PubMed]
8. Veerasingam, S.; Venkatachalapathy, R.; Ramkumar, T. Heavy metals and ecological risk assessment in marine sediments of Chennai, India. *Carpathian J. Earth Environ. Sci.* **2012**, *7*, 111–124.
9. Tosic, R.; Dragicevic, S.; Kostadinov, S.; Dragovic, N. Assessment of Soil Erosion Potential by the Usle Method: Case Study, Republic of Srpska—BiH. *Fresenius Environ. Bull.* **2011**, *20*, 1910–1917.
10. Dotterweich, M. The history of soil erosion and fluvial deposits in small catchments of central Europe: Deciphering the long-term interaction between humans and the environment—A review. *Geomorphology* **2008**, *101*, 192–208. [CrossRef]
11. Yang, D.; Kanae, S.; Oki, T.; Koike, T.; Musiake, K. Global potential soil erosion with reference to land use and climate changes. *Hydrol. Process.* **2003**, *17*, 2913–2928. [CrossRef]
12. Routschek, A.; Schmidt, J.; Kreienkamp, F. Impact of climate change on soil erosion—A high-resolution projection on catchment scale until 2100 in Saxony/Germany. *Catena* **2014**, *121*, 99–109. [CrossRef]
13. Lal, R. Soil degradation by erosion. *Land Degrad. Dev.* **2001**, *12*, 519–539. [CrossRef]
14. Lal, R. Soil Erosion by Wind and Water: Problems and Prospects. In *Soil Erosion Research Methods*; Routledge: London, UK, 2017; pp. 1–10. [CrossRef]
15. Pimentel, D. Soil Erosion: A Food and Environmental Threat. *Environ. Dev. Sustain.* **2006**, *8*, 119–137. [CrossRef]
16. Bhattarai, R.; Dutta, D. Estimation of Soil Erosion and Sediment Yield Using GIS at Catchment Scale. *Water Resour. Manag.* **2006**, *21*, 1635–1647. [CrossRef]
17. Rojas-González, A.M. Soil Erosion Calculation Using Remote Sensing and GIS in Río Grande de Arecibo Watershed, Puerto Rico. In Proceedings of the ASPRS 2008 Annual Conference Bridging the Horizons: New Frontiers in Geospatial Collaboration, Portland, OR, USA, 28 April–2 May 2008.
18. Marsh, W.M.; Grossa, J., Jr. *Environmental Geography: Science, Land Use, and Earth Systems*; John Wiley and Sons: Hoboken, NJ, USA, 1996.
19. Wang, M.; Pan, J.; Zhao, J. Quantitative survey of the soil erosion change based on GIS and RS: Take the Qingcheng area as an example. *Agric. Res. Arid. Areas* **2007**, *25*, 116–121.
20. Nyakatawa, E.; Reddy, K.; Sistani, K. Tillage, cover cropping, and poultry litter effects on selected soil chemical properties. *Soil Tillage Res.* **2001**, *58*, 69–79. [CrossRef]
21. Oldeman, L.R. Global extent of soil degradation. In *Bi-Annual Report 1991-1992/ISRIC*; International Soil Reference and Information Centre (ISRIC): Wageningen, The Netherlands, 1992; pp. 19–36.
22. Blanco-Canqui, H.; Lal, R. Soil and water conservation. In *Principles of Soil Conservation and Management*; Springer: Berlin/Heidelberg, Germany, 2010; pp. 1–19.
23. Butzer, K.W. Accelerated soil erosion: A problem of man-land relationships. *Perspect. Environ.* **1974**, 57–77.
24. Woodward, J.; Foster, I. Erosion and suspended sediment transfer in river catchments: Environmental controls, processes and problems. *Geography* **1997**, *82*, 353–376.
25. Didone, E.J.; Minella, J.P.G.; Merten, G.H. Quantifying soil erosion and sediment yield in a catchment in southern Brazil and implications for land conservation. *J. Soils Sediments* **2015**, *15*, 2334–2346. [CrossRef]
26. Keesstra, S.D.; Bouma, J.; Wallinga, J.; Tittonell, P.; Smith, P.; Cerdà, A.; Montanarella, L.; Quinton, J.N.; Pachepsky, Y.; van der Putten, W.H.; et al. The significance of soils and soil science towards realization of the United Nations Sustainable Development Goals. *Soil* **2016**, *2*, 111–128. [CrossRef]
27. Gholami, V.; Sahour, H.; Amri, M.A.H. Soil erosion modeling using erosion pins and artificial neural networks. *Catena* **2021**, *196*, 104902. [CrossRef]

28. Ashraf, A. Risk modeling of soil erosion under different land use and rainfall conditions in Soan river basin, sub-Himalayan region and mitigation options. *Model. Earth Syst. Environ.* **2019**, *6*, 417–428. [CrossRef]

29. Boardman, J.; Evans, R. The measurement, estimation and monitoring of soil erosion by runoff at the field scale: Challenges and possibilities with particular reference to Britain. *Prog. Phys. Geogr. Earth Environ.* **2020**, *44*, 31–49. [CrossRef]

30. Panagos, P.; Katsoyiannis, A. Soil erosion modelling: The new challenges as the result of policy developments in Europe. *Environ. Res.* **2019**, *172*, 470–474. [CrossRef]

31. Rosas, M.; Gutierrez, R.R. Assessing soil erosion risk at national scale in developing countries: The technical challenges, a proposed methodology, and a case history. *Sci. Total Environ.* **2020**, *703*, 135474. [CrossRef]

32. Bennett, H.H. A Permanent Loss to New England: Soil Erosion Resulting from the Hurricane. *Geogr. Rev.* **1939**, *29*, 196. [CrossRef]

33. SMITH, D.D. Interpretation of soil conservation data for field use. *Agric. Eng.* **1941**, *22*, 173–175.

34. Zingg, A.W. Degree and length of land slope as it affects soil loss in run-off. *Agric. Eng.* **1940**, *21*, 59–64.

35. De Vente, J.; Poesen, J. Predicting soil erosion and sediment yield at the basin scale: Scale issues and semi-quantitative models. *Earth-Sci. Rev.* **2005**, *71*, 95–125. [CrossRef]

36. Jetten, V.; Govers, G.; Hessel, R. Erosion models: Quality of spatial predictions. *Hydrol. Process.* **2003**, *17*, 887–900. [CrossRef]

37. Karydas, C.; Panagos, P.; Gitas, I. A classification of water erosion models according to their geospatial characteristics. *Int. J. Digit. Earth* **2012**, *7*, 229–250. [CrossRef]

38. King, C.; Delpont, G. Spatial assessment of erosion: Contribution of remote sensing, A review. *Remote. Sens. Rev.* **1993**, *7*, 223–232. [CrossRef]

39. Merritt, W.; Letcher, R.; Jakeman, A. A review of erosion and sediment transport models. *Environ. Model. Softw.* **2003**, *18*, 761–799. [CrossRef]

40. Toy, T.J.; Foster, G.R.; Renard, K.G. *Soil Erosion: Processes, Prediction, Measurement, and Control*; John Wiley & Sons: Hoboken, NJ, USA, 2002.

41. Vrieling, A. Satellite remote sensing for water erosion assessment: A review. *Catena* **2006**, *65*, 2–18. [CrossRef]

42. Zhang, L.; O'Neill, A.L.; Lacey, S. Modelling approaches to the prediction of soil erosion in catchments. *Environ. Softw.* **1996**, *11*, 123–133. [CrossRef]

43. Morgan, R.; Nearing, M. Model development: A user's perspective. In *Handbook of Erosion Modelling*; Blackwell Publishing Ltd.: Hoboken, NJ, USA, 2011; pp. 9–32.

44. Boardman, J. Soil erosion science: Reflections on the limitations of current approaches. *Catena* **2006**, *68*, 73–86. [CrossRef]

45. Morgan, R.; Quinton, J.; Smith, R.; Govers, G.; Poesen, J.; Auerswald, K.; Chisci, G.; Torri, D.; Styczen, M. The European Soil Erosion Model (EUROSEM): A dynamic approach for predicting sediment transport from fields and small catchments. *Earth Surf. Process. Landf.* **1998**, *23*, 527–544. [CrossRef]

46. De Roo, A.; Wesseling, C.; Ritsema, C. LISEM: A single-event physically based hydrological and soil erosion model for drainage basins. I: Theory, input and output. *Hydrol. Process.* **1996**, *10*, 1107–1117. [CrossRef]

47. Kirkby, M.J.; Irvine, B.J.; Jones, R.J.A.; Govers, G.; Team, P. The PESERA coarse scale erosion model for Europe. I.—Model rationale and implementation. *Eur. J. Soil Sci.* **2008**, *59*, 1293–1306. [CrossRef]

48. Bagarello, V.; Di Piazza, G.; Ferro, V.; Giordano, G. Predicting unit plot soil loss in Sicily, south Italy. *Hydrol. Process. Int. J.* **2008**, *22*, 586–595. [CrossRef]

49. Bagarello, V.; Ferro, V.; Pampalone, V. A new version of the USLE-MM for predicting bare plot soil loss at the Sparacia (South Italy) experimental site. *Hydrol. Process.* **2015**, *29*, 4210–4219. [CrossRef]

50. Bagarello, V.; Ferro, V. Plot-scale measurement of soil erosion at the experimental area of Sparacia (southern Italy). *Hydrol. Process.* **2004**, *18*, 141–157. [CrossRef]

51. Di Stefano, C.; Ferro, V.; Pampalone, V. Testing the USLE-M Family of Models at the Sparacia Experimental Site in South Italy. *J. Hydrol. Eng.* **2017**, *22*, 05017012. [CrossRef]

52. Kinnell, P.I.A. Applying the QREI30index within the USLE modelling environment. *Hydrol. Process.* **2014**, *28*, 591–598. [CrossRef]

53. Cao, L.; Zhang, K.; Dai, H.; Liang, Y. Modeling Interrill Erosion on Unpaved Roads in the Loess Plateau of China. *Land Degrad. Dev.* **2016**, *26*, 825–832. [CrossRef]

54. Gessesse, B.; Bewket, W.; Bräuning, A. Model-Based Characterization and Monitoring of Runoff and Soil Erosion in Response to Land Use/land Cover Changes in the Modjo Watershed, Ethiopia. *Land Degrad. Dev.* **2014**, *26*, 711–724. [CrossRef]

55. Diodato, N.; Borrelli, P.; Fiener, P.; Bellocchi, G.; Romano, N. Discovering historical rainfall erosivity with a parsimonious approach: A case study in Western Germany. *J. Hydrol.* **2017**, *544*, 1–9. [CrossRef]

56. Doetterl, S.; Van Oost, K.; Six, J. Towards constraining the magnitude of global agricultural sediment and soil organic carbon fluxes. *Earth Surf. Process. Landf.* **2011**, *37*, 642–655. [CrossRef]

57. Van Oost, K.; Quine, T.A.; Govers, G.; De Gryze, S.; Six, J.; Harden, J.W.; Ritchie, J.C.; McCarty, G.W.; Heckrath, G.; Kosmas, C.; et al. The Impact of Agricultural Soil Erosion on the Global Carbon Cycle. *Science* **2007**, *318*, 626–629. [CrossRef]

58. Di Stefano, C.; Ferro, V.; Porto, P. Linking Sediment Yield and Caesium-137 Spatial Distribution at Basin Scale. *J. Agric. Eng. Res.* **1999**, *74*, 41–62. [CrossRef]

59. Ferro, V. Further remarks on a distributed approach to sediment delivery. *Hydrol. Sci. J.* **1997**, *42*, 633–647. [CrossRef]

60. Renard, K.G.; Foster, G.R.; Weesies, G.A.; Porter, J.P. RUSLE: Revised universal soil loss equation. *J. Soil Water Conserv.* **1991**, *46*, 30–33.

61. Risse, L.M.; Nearing, M.A.; Laflen, J.M.; Nicks, A.D. Error Assessment in the Universal Soil Loss Equation. *Soil Sci. Soc. Am. J.* **1993**, *57*, 825–833. [CrossRef]

62. Ganasri, B.P.; Ramesh, H. Assessment of soil erosion by RUSLE model using remote sensing and GIS—A case study of Nethravathi Basin. *Geosci. Front.* **2016**, *7*, 953–961. [CrossRef]

63. Coen, G.; Tatarko, J.; Martin, T.; Cannon, K.; Goddard, T.; Sweetland, N. A method for using WEPS to map wind erosion risk of Alberta soils. *Environ. Model. Softw.* **2004**, *19*, 185–189. [CrossRef]

64. Panagos, P.; Meusburger, K.; Van Liedekerke, M.; Alewell, C.; Hiederer, R.; Montanarella, L. Assessing soil erosion in Europe based on data collected through a European network. *Soil Sci. Plant Nutr.* **2014**, *60*, 15–29. [CrossRef]

65. Wall, G.; Coote, D.; Pringle, E.; Shelton, I. *Revised Universal Soil Loss Equation for Application in Canada: A Handbook for Estimating Soil Loss from Water Erosion in Canada*; Agriculture and Agri-Food Canada, Research Branch: Ottawa, ON, Canada, 2002.

66. Teng, H.; Rossel, R.A.V.; Shi, Z.; Behrens, T.; Chappell, A.; Bui, E. Assimilating satellite imagery and visible–near infrared spectroscopy to model and map soil loss by water erosion in Australia. *Environ. Model. Softw.* **2016**, *77*, 156–167. [CrossRef]

67. Wang, X.; Zhao, X.; Zhang, Z.; Yi, L.; Zuo, L.; Wen, Q.; Liu, F.; Xu, J.; Hu, S.; Liu, B. Assessment of soil erosion change and its relationships with land use/cover change in China from the end of the 1980s to 2010. *Catena* **2016**, *137*, 256–268. [CrossRef]

68. Wischmeier, W.H.; Smith, D.D. *Predicting Rainfall Erosion Losses: A Guide to Conservation Planning*; Department of Agriculture, Science and Education Administration: Corvallis, OR, USA, 1978.

69. Angima, S.; Stott, D.; O'Neill, M.; Ong, C.; Weesies, G. Soil erosion prediction using RUSLE for central Kenyan highland conditions. *Agric. Ecosyst. Environ.* **2003**, *97*, 295–308. [CrossRef]

70. Garde, R.; Kathyari, U. Erosion Prediction Models for Large Catchments. In Proceedings of the International Symposium on Water Erosion, Sedimentation, and Resource Conservation, Dehradun, India, 9–13 October 1990; pp. 89–102.

71. Millward, A.A.; Mersey, J.E. Adapting the RUSLE to model soil erosion potential in a mountainous tropical watershed. *Catena* **1999**, *38*, 109–129. [CrossRef]

72. Meyer, L.W.; Wischmeier, W.H. Mathematical simulation of the process of soil erosion by water. *Trans. ASAE* **1969**, *12*, 754–0758.

73. Renard, K.G. *Predicting Soil Erosion by Water: A Guide to Conservation Planning with the Revised Universal Soil Loss Equation (RUSLE)*; United States Government Printing: Washington, DC, USA, 1997.

74. Williams, J. Sediment delivery ratios determined with sediment and runoff models. *IAHS Publ.* **1977**, *122*, 168–179.

75. Woznicki, S.A.; Nejadhashemi, A.P. Spatial and Temporal Variabilities of Sediment Delivery Ratio. *Water Resour. Manag.* **2013**, *27*, 2483–2499. [CrossRef]

76. Weifeng, Z.; Bingfang, W. Assessment of soil erosion and sediment delivery ratio using remote sensing and GIS: A case study of upstream Chaobaihe River catchment, north China. *Int. J. Sediment Res.* **2008**, *23*, 167–173. [CrossRef]

77. Renfro, G.W. Use of erosion equations and sediment delivery ratios for predicting sediment yield. In *Proceedings of the Sediment Yield Workshop: Present and Prospective Technology for Predicting Sediment Yield and Sources*; U.S. Department of Agriculture: Washington, DC, USA, 1972; pp. 33–45.

78. Ebisemiju, F.S. Sediment delivery ratio prediction equations for short catchment slopes in a humid tropical environment. *J. Hydrol.* **1990**, *114*, 191–208. [CrossRef]

79. Wu, L.; He, Y.; Ma, X. Using five long time series hydrometeorological data to calibrate a dynamic sediment delivery ratio algorithm for multi-scale sediment yield predictions. *Environ. Sci. Pollut. Res.* **2020**, *27*, 16377–16392. [CrossRef] [PubMed]

80. Saygın, S.D.; Ozcan, A.U.; Basaran, M.; Timur, O.B.; Dolarslan, M.; Yılman, F.E.; Erpul, G. The combined RUSLE/SDR approach integrated with GIS and geostatistics to estimate annual sediment flux rates in the semi-arid catchment, Turkey. *Environ. Earth Sci.* **2013**, *71*, 1605–1618. [CrossRef]

81. Ebrahimzadeh, S.; Motagh, M.; Mahboub, V.; Harijani, F.M. An improved RUSLE/SDR model for the evaluation of soil erosion. *Environ. Earth Sci.* **2018**, *77*, 454. [CrossRef]

82. Bhattacharya, R.K.; Das Chatterjee, N.; Das, K. Estimation of erosion susceptibility and sediment yield in ephemeral channel using RUSLE and SDR model: Tropical Plateau Fringe Region, India. In *Gully Erosion Studies from India and Surrounding Regions*; Springer: Berlin/Heidelberg, Germany, 2019; pp. 163–185. [CrossRef]

83. de Vente, J.; Poesen, J.; Verstraeten, G.; Van Rompaey, A.; Govers, G. Spatially distributed modelling of soil erosion and sediment yield at regional scales in Spain. *Glob. Planet. Chang.* **2008**, *60*, 393–415. [CrossRef]

84. Kamaludin, H.; Lihan, T.; Ali Rahman, Z.; Mustapha, M.; Idris, W.; Rahim, S. Integration of remote sensing, RUSLE and GIS to model potential soil loss and sediment yield (SY). *Hydrol. Earth Syst. Sci. Discuss.* **2013**, *10*, 4567–4596.

85. Cerdan, O.; Govers, G.; Le Bissonnais, Y.; Van Oost, K.; Poesen, J.; Saby, N.; Gobin, A.; Vacca, A.; Quinton, J.; Auerswald, K.; et al. Rates and spatial variations of soil erosion in Europe: A study based on erosion plot data. *Geomorphology* **2010**, *122*, 167–177. [CrossRef]

86. Ali, S.; Cheema, M.J.M.; Waqas, M.M.; Waseem, M.; Leta, M.K.; Qamar, M.U.; Awan, U.K.; Bilal, M.; Rahman, M.H.U. Flood Mitigation in the Transboundary Chenab River Basin: A Basin-Wise Approach from Flood Forecasting to Management. *Remote Sens.* **2021**, *13*, 3916. [CrossRef]

87. Safari, Z.; Rahimi, S.; Ahmed, K.; Sharafati, A.; Ziarh, G.; Shahid, S.; Ismail, T.; Al-Ansari, N.; Chung, E.-S.; Wang, X. Estimation of Spatial and Seasonal Variability of Soil Erosion in a Cold Arid River Basin in Hindu Kush Mountainous Region Using Remote Sensing. *Sustainability* **2021**, *13*, 1549. [CrossRef]

88. Ali, S.; Cheema, M.; Waqas, M.; Waseem, M.; Awan, U.; Khaliq, T. Changes in Snow Cover Dynamics over the Indus Basin: Evidences from 2008 to 2018 MODIS NDSI Trends Analysis. *Remote. Sens.* **2020**, *12*, 2782. [CrossRef]
89. Ali, S.; Cheema, M.J.M.; Bakhsh, A.; Khaliq, T. Near Real Time Flood Forecasting in the Transboundary Chenab River Using Global Satellite Mapping of Precipitation. *Pak. J. Agric. Sci.* **2020**, *57*, 1327–1335.
90. Williams, J. Sediment routing for agricultural watersheds. *JAWRA J. Am. Water Resour. Assoc.* **1975**, *11*, 965–974. [CrossRef]
91. Evans, M.; Lindsay, J. High resolution quantification of gully erosion in upland peatlands at the landscape scale. *Earth Surf. Process. Landf.* **2010**, *35*, 876–886. [CrossRef]
92. Onnen, N.; Heckrath, G.; Stevens, A.; Olsen, P.; Greve, M.B.; Pullens, J.W.; Kronvang, B.; Van Oost, K. Distributed water erosion modelling at fine spatial resolution across Denmark. *Geomorphology* **2019**, *342*, 150–162. [CrossRef]
93. Ismail, J.; Ravichandran, S. RUSLE2 Model Application for Soil Erosion Assessment Using Remote Sensing and GIS. *Water Resour. Manag.* **2007**, *22*, 83–102. [CrossRef]
94. Lu, D.; Li, G.; Valladares, G.S.; Batistella, M. Mapping soil erosion risk in Rondonia, Brazilian Amazonia: Using RUSLE, remote sensing and GIS. *Land Degrad. Dev.* **2004**, *15*, 499–512. [CrossRef]
95. Aiello, A.; Adamo, M.; Canora, F. Remote sensing and GIS to asess soil erosion with RUSLE3D and USPED at river basin scale in southern Italy. *Catena* **2015**, *131*, 174–185. [CrossRef]
96. Alkharabsheh, M.; Alexandridis, T.; Bilas, G.; Misopolinos, N.; Silleos, N. Impact of Land Cover Change on Soil Erosion Hazard in Northern Jordan Using Remote Sensing and GIS. *Procedia Environ. Sci.* **2013**, *19*, 912–921. [CrossRef]
97. Kouli, M.; Soupios, P.; Vallianatos, F. Soil erosion prediction using the Revised Universal Soil Loss Equation (RUSLE) in a GIS framework, Chania, Northwestern Crete, Greece. *Environ. Geol.* **2009**, *57*, 483–497. [CrossRef]
98. Rao, S.V.N.; Rao, M.V.; Ramasastri, K.S.; Singh, R.N.P. A Study of Sedimentation in Chenab Basin in Western Himalayas. *Hydrol. Res.* **1997**, *28*, 201–216. [CrossRef]
99. Amin, M.; Romshoo, S.A. Comparative assessment of soil erosion modelling approaches in a Himalayan watershed. *Model. Earth Syst. Environ.* **2018**, *5*, 175–192. [CrossRef]
100. Romshoo, S.A.; Altaf, S.; Amin, M.; Ameen, U. Sediment yield estimation for developing soil conservation strategies in GIS environment for the mountainous Marusudar catchment, Chenab basin, J&K, India. *J. Himal. Ecol. Sustain. Dev.* **2017**, *12*, 16–32.
101. Singh, P.; Ramasastri, K.S.; Kumar, N. Topographical Influence on Precipitation Distribution in Different Ranges of Western Himalayas. *Hydrol. Res.* **1995**, *26*, 259–284. [CrossRef]
102. Leta, M.; Demissie, T.; Tränckner, J. Modeling and Prediction of Land Use Land Cover Change Dynamics Based on Land Change Modeler (LCM) in Nashe Watershed, Upper Blue Nile Basin, Ethiopia. *Sustainability* **2021**, *13*, 3740. [CrossRef]
103. Singh, S.K.; Srivastava, P.K.; Gupta, M.; Thakur, J.K.; Mukherjee, S. Appraisal of land use/land cover of mangrove forest ecosystem using support vector machine. *Environ. Earth Sci.* **2013**, *71*, 2245–2255. [CrossRef]
104. Koneti, S.; Sunkara, S.L.; Roy, P.S. Hydrological Modeling with Respect to Impact of Land-Use and Land-Cover Change on the Runoff Dynamics in Godavari River Basin Using the HEC-HMS Model. *ISPRS Int. J. Geo-Inf.* **2018**, *7*, 206. [CrossRef]
105. Dwarakish, G.; Ganasri, B. Impact of land use change on hydrological systems: A review of current modeling approaches. *Cogent Geosci.* **2015**, *1*, 1115691. [CrossRef]
106. Leta, M.K.; Demissie, T.A.; Tränckner, J. Hydrological Responses of Watershed to Historical and Future Land Use Land Cover Change Dynamics of Nashe Watershed, Ethiopia. *Water* **2021**, *13*, 2372. [CrossRef]
107. ESA (European Space Agency). Land Cover CCI Product User Guide Version 2. Tech. Rep. 2017. Available online: Maps.elie.ucl. ac.be/CCI/viewer/download/ESACCI-LC-Ph2-PUGv2_2.0.pdf.2017 (accessed on 19 November 2008).
108. Srivastava, P.K.; Han, D.; Rico-Ramirez, M.A.; O'Neill, P.; Islam, T.; Gupta, M. Assessment of SMOS soil moisture retrieval parameters using tau–omega algorithms for soil moisture deficit estimation. *J. Hydrol.* **2014**, *519*, 574–587. [CrossRef]
109. Yang, D.; Gao, B.; Jiao, Y.; Lei, H.; Zhang, Y.; Yang, H.; Cong, Z. A distributed scheme developed for eco-hydrological modeling in the upper Heihe River. *Sci. China Earth Sci.* **2015**, *58*, 36–45. [CrossRef]
110. Renard, K.; Foster, G.; Weesies, G.; McCool, D.; Yoder, D. Predicting soil erosion by water: A guide to conservation planning with the Revised Universal Soil Loss Equation (RUSLE). *Agric. Handb.* **1996**, *703*, 25–28.
111. Morgan, R.P.C.; Quinton, J.; Rickson, R.J. Modelling Methodology for Soil Erosion Assessment and Soil Conservation Design: The EUROSEM Approach. *Outlook Agric.* **1994**, *23*, 5–9. [CrossRef]
112. Jung, P.-K.; Ko, M.-H.; Im, J.-N.; Um, K.-T.; Choi, D.-U. Rainfall erosion factor for estimating soil loss. *Korean J. Soil Sci. Fertil.* **1983**, *16*, 112–118.
113. Sharpley, A.N.; Williams, J.R. *EPIC. Erosion/Productivity Impact Calculator: 1. Model Documentation. 2. User Manual*; EPIC: Cary, NC, USA, 1990.
114. Beskow, S.; de Mello, C.; Norton, L.; Curi, N.; Viola, M.R.; Avanzi, J. Soil erosion prediction in the Grande River Basin, Brazil using distributed modeling. *Catena* **2009**, *79*, 49–59. [CrossRef]
115. Pal, S.C.; Shit, M. Application of RUSLE model for soil loss estimation of Jaipanda watershed, West Bengal. *Spat. Inf. Res.* **2017**, *25*, 399–409. [CrossRef]
116. Pandey, A.; Chowdary, V.M.; Mal, B.C. Identification of critical erosion prone areas in the small agricultural watershed using USLE, GIS and remote sensing. *Water Resour. Manag.* **2007**, *21*, 729–746. [CrossRef]
117. Karaburun, A. Estimation of C factor for soil erosion modeling using NDVI in Buyukcekmece watershed. *Ozean J. Appl. Sci.* **2010**, *3*, 77–85.

118. Kayet, N.; Pathak, K.; Chakrabarty, A.; Sahoo, S. Urban heat island explored by co-relationship between land surface temperature vs multiple vegetation indices. *Spat. Inf. Res.* **2016**, *24*, 515–529. [CrossRef]

119. Julien, P.Y. *Erosion and Sedimentation*; Cambridge University Press: Cambridge, UK, 2010.

120. Van Rompaey, A.J.J.; Verstraeten, G.; Van Oost, K.; Govers, G.; Poesen, J. Modelling mean annual sediment yield using a distributed approach. *Earth Surf. Process. Landf.* **2001**, *26*, 1221–1236. [CrossRef]

121. USDA, S. *National Engineering Handbook*; Section 4: Hydrology; U.S. Department of Agriculture: Washington, DC, USA, 1972.

122. Klaghofer, E.; Summer, W.; Villeneuve, J. Some remarks on the determination of the sediment delivery ratio. *Int. Assoc. Hydrol. Sci. Publ.* **1992**, *209*, 113–118.

123. Roehl, J. *Sediment Source Areas, Delivery Ratios and Influencing Morphological Factors*; Internation Association for Scientific Hydrology Commission of Land Erosion: London, UK, 1962.

124. Vanoni, V.A. Sediment Deposition Engineering. In *ASCE Manuals and Reports on Engineering Practices, No. 54*; American Society of Civil Engineers: Reston, VA, USA, 1975.

125. Bazzoffi, P.; Baldassarre, G.; Vacca, S. Validation of PISA2 model for automatic assessment of reservoir sedimentation. In *Proceedings of the International Conference on Reservoir Sedimentation*; Albertson, M., Ed.; Colorado State University: Fort Collins, CO, USA, 1996; pp. 519–528.

126. Williams, J.R.; Berndt, H.D. Sediment Yield Prediction Based on Watershed Hydrology. *Trans. ASAE* **1977**, *20*, 1100–1104. [CrossRef]

127. Panagos, P.; Borrelli, P.; Poesen, J.; Meusburger, K.; Ballabio, C.; Lugato, E.; Montanarella, L.; Alewell, C. Reply to "The new assessment of soil loss by water erosion in Europe. Panagos P. et al., 2015 Environ. Sci. Policy 54, 438–447—A response" by Evans and Boardman [Environ. Sci. Policy 58, 11–15]. *Environ. Sci. Policy* **2016**, *59*, 53–57. [CrossRef]

128. Shin, G. The Analysis of Soil Erosion Analysis in Watershed using GIS. Ph. D. Thesis, Department of Civil Engineering, Gang-won National University, Gangwon-do, Korea, 1999.

129. Singh, G.; Babu, R.; Narain, P.; Bhushan, L.; Abrol, I. Soil erosion rates in India. *J. Soil Water Conserv.* **1992**, *47*, 97–99.

130. Maqsoom, A.; Aslam, B.; Hassan, U.; Kazmi, Z.A.; Sodangi, M.; Tufail, R.F.; Farooq, D. Geospatial Assessment of Soil Erosion Intensity and Sediment Yield Using the Revised Universal Soil Loss Equation (RUSLE) Model. *ISPRS Int. J. Geo-Inf.* **2020**, *9*, 356. [CrossRef]

131. Walling, D. Erosion and sediment yield research—Some recent perspectives. *J. Hydrol.* **1988**, *100*, 113–141. [CrossRef]

132. Morgan, R.P.C. *Soil Erosion and Conservation*; John Wiley & Sons: Hoboken, NJ, USA, 2009.

133. Bai, L.; Shi, C.; Li, L.; Yang, Y.; Wu, J. Accuracy of CHIRPS Satellite-Rainfall Products over Mainland China. *Remote Sens.* **2018**, *10*, 362. [CrossRef]

134. Haan, C.T.; Barfield, B.J.; Hayes, J.C. *Design Hydrology and Sedimentology for Small Catchments*; Elsevier: Amsterdam, The Netherlands, 1994.

135. Ullah, S.; Ali, A.; Iqbal, M.; Javid, M.; Imran, M. Geospatial assessment of soil erosion intensity and sediment yield: A case study of Potohar Region, Pakistan. *Environ. Earth Sci.* **2018**, *77*, 705. [CrossRef]

136. Gajbhiye, S.; Mishra, S.K.; Pandey, A. Simplified sediment yield index model incorporating parameter curve number. *Arab. J. Geosci.* **2014**, *8*, 1993–2004. [CrossRef]

137. Mutua, B.M.; Klik, A. Estimating spatial sediment delivery ratio on a large rural catchment. *J. Spat. Hydrol.* **2006**, *6*, 1.

138. Lee, S.E.; Kang, S.H. Geographic information system-coupling sediment delivery distributed modeling based on observed data. *Water Sci. Technol.* **2014**, *70*, 495–501. [CrossRef] [PubMed]

139. Lu, H.; Moran, C.; Prosser, I.P. Modelling sediment delivery ratio over the Murray Darling Basin. *Environ. Model. Softw.* **2006**, *21*, 1297–1308. [CrossRef]

140. Diodato, N.; Grauso, S. An improved correlation model for sediment delivery ratio assessment. *Environ. Earth Sci.* **2009**, *59*, 223–231. [CrossRef]

141. Lu, H.; Moran, C.; Sivapalan, M. A theoretical exploration of catchment-scale sediment delivery. *Water Resour. Res.* **2005**, *41*. [CrossRef]

142. Heckmann, T.; Vericat, D. Computing spatially distributed sediment delivery ratios: Inferring functional sediment connectivity from repeat high-resolution digital elevation models. *Earth Surf. Process. Landf.* **2018**, *43*, 1547–1554. [CrossRef]

143. Park, Y.; Kim, J.; Kim, N.; Kim, K.-S.; Choi, J.; Lim, K.J. Analysis of sediment yields at watershed scale using area/slope-based sediment delivery ratio in SATEEC. *J. Korean Soc. Water Environ.* **2007**, *23*, 650–658.

144. Pelletier, J.D. A spatially distributed model for the long-term suspended sediment discharge and delivery ratio of drainage basins. *J. Geophys. Res. Earth Surf.* **2012**, *117*. [CrossRef]

145. Ferro, V.; Minacapilli, M. Sediment delivery processes at basin scale. *Hydrol. Sci. J.* **1995**, *40*, 703–717. [CrossRef]

146. Brasington, J.; Richards, K. Interactions between model predictions, parameters and DTM scales for topmodel. *Comput. Geosci.* **1998**, *24*, 299–314. [CrossRef]

147. Gao, J. Impact of sampling intervals on the reliability of topographic variables mapped from grid DEMs at a micro-scale. *Int. J. Geogr. Inf. Sci.* **1998**, *12*, 875–890. [CrossRef]

148. Zhang, X.; Drake, N.A.; Wainwright, J.; Mulligan, M. Comparison of slope estimates from low resolution DEMs: Scaling issues and a fractal method for their solution. *Earth Surf. Process. Landf.* **1999**, *24*, 763–779. [CrossRef]

149. Rajbanshi, J.; Bhattacharya, S. Assessment of soil erosion, sediment yield and basin specific controlling factors using RUSLE-SDR and PLSR approach in Konar river basin, India. *J. Hydrol.* **2020**, *587*, 124935. [CrossRef]
150. Gelagay, H.S. RUSLE and SDR Model Based Sediment Yield Assessment in a GIS and Remote Sensing Environment; A Case Study of Koga Watershed, Upper Blue Nile Basin, Ethiopia. *Gelagay Hydrol. Curr. Res.* **2016**, *7*, 2. [CrossRef]
151. Thomas, J.; Joseph, S.; Thrivikramji, K.P. Assessment of soil erosion in a monsoon-dominated mountain river basin in India using RUSLE-SDR and AHP. *Hydrol. Sci. J.* **2018**, *63*, 542–560. [CrossRef]
152. Behera, M.; Sena, D.R.; Mandal, U.; Kashyap, P.S.; Dash, S.S. Integrated GIS-based RUSLE approach for quantification of potential soil erosion under future climate change scenarios. *Environ. Monit. Assess.* **2020**, *192*, 1–18. [CrossRef]
153. Swarnkar, S.; Malini, A.; Tripathi, S.; Sinha, R. Assessment of uncertainties in soil erosion and sediment yield estimates at ungauged basins: An application to the Garra River basin, India. *Hydrol. Earth Syst. Sci.* **2018**, *22*, 2471–2485. [CrossRef]
154. Gashaw, T.; Tulu, T.; Argaw, M.; Worqlul, A.W. Modeling the impacts of land use–land cover changes on soil erosion and sediment yield in the Andassa watershed, upper Blue Nile basin, Ethiopia. *Environ. Earth Sci.* **2019**, *78*, 1–22. [CrossRef]
155. Anees, M.T.; Abdullah, K.; Nawawi, M.N.M.; Norulaini, N.A.N.; Syakir, M.I.; Omar, A.K.M. Soil erosion analysis by RUSLE and sediment yield models using remote sensing and GIS in Kelantan state, Peninsular Malaysia. *Soil Res.* **2018**, *56*, 356. [CrossRef]
156. Fistikoglu, O.; Harmancioglu, N.B. Integration of GIS with USLE in Assessment of Soil Erosion. *Water Resour. Manag.* **2002**, *16*, 447–467. [CrossRef]
157. Nasir, A.; Uchida, K.; Ashraf, M. Estimation of soil erosion by using RUSLE and GIS for small mountainous watersheds in Pakistan. *Pak. J. Water Resour.* **2008**, *10*, 11–21.

Article

Sediment Influx and Its Drivers in Farmers' Managed Irrigation Schemes in Ethiopia

Zerihun Anbesa Gurmu [1,2,*], Henk Ritzema [2], Charlotte de Fraiture [3], Michel Riksen [4] and Mekonen Ayana [5]

1 Faculty of Water Resources and Irrigation Engineering, Arba Minch University, P.O. Box 21, Arba Minch 4400, Ethiopia

2 Water Resources Management Group, Wageningen University, Droevendaalsesteeg 3a, 6708 PB Wageningen, The Netherlands; henk.ritzema@wur.nl

3 Department of Water Science and Engineering, IHE Delft Institute for Water Education, Westvest 7, 2611 AX Delft, The Netherlands; c.defraiture@un-ihe.org

4 Soil Physics and Land Management Group, Wageningen University, Droevendaalsesteeg 3a, 6708 PB Wageningen, The Netherlands; michel.riksen@wur.nl

5 Department of Water Resources Engineering, Adama Science and Technology University, P.O. Box 1888, Adama 2118, Ethiopia; mekonen.ayana24@gmail.com

* Correspondence: zerihun.gurmu@wur.nl

Abstract: Excessive soil erosion hampers the functioning of many irrigation schemes throughout sub-Saharan Africa, increasing management difficulties and operation and maintenance costs. River water is often considered the main source of sedimentation, while overland sediment inflow is overlooked. From 2016 to 2018, participatory research was conducted to assess sediment influx in two irrigation schemes in Ethiopia. Sediment influx was simulated using the revised universal soil loss equation (RUSLE) and compared to the amount of sediment removed during desilting campaigns. The sediment deposition rate was 308 m^3/km and 1087 m^3/km, respectively, for the Arata-Chufa and Ketar schemes. Spatial soil losses amounts to up to 18 t/ha/yr for the Arata-Chufa scheme and 41 t/ha/yr for the Ketar scheme. Overland sediment inflow contribution was significantly high in the Ketar scheme accounting for 77% of the deposited sediment, while only 4% of the sedimentation at the Arata-Chufa scheme came from overland flow. Feeder canal length and the absence of canal banks increased the sedimentation rate, however, this was overlooked by the stakeholders. We conclude that overland sediment inflow is an often neglected component of canal sedimentation, and this is a major cause of excessive sedimentation and management problems in numerous irrigation schemes in sub-Saharan Africa.

Keywords: irrigation; sediment; overland flow; soil loss

Citation: Gurmu, Z.A.; Ritzema, H.; Fraiture, C.d.; Riksen, M.; Ayana, M. Sediment Influx and Its Drivers in Farmers' Managed Irrigation Schemes in Ethiopia. *Water* **2021**, *13*, 1747. https://doi.org/10.3390/w13131747

Academic Editor: Achim A. Beylich

Received: 30 April 2021
Accepted: 20 June 2021
Published: 24 June 2021

Publisher's Note: MDPI stays neutral with regard to jurisdictional claims in published maps and institutional affiliations.

1. Introduction

Excessive sediment influx hampers the function of many water resource systems and irrigation infrastructures in sub-Saharan Africa, causing storage capacity reductions, opportunity costs and safety hazards [1–7]. The impact of excessive sedimentation is especially high in countries such as Ethiopia, where overland soil erosion is severe and limited resources are available to address the problem [8–10]. Soil erosion is a major factor limiting agriculture due to the loss of fertile topsoil. It has a prolonged effect on the agricultural sector as the rate of soil loss exceeds the soil formation rate [11].

Soil erosion also affects the overall performance of irrigation schemes. Due to excessive sedimentation, many irrigation schemes have been abandoned or operate far below full capacity [12,13]. In Ethiopia, most irrigation systems are the river diversion type. However, the country's rivers carry huge sediment loads, and therefore are a major source of sedimentation. Although soil erosion from the upland catchment is the ultimate source of sedimentation in many irrigation schemes, the specific source of sedimentation varies with the mechanism through which the sediment enters an irrigation scheme. An irrigation

229

scheme can be threatened by sediment that comes from a river and an overland flow. River sediment enters an irrigation scheme via an intake structure. For example, Gurmu et al. [14] found that river sediment contributed more than 95% of the total sediment deposition in the studied irrigation schemes. Nonetheless, overland erosion flow can also contribute large quantities of sediment. The overland sediment inflow from onsite soil erosion of the catchment area after the intake structure (upland of the main canal) happens when the generated soil loss joins the canal after the intake structures. In some schemes, overland flow is the only source of sedimentation. The Bebeks irrigation scheme, for instance, is threatened only by overland sediment inflow [15]. The scheme is irrigated by entirely sediment-free spring water, nonetheless it performs far below capacity, mainly due to the sediment that entirely comes from an overland flow.

While many stakeholders recognize upstream erosion as a major driver of sedimentation in irrigation canals, most focus on erosion occurring upstream of the intake [16]. However, much of the overland sediment inflow emanates from the catchment upland of the main canal of the scheme itself. Moreover, deposition from overland flow is typically concentrated in the main canals, as secondary and tertiary canals tend to be built at higher elevations relative to field plots, with canals laid along the contour.

A lack of resources for operation and maintenance aggravates problems of excessive sedimentation [17], as the physical infrastructure of many schemes is deteriorated. In farmer-led schemes, farmers apply tacit knowledge to temporarily reduce the quantity of sediment entering their irrigation schemes, for example, by delaying water abstraction when river sediment loads are particularly heavy [16] and diverting surface runoff to prevent it from entering the canal (Figure 1). To clear excessive sedimentation, they organize seasonal or annual desilting campaigns, which are labor-intensive and require participation of many farmers over several days. For example, in one irrigation scheme serving 430 ha with a main canal length of 12 km, some 3118 labor days were required per campaign to remove the accumulated sediment [16]. Of the total time required for crop cultivation, farmers were found to invest one-fourth of their time in sediment management activities [16]. However, even with this management, farmers have been unable to adequately and sustainably deal with problems of excessive sedimentation.

Figure 1. Farmers at the Ketar irrigation scheme diverting surface runoff to prevent sediment from entering the main canal, 25 August 2018.

Sustainable sedimentation management requires identification of sedimentation sources and quantification of their respective contributions. Yet, most studies on sediment transport in irrigation schemes deal mainly with river sediment. Despite taking a greater share of overall sedimentation quantity in the irrigation schemes, little is known about the contribution of overland erosion flow to sedimentation problems. Therefore, in the current research we quantified soil loss and sediment yield and compared it with the sedimentation volume measured in two small-scale irrigation schemes in the Great Rift Valley Basin of Ethiopia—one of the River Basins in the country that exhibit severe soil losses.

2. Materials and Methods

2.1. Location of the Study

Two representative small-scale irrigation schemes, namely Arata-Chufa and Ketar from Ethiopia, were selected for the study. Both are gravity type river diversion schemes and both are affected by river and overland sediment inflow. Furthermore, both schemes are operated and maintained by farmers, and were in proper use at the time of the research. Farmers devote time and labor to keep the schemes in working order, despite problems of excessive sediment load and deposition. However, both schemes have differences in the sources and quantity of sedimentation, command area size, type and layout and management structure. Figure 2 presents the location of the two schemes in the Great Rift Valley Basin of Central Ethiopia, on the lower reach of the Ketar River, a few kilometers before it joins Lake Ziway. Geographically, Arata-Chufa is located at 7°59′ N and 39°02′ E with an average elevation of 1740 m above mean sea level. Ketar was located at 7°49′ N and 39°02′ E at a mean elevation of 2294 m above mean sea level. The Arata-Chufa scheme covers 100 ha and serves 324 beneficiaries. The Ketar scheme covers 430 ha and serves 1074 beneficiaries.

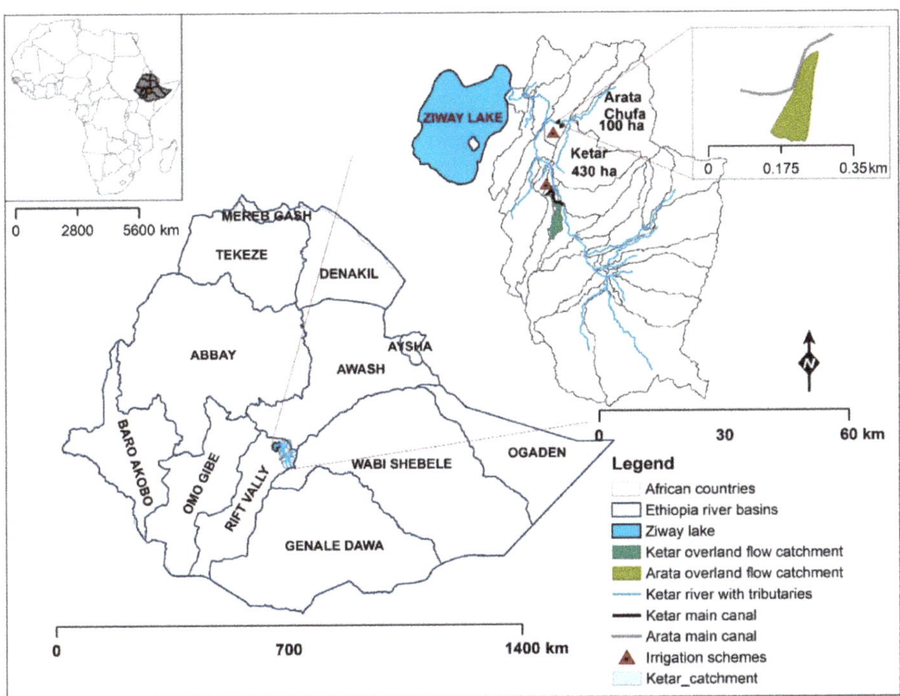

Figure 2. Location of the Arata-Chufa and Ketar irrigation schemes and the catchments contributing overland sediment inflow.

2.2. Field Data Collection

Field data collection began with an inventory of the schemes, to get acquainted with the canal layout and to understand local conditions, sediment hotspots and canal desilting periods. Farmers reported that desilting campaigns took two to three weeks, with the work conducted only on two to three days in each of those weeks. Canal cleaning and repair activities were undertaken at the end of the rainy season, before the start of the new irrigation season. The summer (wet) season usually ceases in late August. Sediment cleaning activities started in the last week of August and were completed in early September. On average, sediment cleaning took 3 days at Arata-Chufa and 5.5 days at Ketar.

We measured the volume of sediment deposited in the canal and removed by the farmers in two years: 2017 and 2018. The volume of sediment removed in the year before the fieldwork, 2016, was estimated based on the flood marks on the sides of the canal with the participation of farmers. Most canal sections were lined with concrete, which meant that canal cross-sections were relatively uniform. For unlined canal sections, irregularities in canal depth, width and shape were considered in measuring and calculating sediment volumes. Canal transition and culvert sections were measured separately.

2.3. Soil Erosion Modeling

There are many empirical models for predicting soil losses and the corresponding sediment yields. However, their scope of application is limited, as they were developed using site-specific empirical data [18,19]. To deal with this shortcoming, numerical and physically based distributed models have been developed. These, however, require large amounts of input data for calibration and simulation [20] and show limited accuracy in data-scarce conditions [19]. Recent advancements in GIS and remote sensing have enabled empirical models to predict soil erosion cell by cell.

Since our study area is characterized by data scarcity, we modeled soil erosion using the revised universal soil loss equation (RUSLE) developed by Wishmeier and Smith [21] coupled with GIS and remote sensing. Due to its simplicity, RUSLE has been widely applied globally and proven to be of value in the Ethiopian highlands [19,22]. Figure 3 presents our conceptual framework, in which RUSLE was used to identify the main upland sediment sources and to quantify soil loss and sediment yield in the main canals of the schemes under investigation from overland flow sources.

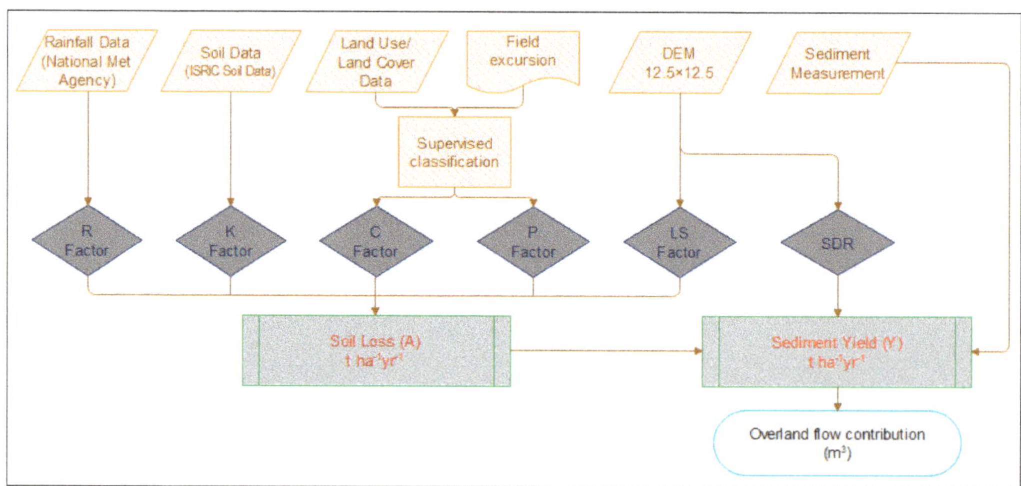

Figure 3. Conceptual framework for quantifying soil loss and sediment yield in the canals of the irrigation schemes from overland flow sources.

Usually, the irrigation schemes are closed during the wet season (June to August) and irrigation is resumed after the farmers cleaned their scheme. During the wet season, the sediment enters the canal from the onsite soil erosion of the catchment area upland of the main canal. The volume of the sediment removed by the farmers incorporated both river and overland sediment inflow. We compared the sediment yield computed by RUSLE to the volume of sediment removed by farmers from the canals in their desilting campaigns to estimate the relative contribution of overland sediment inflow to total sediment deposition in the schemes. We conducted transect walks and participatory erosion mapping to identify erosion hotspots and major gully formations. Note that although the RUSLE model is limited in predicting gully erosion, major gully formations were absent in the study area.

Empirically RUSLE is expressed as follows:

$$A = R \times K \times LS \times C \times P \tag{1}$$

where

A is the mean annual soil loss (t/ha/yr),
R is the rainfall erosivity factor (MJ mm/ha h yr),
K is the soil erodibility factor (t ha h /ha MJ mm),
LS is the slope length and steepness factor (dimensionless),
C is the land cover and management factor (dimensionless, ranges from zero to one),
P is the support practices factor (dimensionless, ranges from zero to one).

A 12.5 m × 12.5 m digital elevation model (DEM) was used to delineate the catchment contributing overland sediment flow to the canals. First, a larger catchment was delineated taking outlet points in the river a bit downstream to the schemes. Then, many sub-catchments were redelineated considering numerous outlet points in the main canal and the subcatchments were merged together. Using this method, the catchment contributing overland sediment flow to the Arata-Chufa scheme was delineated as 1.14 ha and it was delineated as 1082 ha for the Ketar scheme.

2.3.1. Rainfall Erosivity

Rainfall erosivity (the R factor) measures the ability of the impact of a raindrop to detach a soil particle. It is determined based on rainfall kinetic energy and 30-min rainfall intensity records. However, such rainfall measurements were hardly available for the study area. We thus estimated the R factor, following Hurni [23], based on the mean annual precipitation as follows:

$$R = 0.562 \times P - 8.12 \tag{2}$$

where P is the mean annual rainfall.

For the Arata-Chufa scheme, we used mean annual precipitation for 1987–2017 from Arata station records (Figure 4). For the Ketar scheme, nine meteorological stations were nearby. Rainfall interpolation mapping indicated that only the Ketar-Genet station was sufficiently representative of the rainfall characteristics of the catchment of interest. We therefore computed the rainfall erosivity factor using the mean annual precipitation data from the Ketar-Genet station for 1978–2014 (Figure 4).

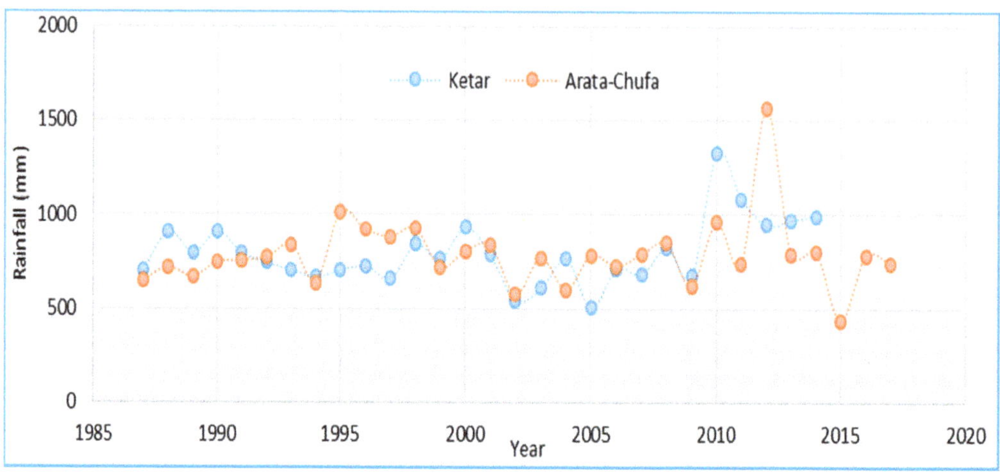

Figure 4. Mean annual rainfall for the Ketar-Genet station (7°82′ N and 39°029′ E, altitude 2314 m) and Arata station (7°83′ N and 39°1′ E, altitude 2400 m).

2.3.2. Soil Erodibility

Soil erodibility (the K factor) represents the resistivity of soil particles to the impact of a raindrop. K is determined based on soil physical and chemical properties, such as the percentage of silt, clay and sand, organic carbon content and soil structure and permeability [19]. Data scarcity was again an obstacle in the study area. Previous authors [19,24] estimated K values based on observed soil color, as suggested by Hurni [23]. Williams [25] estimated K as a function of the percentage of silt, clay and sand and the organic carbon content of the topsoil. We explored different soil databases, including those of the Ethiopian Ministry of Water, Irrigation and Energy, the Ministry of Agriculture and Natural Resources and the Food and Agricultural Organization of the United Nations (FAO). Ultimately, we used data from the International Soil Reference and Information Centre (ISRIC), as it had better resolution (1 km × 1 km) than the other sources. The following function was used to generate a K factor raster map for the catchments:

$$K = f_{csand} \times f_{cl-si} \times f_{org} \times f_{hisand} \tag{3}$$

where

f_{csand} is the function of coarse sand content,
f_{cl-si} is the function of the clay-to-silt ratio,
f_{org} is the function of the organic carbon content,
f_{hisand} is the function for high sand content.

Raster files for the above functions were processed in ArcGIS, using the data retrieved from the ISRIC soil database (Figure 5), as follows:

$$f_{csand} = \left[0.2 + 0.3 \times \left(-0.256 \times m_s \times \left(1 - \frac{m_{silt}}{100} \right) \right) \right] \tag{4}$$

$$f_{cl-si} = \left[\frac{m_{silt}}{m_c + m_{silt}} \right]^{0.3} \tag{5}$$

$$f_{org} = \left[1 - \frac{0.25 \times orgC}{orgC + exp(3.72 - 2.95 \times orgC)} \right] \tag{6}$$

$$f_{hisand} = \left[1 - \frac{0.7 \times \left(1 - \frac{m_s}{100}\right)}{\left(1 - \frac{m_s}{100}\right) + exp\left\langle -5.51 + 22.9 \times \left(1 - \frac{m_s}{100}\right)\right\rangle}\right] \tag{7}$$

where

m_s is the sand content (%),
m_{silt} is the silt content (%),
m_c is the clay content (%),
$orgC$ is the organic carbon content (%).

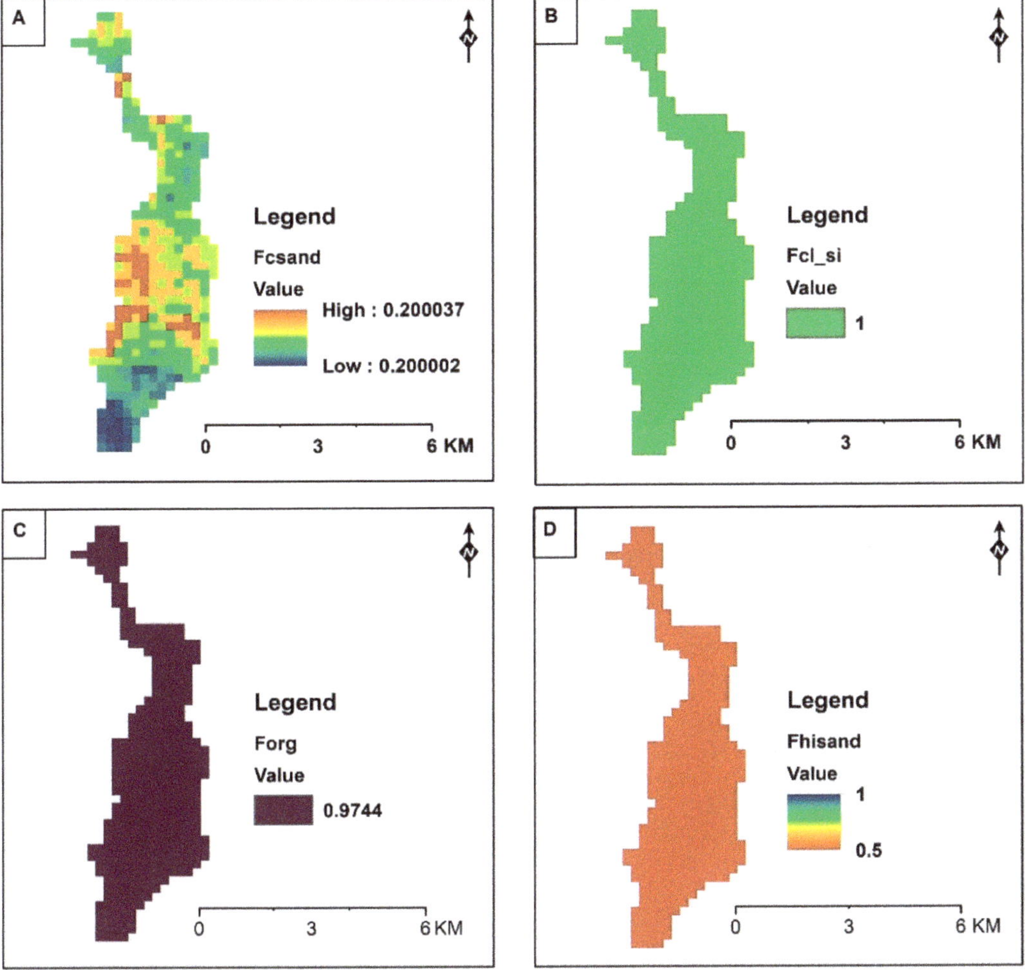

Figure 5. Physical and chemical properties of the soil in the study area. (**A**): Coarse sand content; (**B**): clay-to-silt ratio; (**C**): organic carbon content; (**D**): high sand content. Source: Data from the International Soil Reference and Information Centre (ISRIC). Available online at: https://data.isric.org/geonetwork/srv/eng/catalog.search#/home (accessed on 3 October 2019).

2.3.3. Slope Length and Steepness

Slope length and steepness (the LS factor) represents the rate of soil loss per unit area of land from a field of length 22.13 m and a uniform 9% slope steepness [21]. LS is thus a topographic factor that reflects the sediment transport capacity of surface runoff [26]. The slope length (L) is the distance from the beginning of surface runoff to a point where either a change in slope occurs or the flow concentrates in depressions [21]. The approach initially introduced by Wischmeier and Smith [21] to estimate LS did not fully account for the effects of uphill slope and vegetation cover [27,28]. Compared to the other erosion parameters, estimation of LS is more controversial for catchments with complex topography [27,28]. This is because downhill erosion is determined not only by the erosive power of rainfall and the erodibility of a particular soil, but also by upslope flow accumulation due to uphill topography and land use types and vegetation cover [27–29].

To calculate slope length (L) of a complex, three-dimensional terrain, many studies (e.g., [19,24,26–28,30,31]) adopt a grid-based approach based on the upslope contributing area. The current study used such an approach, as follows:

$$LS = \left(\frac{A_s}{22.13} \right)^m \left(\frac{\sin \beta}{0.0896} \right)^n \tag{8}$$

where A_s is the upslope contributing area and β is the slope angle.

Equation (5) was used in a GIS environment to generate an LS factor map of the area contributing overland runoff flow to the main canals under study. For this purpose, a 12.5 m × 12.5 m DEM was employed to derive the slope angle to compute the topographic factor.

$$LS = \left(\text{Flow accumulation} \times \frac{\text{Cell size}}{22.13} \right)^{0.4} \times \left(\frac{\sin \text{slope}}{0.0896} \right)^{1.3} \tag{9}$$

2.3.4. Land Cover and Management

Land cover and management (the C factor) considers the effect of land cover, soil biomass and farming practices on the rate of soil loss [18,32]. The C factor is the ratio of soil loss with a specific surface cover to the corresponding soil loss from a bare fallow area [19–21]. For this study, we mapped the C factor in conformance with land use and land cover maps obtained from the Ethiopian Ministry of Water, Irrigation and Energy and the Ministry of Agriculture and Natural Resources. As the temporal and spatial scale of these maps did not accurately represent real-time land use and land cover conditions in the study area, we minimized uncertainty in C value determination [33,34] with supplementation of land use and land cover data gathered during the fieldwork. The development of a C factor map was supported by supervised classification of locally collected land use data, following recommendations from different studies. For agricultural land use types, C values were derived based on the type of farming and slope of the area (Table 1).

Table 1. Land cover and management (C factor) and support practices (P factor) values used to compute soil loss with the revised universal soil loss equation (RUSLE).

Land Use/Cover	Description	Slope (%)	C	P	References
Cropland	Areas intensively cultivated to grain crops with contour planting and no soil and water conservation measures	0–7 7–11.3 11.3–17.6 17.6–26.8 >26.8	0.17 0.20 0.30 0.34 0.4	0.65 0.70 0.75 0.80 0.90	
Bare soil	Land surface without vegetation cover		0.4	0.65	
Closed shrub	Mixed shrub and grassland, with 50–70% of land area covered		0.1	0.8	[19,21,32–36]
Open shrub	Mixed shrub and grassland, with fair to good cover		0.12	0.75	
Open grassland	Fair to good grass cover (closed grazing)		0.15	0.7	
Sparse forest	Open forest with grassland, with fair to good cover		0.03	0.85	

2.3.5. Support Practices

Support practices (the P factor) represents the effect of specific land management practices in reducing runoff and resultant soil losses compared to a situation without those practices with upslope or downslope cultivation [19,21]. The P factor accounts for the effect of structural and non-structural erosion control measures on soil loss. Taye et al. [34] established p values for agricultural and range lands with various soil and water conservation measures in Northern Ethiopia. For the current study in Central Ethiopia, we determined p values based on recommendations from the literature (Table 1).

2.4. Sediment Yield

The volume of sediment that ended up in the cross-section of the main canals was computed as a function of the gross soil loss from the catchment contributing surface runoff and the sediment delivery ratio (SDR). Haregeweyn et al. [37], Nyssen et al. [35] and Williams and Berndt [38] developed SDR as a function of catchment physiography, sediment particle size, runoff rate and land use or cover types. The attempt to develop SDR for Ethiopian highlands by Haregeweyn et al. [37] was reportedly unsuccessful. Jain et al. [39] computed SDR based on the relationship between suspended sediment and discharge. In a similar study, Haregeweyn et al. [19], following Nyssen et al. [35], computed SDR based on land use types with or without soil and water conservation practices and they used a SDR of 30% for agricultural land and 25% for non-agricultural land. Bhattarai and Dutta [40] derived SDR from overland flow travel time, which is dependent on the terrain and land cover characteristics.

We used the approach suggested by Williams and Berndt [38], computing the SDR for the study area as follows:

$$SDR = 0.627 \times SLP^{0.403} \tag{10}$$

where SLP is the slope of the main stream channel (‰).

This method has been found to yield reasonable estimates of sediment yield in data-scarce regions [20,41]. As for many empirical equations, this method may not result in an accurate estimate of SDR. Nonetheless, due to limited data availability in the study area, using another option of SDR would still result in the same uncertainty. To minimize the uncertainty, we compared the estimated SDR value computed using this approach with the findings of other studies reported in the country.

We computed the RUSLE factors for the two irrigation schemes under study and used Map Algebra in ArcGIS to quantify the corresponding soil loss and sediment yield. Various statistical analyses were performed to classify the catchment based on soil erosion rates.

3. Results

3.1. Raster Maps of RUSLE Factors

Raster maps depicting the RUSLE parameters were created for each scheme, Arata-Chufa (Figure 6) and Ketar (Figure 7). These maps show the spatial distribution of rainfall erosivity (Figures 6A and 7A), soil erodibility (Figures 6B and 7B), topography (Figures 6C and 7C), land cover and management (Figures 6D and 7D) and support practices (Figures 6E and 7E). Figures 6F and 7F present the land cover map of the catchment used to develop the RUSLE parameters for the Arata-Chufa and the Ketar schemes, respectively.

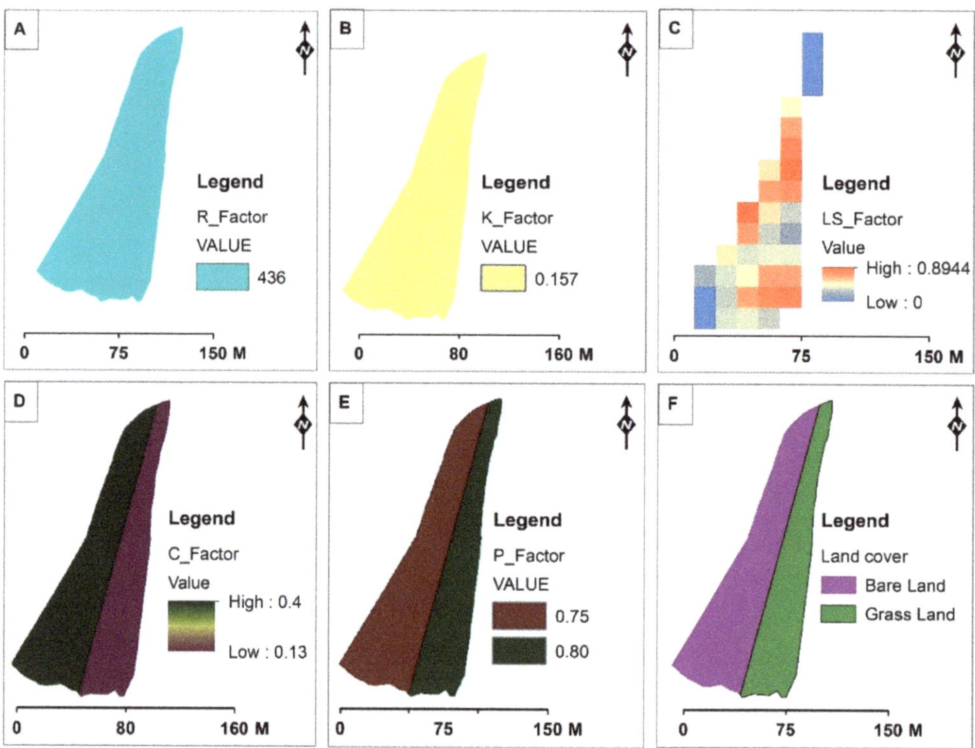

Figure 6. Raster maps of catchment contributing overland sediment inflow to the main canal of the Arata-Chufa scheme. (**A–E**) depict factors of the revised universal soil loss equation. (**A**): Mean annual soil loss (R factor); (**B**): soil erodibility (K factor); (**C**): slope length and steepness (LS factor); (**D**): land cover and management (C factor); (**E**): support practices (P factor). (**F**): Maps land cover in the study area.

R is uniform for the whole catchment as the mean annual precipitation from a single station used to estimate the rainfall erosivity factor. Note that the catchments were quite small, which limits spatial rainfall variability.

Pellic vertisols were the dominant soil types in the study area. These have a soil erodibility (K factor) of about 0.15 for black cotton soil, estimated from the easily identifiable soil color [22]. Using the ISRIC soil database, we estimated the K factor as 0.157 for Arata-Chufa and as 0.195 for Ketar. These values were largely in line with the estimated values based on soil color.

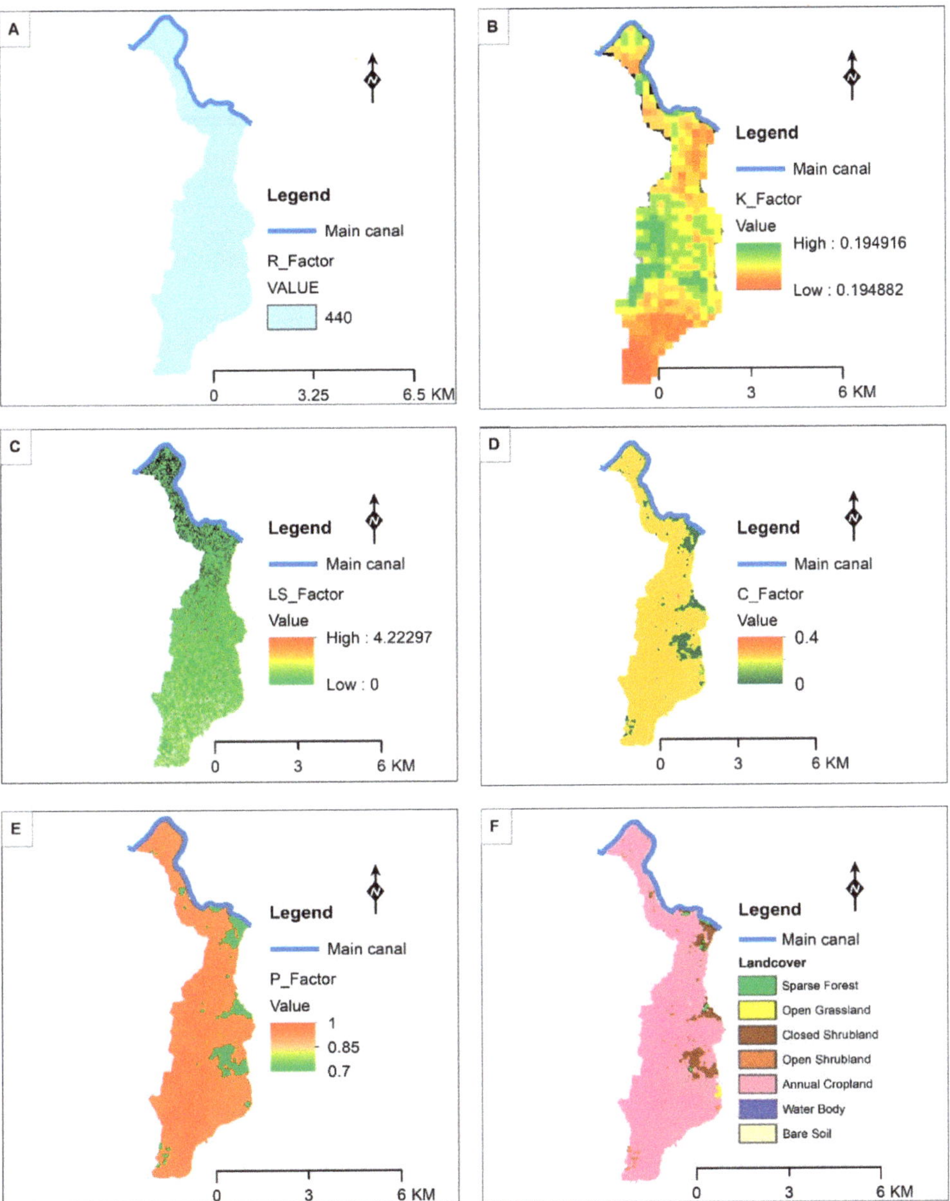

Figure 7. Raster maps of catchment contributing overland sediment inflow to the main canal of the Ketar scheme. (**A–E**) depict factors of the revised universal soil loss equation. (**A**): Mean annual soil loss (R factor); (**B**): soil erodibility (K factor); (**C**): slope length and steepness (LS factor); (**D**): land cover and management (C factor); (**E**): support practices (P factor). (**F**): Maps land cover in the study area.

The complexity of the terrain affects the computation of the LS factor or slope length and steepness. The Arata-Chufa catchment exhibited moderate topographic variability, with elevations ranging from 1725 to 1730 m above mean sea level. The elevation gradient of the Ketar catchment was larger, with elevations ranging from 2258 m above mean sea level close to the main canal to 2488 m above mean sea level at the upstream escarpment of the catchment.

At the Arata-Chufa scheme, sedimentation from surface runoff came mainly from a gravel road that crossed the main canal and an open area of grazing land between the main canal and this gravel road. At the Ketar scheme, various land cover and land use types contributed to the overland sediment flow. Particularly, a rainfed cropland upland of the main canal was the origin of most of the sediment, though there were also mixed grasslands, shrub and open forest in the catchment, with bare areas in between. These characteristics were considered in determining the C factor for the study area. C values ranged from 0.13 to 0.40 for Arata-Chufa and from 0 to 0.4 for Ketar.

No large-scale interventions have been implemented to reduce soil erosion. However, farmers use contour farming and a few have constructed soil bunds at the boundaries of their field plots, particularly at the Ketar irrigation scheme. Moreover, farmers leave biomass on the land after harvesting until the following plowing season. All of these practices help to reduce soil erosion and thus were considered in determining the P factor for the catchments. P values ranged from 0.75 to 0.80 for the Arata-Chufa scheme and from 0.65 to 1.00 for the Ketar scheme.

3.2. Estimation of Soil Loss Rate

The pixel-by-pixel estimate of soil loss rates for the catchment of the Arata-Chufa scheme varies from 18 t/ha/yr for bare land (the gravel road) in the upstream part of the catchment to zero for the largely grass-covered zone in the lower catchment, close to the main canal (Figure 8A). Mean annual soil loss for the catchment was estimated at 8.9 t/ha/yr, whereas the mean annual sediment yield to the Arata-Chufa main canal from the corresponding catchment was 2.32 t/ha/yr. Sediment yields varied across the catchment, ranging from zero to 4.3 t/ha/yr (Figure 8B).

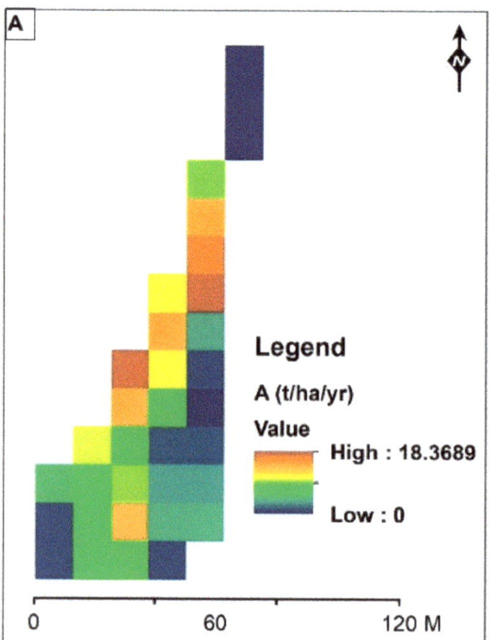

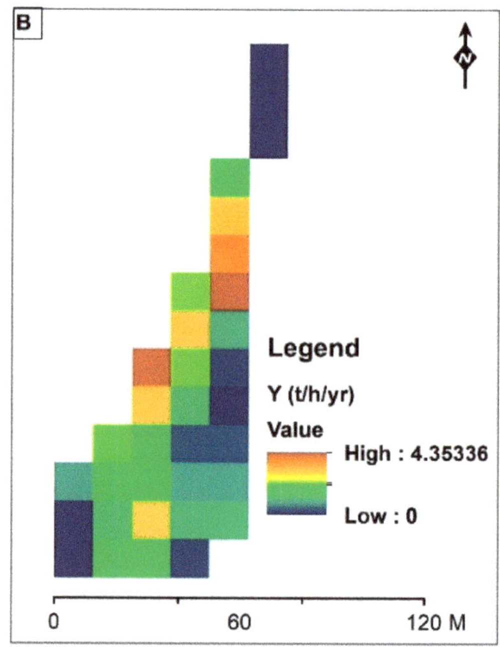

Figure 8. Annual soil loss (**A**) and sediment yield (**B**) in the catchment contributing overland sediment inflow to the main canal of the Arata-Chufa irrigation scheme.

The grid-based soil loss modeling for the catchment at the Ketar scheme shows annual soil losses ranging from 0, in the lower reach of the catchment, to 41 t/ha/yr (Figure 9A). Particularly high soil loss rates were registered along the steep, narrow drainage channels extending upland from the main canal. Mean annual soil loss of the catchment was estimated at 18.5 t/ha/yr, whereas sediment yield to the catchment contributing sediment to the Ketar main canal ranged from 0 to 6.2 t/ha/yr (Figure 9B).

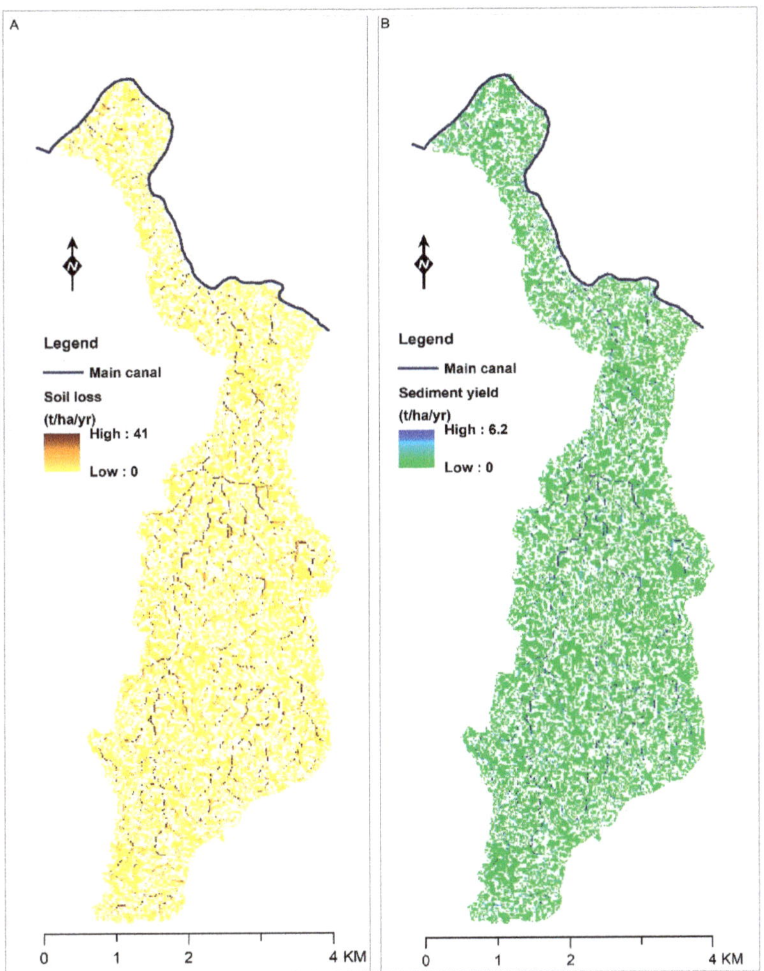

Figure 9. Annual soil loss (**A**) and sediment yield (**B**) in the catchment contributing overland sediment inflow to the main canal of the Ketar irrigation scheme.

3.3. Field Measurement of Sedimentation in the Schemes

Sedimentation, both river sediment and overland sediment inflow, in the main canals of the schemes was measured at the end of the wet season. At the Arata-Chufa scheme, sedimentation averaged 181 m^3/yr. To remove this volume of sediment, some 256 farmers worked 4.5 h a day for 5.5 days, together removing 0.22 m^3 of sediment per day (Figure 10, Figure A1, Table A1).

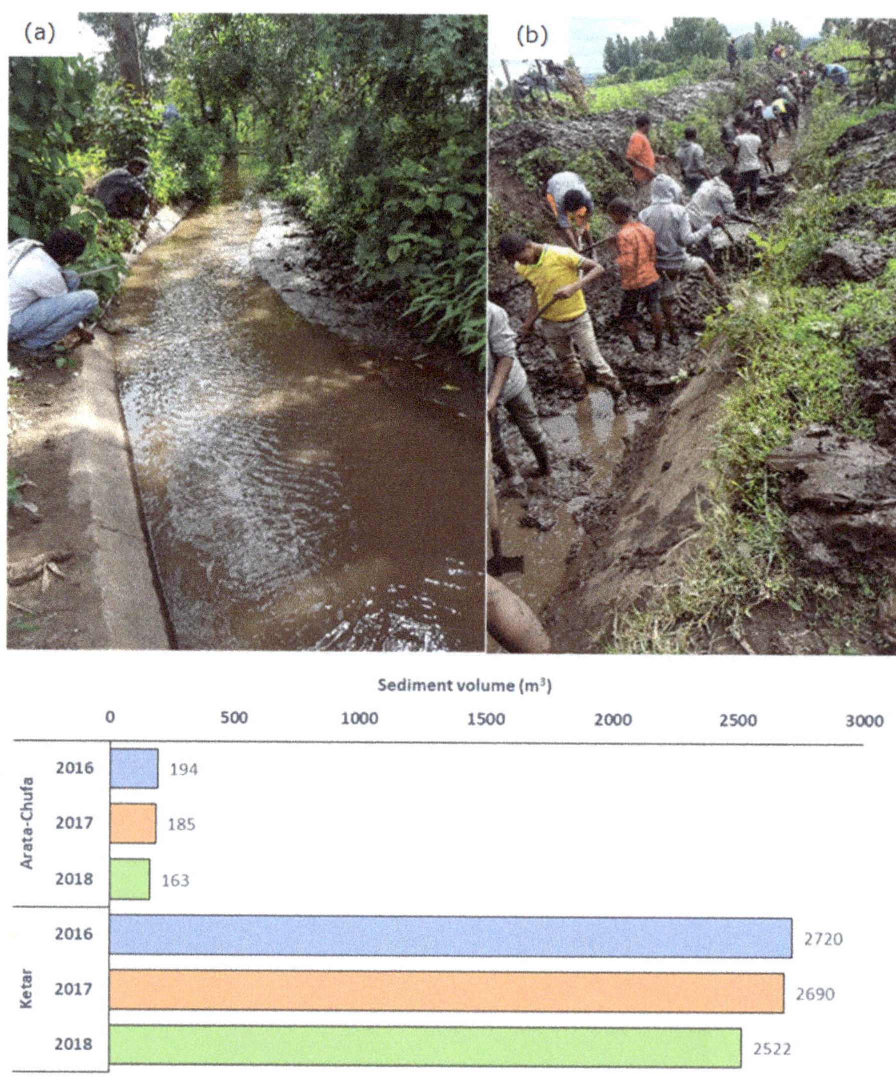

Figure 10. Volume of sediment removed from the main canal of the (**a**) Arata-Chufa and (**b**) Ketar irrigation scheme.

At Ketar, much of the sediment was deposited over only 20% of the main canal (2433 m). This critical section was 4.5 km from the intake and had a milder longitudinal bed slope (0.130‰) compared to the other sections of the main canal. On average, 2644 m^3 of sediment per year was removed from this section of the main canal (see Figure 10). Totally 3118 farmers participated in the desilting campaigns, together removing 0.83 m^3 of sediment over three 5-h working days (Figure 10, Figure A1, Table A1).

Comparison of the volumes of sediment measured in 2017 and 2018 to the sediment volumes estimated for the year prior to the fieldwork (2016) indicate a decrease in sediment volumes from 2016 to 2018, by 10.3% and 4.2%, respectively, for the Arata-Chufa and Ketar schemes. There is a strong correlation between the sediment volume in 2016 and the mean of the sediment volumes in 2017 and 2018, with the correlation being 0.76 for Arata-Chufa and 0.83 for Ketar (Figure 11).

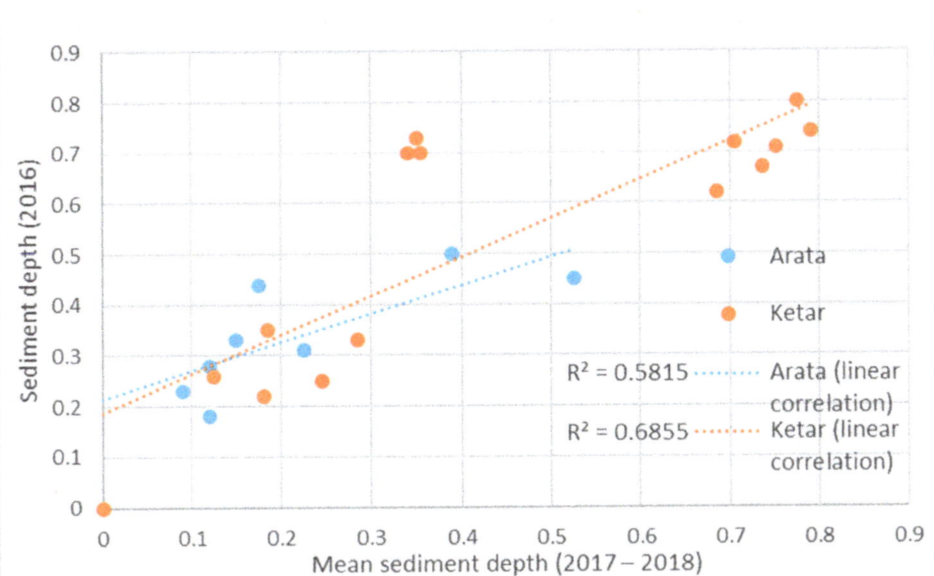

Figure 11. Correlation between the mean of the sediment volumes in 2017 and 2018 and the sediment volume estimated for 2016 using flood and sediment marks with farmer participation.

3.4. Overland Sediment Inflow Contribution

Overland flow sediment inflow concerns the part of the sediment that comes from the erosion of the catchment area upland of the main canal after the diversion structure and does not enter the scheme via regular intake structures. The onsite overland flow sediment enters the schemes along the main canal lateral. The contribution of overland sediment inflow is estimated by comparing the sediment yield modeled using RUSLE with the gross sediment volume removed from the schemes. The irrigation season runs from September to May (dry season) after dredging the deposited sediment that comes from river and overland flow. During the fieldwork at Arata-Chufa, we observed sediment inflow from surface runoff, despite the small size of the sediment-contributing catchment. Our erosion models indicate that the gross soil loss from this catchment was 10 t/yr. The corresponding sediment yield to the Arata-Chufa main canal was estimated as 2.6 t/yr (Table 2).

Table 2. Annual soil loss, sediment yield to contributing catchment and quantity of sediment dredged from the main canal of the Arata-Chufa and Ketar schemes.

Soil Loss (A)		Sediment Yield (Y)		Measured Dredged Sediment
Rate	Gross	Rate	Gross	Gross (2016–2018)
(m³/ha/yr)	(m³/yr)	(m³/ha/yr)	(m³/yr)	(m³/yr)
		Arata-Chufa irrigation scheme		
25.2	28.7	6.6	7.52	181
		Ketar irrigation scheme		
52.4	56,697	9.5	2042	2644

The Ketar scheme experienced higher soil loss from the catchment and correspondingly large sediment inflow to the main canal. Gross annual soil loss was estimated as 20,017 t, and the corresponding sediment yield to the main canal of the scheme was estimated as 720 t, with a mean annual sediment yield of 3.44 t/ha (see Table 2).

The Arata-Chufa scheme was affected mainly by sediment delivered by the river water feeding the scheme. Most erosion surface flow was conveyed into the river by a channel along the gravel road, which crossed the main canal (Figure 12). Measurement of sediment volumes in the main canal and soil erosion modeling indicate that surface runoff contributed about 4.3% (7.5 m^3) of the total volume of sediment deposited in the main canal.

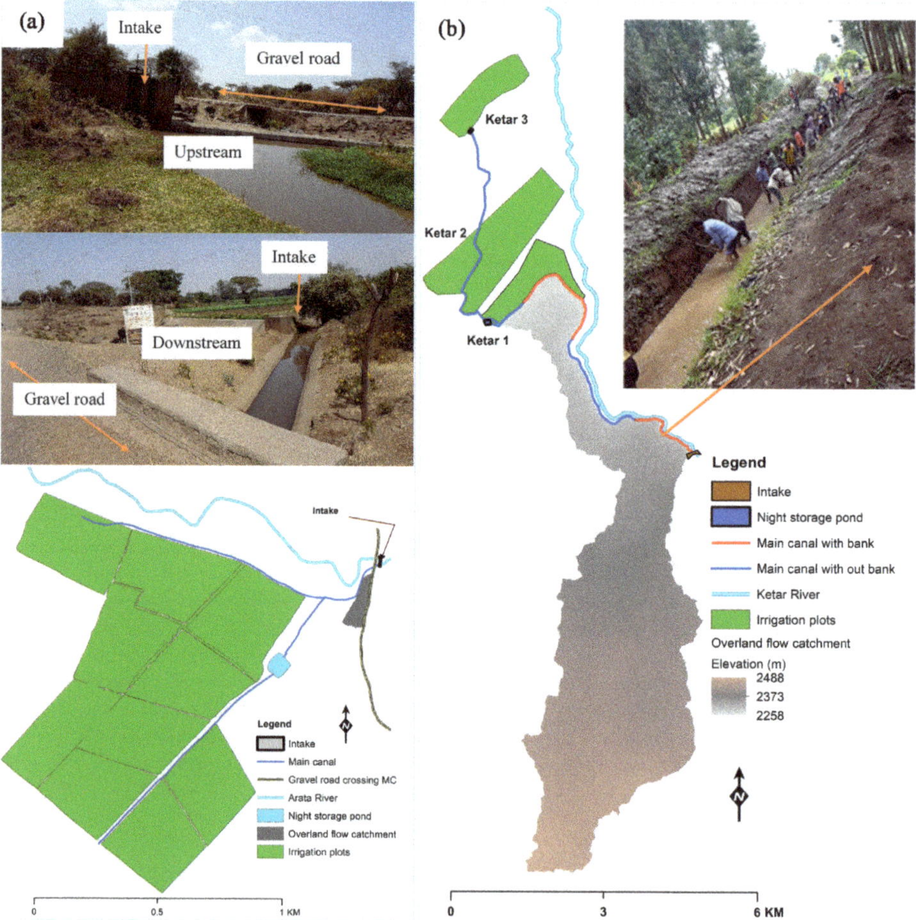

Figure 12. Layout of irrigation schemes and overland sediment flow in contributing catchment: (**a**) Arata-Chufa and (**b**) Ketar. At Ketar, the main canal segments labeled 'with bank' have a ridge embankment that helps protect the canal against overland sediment inflow.

The Ketar scheme main canal travelled some 4.5 km as a headrace canal from the intake to the field plots through various land use types, though mostly croplands (see Figure 12). Moreover, there was a lack of land conservation activities and the main canal was highly deteriorated due to years of use and a lack of maintenance. These factors contributed to overland sedimentation inflow to the main canal. Another factor, however, was the ridges, which had been formed alongside the main canal from sediment removed over years of desilting campaigns. These ridges played an important role in reducing sediment inflow to the canal. Nonetheless, sediment yield analyses show a large contribution of overland sediment inflow to the total volume of sediment deposited in the Ketar main canal. Specifically, overland flows accounted for some 77% (2042 m^3) of the gross volume of sediment deposited in the Ketar main canal.

The Arata-Chufa scheme had a shorter feeder canal. Here, the contribution of overland sediment flow into the main canal was found to be minimal (Table 3). Notwithstanding this, the main canal of the scheme became fully silted-up at the end of the cropping season, that is, within a three to four months period. For such a scheme, therefore, overland sediment will likely not be a priority concern. For the Ketar irrigation scheme, however, the volume of overland sediment inflow per unit of main canal was high (167 m^3/km) (Table 3). Explanations for this high overland sediment inflow include the long length of the main canal from intake to the first irrigation plot (4.5 km) and a lack of protection of the main canal from overland sediment inflow.

Table 3. Overland sediment inflow to the schemes per unit of irrigable land, per length of main canal and per user.

Per unit of Irrigable Land	Per Length of Main Canal	Per User
(m^3/ha)	(m^3/km)	(m^3/farmer)
Arata-Chufa irrigation scheme		
0.08	5.76	0.02
Ketar irrigation scheme		
4.74	167.05	1.90

3.5. Soil Loss Severity Analysis

While sedimentation of the Arata-Chufa scheme was found to be due primarily to the entry of sediment-laden river water, with the contribution of overland sediment flow relatively low, it is noteworthy that 92% of overland sediment inflow to the Arata-Chufa main canal came from the gravel road that crossed the main canal (Table 4, Figure 12).

Table 4. Severity classes of soil erosion loss for the area contributing sediment to the main canal of the Arata-Chufa and Ketar irrigation schemes. The severity classes are adapted from [19].

Erosion Severity Classes	Range of Soil Loss	Area	Percentage of Total Area	Mean Annual Soil Loss	Total Annual Soil Loss	Percentage of Total Soil Loss
	(t/ha/yr)	(ha)	(%)	(t/ha/yr)	(t/ha/yr)	%
Arata-Chufa irrigation scheme						
Very slight	0–5	0.29	25.44	3.12	0.90	8.42
Slight	5–15	0.75	65.79	10.78	8.09	75.21
Moderate	15–30	0.1	8.77	17.60	1.76	16.37
Severe	30–50	-	-	-	-	-
Very severe	>50	-	-	-	-	-
Total		1.14			10.75	
Ketar irrigation scheme						
Very slight	0–5	1067.70	98.65	0.4	17055.00	70.58
Slight	5–15	10.93	1.01	9.2	3931.00	16.27
Moderate	15–30	3.53	0.33	21.3	2952.00	12.22
Severe	30–50	0.15	0.01	37.8	227.00	0.94
Very severe	>50	-	-	-	-	-
Total		1082			24,165	

At the Ketar scheme, our soil erosion risk analysis indicates that 12% and 1% of sediment deposition in the main canal originated, respectively, from lands classified as 'moderately' and 'severely' at risk from soil erosion (Table 4). These classes are considered top priority when implementing structural and non-structural soil and water conservation measures. However, the total area of the catchment experiencing moderate to severe erosion rates was quite small compared to the entire catchment size. Indeed, the area exhibiting the highest erosion rates accounted for only about 0.4% of the total catchment area. Thus, to sustainably reduce excessive sedimentation, soil and water conservation activities should be implemented addressing the entire catchment.

3.6. Uncertainty in the RUSLE Model

Due to nonlinear spatiotemporal variability of parameters, the RUSLE model is sensitive to input variable uncertainties and the modeling results should be verified using local measurement data [42]. In particular, the model is highly sensitive to the LS factor (slope length and steepness) [43–45]. Moreover, the model cannot predict gully erosion. We used local data to minimize uncertainty in the input parameters and therefore in the model outcomes. Absence of gullies and an overall less complex catchment points to a general reliability of the sediment yield predictions for the Arata-Chufa scheme. At the Ketar scheme, land dynamics were more complex. Nonetheless, considering river sediment and total sediment inflows, the sediment yield volumes estimated by the RUSLE model were in a reasonable range.

4. Discussion

The annual soil losses estimated in this study are reasonably close to those reported by other authors from studies in the country. However, our mean annual soil loss estimate (18.5 t/ha/yr) is lower than the national-level estimate of 29.9 t/ha/yr by Haregeweyn et al. [46] and figures reported for North and North-Western Ethiopia, that is, 27.5 t/ha/yr [19], 47.4 t/ha/yr [24], 42.67 t/ha/yr [47], 84 t/ha/yr [48], 30.6 t/ha/yr [49] and 37 t/ha/yr [50]. In a nationwide study, Sonneveld et al. [51] reported that mean annual soil losses varied from 0 in the east and south to greater than 100 t/ha/yr in the northern and north-western escarpment. Kebede et al. [52] conducted a study in the Cheleleka watershed of the Central Rift Valley Basin of Ethiopia, where the current study area was also located. They reported annual soil losses in the range of 2.5–86 t/ha. The current study's mean annual soil loss estimate (18.5 t/ha/yr) is within this range and close to the 18.2 t/ha/yr estimated by Hui et al. [53].

There is high uncertainty associated with the values estimated using the revised universal soil loss equation (RUSLE) model. To reduce the associated uncertainties, we verified the RUSLE input parameters against the data collected during fieldwork. For instance, the absence or presence of soil and water conservation activities, types of crop, length of the growing period, post-harvest activities, soil type, land use type and absence or presence of gullies were carefully analyzed while determining the RUSLE input parameters. The empirical equation (Equation (10)), used to estimate the sediment delivery ratio (SDR), is also subjected to uncertainty. The estimated value of SDR was 26% for the Arata-Chufa and 18% for the Ketar scheme. The estimated SDR values by the current study are close to 30% for agricultural land and 25% for non-agricultural land estimated by Nyssen et al. [35] as used by Haregeweyn et al. [37]. One reason why our estimated mean annual soil loss is lower than the values reported by other authors for North and North-Western Ethiopia, could be the complexity of the terrain. As noted, topographic complexity plays a substantial role in the estimation of LS, which is a highly sensitive RUSLE parameter [54]. The current study area had moderate topographic complexity, while North and North-Western Ethiopia are well known for their rugged terrain and steep mountains.

Nearly 80% and 26% of the catchments at the Arata-Chufa and Ketar schemes, respectively, exhibited soil loss rates greater than the tolerable limits of 7.2 t/ha/yr [55] and 10 t/ha/yr [22]. Determination of appropriate tolerable limits is further dependent on local conditions, soil depth, rate of soil formation, terrain and rainfall characteristics. Findings from the current study indicate a need to implement conservation measures before it is too late and degradation becomes irreversible. In most places in the study area, soil was being lost at a rate faster than soil formation, which ranges from 2 to 22 t/ha/yr in Ethiopia [56]. Soil losses greater than 10 t/ha/yr are irreversible within a time span of 50–100 years [57]. Land degradation and a lack of conservation measures, particularly on croplands, contributed to high sedimentation rates in the study area. Soil loss in the study area had multiple effects. Among others, it caused deterioration of irrigation infrastructure and soil fertility loss. Many water conveyance and distribution structures had become dysfunctional due to excessive sedimentation and therefore could not deliver the required services. Water shortages, especially late in the irrigation season, were a major problem due to diminished canal capacities, leakages and malfunctioning water distribution structures. Excessive sedimentation also placed a heavy work burden on farmers, to keep the schemes operational. Reduced agricultural productivity due to a loss of nutrients in topsoil was another undesirable effect of soil erosion faced by farmers in the study area. Irrigated fields tended to be farmed under rainfed conditions during the wet season, which also led to an increased risk of soil loss.

The main determinant of the volume of overland sediment inflow appeared to be the layout of the irrigation scheme and upland land cover and land use. From the participatory mapping and transect walk during the fieldwork, we observed that the main canal of the Arata-Chufa scheme was mostly protected against potential overland sediment inflow. Moreover, the main canal extended only some 400 m before it reached the field plots. This short trajectory was of paramount importance in reducing sediment deposition from overland flow. Moreover, sedimentation from surface runoff came from a limited area, particularly, the gravel road that crossed the main canal downstream of the intake and the open area of grazing land between the main canal and the gravel road. The risk of overland sediment inflow at Ketar was substantially higher, as the canal traversed some 4.5 km from the intake to the field plots, through various land uses and land covers. Most of the sediment deposited into the main canal of this scheme originated from the rainfed croplands upland from the main canal.

We computed overland sediment yield into the canals by systematically delineating and classifying the catchments into subcatchments. This included subcatchments where banks protected the main canal against surface runoff and subcatchments without such canal banks, with the latter being more vulnerable to overland sediment flow into the main canal. Across the entire Ketar catchment, which covers 1082 ha, only 215 ha was found to directly contribute overland sediment flow to the Ketar main canal (12.1 km). Furthermore, over more than 30 years of desilting campaigns, Ketar farmers had dumped the sediment removed from the canal alongside the canal, forming a ridge that served to protect some parts of it from overland sediment inflow. However, this sediment ridge had grown to such a height that further sediment dredging activities were nearly impossible. Thus, the farmers were planning to organize a campaign to excavate the sediment accumulated on the banks, to make canal cleaning easier. Considering that with the protection of these and naturally occurring ridges, overland sediment flow still contributed nearly 77% of the total sediment deposited in the main canal, it is recommended that such excavation be done in tandem with construction of canal banks to prevent surface runoff inflow. This would help farmers sustainably address sedimentation problems, and save labor that would otherwise need to be invested in desilting campaigns.

Data scarcity is often a challenge in understanding processes of sedimentation in irrigation schemes and in designing sustainable measures to address excessive sedimentation. Annual sediment deposition in irrigation canals varies depending on many factors, including rainfall intensity and conservation measures to reduce soil loss. The sediment volumes measured in the current study correlated well with the volumes of sediment estimated with the participation of farmers based on flood and sediment marks on the walls of the canals. This is an important finding, as resource limitations often challenge collection of real-time data. Our correlation analysis reveals that a participatory approach can provide a source of reasonable data for conservation measures to deal with problems of excessive sedimentation.

5. Conclusions

We measured sedimentation volumes in two irrigation schemes in the Great Rift Valley Basin of Ethiopia in two successive years, 2017 and 2018, and estimated volumes for the year prior to the fieldwork, 2016, based on flood and sediment marks with farmers' support. Sediment inflow to the irrigation scheme main canals from overland flow was modelled using RUSLE. Erosion risk maps were prepared to predict the possible implementation of soil and water conservation measures to reduce soil losses. At Arata-Chufa, 4.3% of sedimentation in the canal was found to come from overland flow, while in Ketar this rate was 77%.

Our soil erosion severity map indicates low to moderate erosion rates in most of the areas under study. Some 84% of the Arata-Chufa catchment and 87% of the Ketar catchment, respectively, demonstrated slight to very slight soil erosion. Areas that exhibited a severe risk of erosion were found along surface drainage channels. Prioritizing soil and water conservation measures in the areas with severe erosion risk would not significantly reduce sediment inflow into the canals, as these covered only a small part of the catchment. Addressing the whole catchment when implementing conservation measures or protecting the main canal from surface runoff by constructing canal banks would be of greater help in significantly and sustainably reducing sedimentation, particularly in the Ketar main canal. Land degradation and a lack of soil conservation measures worsened soil erosion in this study area. In the Ketar scheme, excessive sediment inflow with surface runoff was aggravated by deterioration of the canal, the absence of canal banks and the long distance between the intake and field plots. As a result, water availability diminished as the irrigation season progressed. Moreover, water conveyance and distribution structures became damaged and operation and maintenance costs increased.

Farmers were found to be generally unaware of the source of sedimentation in their schemes. Identifying these sources and quantifying their contributions provides a crucial starting point for sustainably addressing sedimentation problems. In the Ketar scheme, the overland sediment inflow was found to be huge. This points to the importance of considering overland sediment inflows when rehabilitating irrigation schemes or designing new schemes, to attain optimum conveyance of water and sediment.

Based on these results, three key recommendations are proposed. First, as sources of sedimentation differ for every scheme, identification and quantification of these sources and areas with higher sediment contributions should be the starting point in addressing problems of excessive sedimentation. Second, collaborating with farmers can help engineers and researchers to acquaint with the system and also to provide reasonable data within a short period of time. Third, reduced costs to clean irrigation canals should be included as a direct benefit of soil conservation plans, in addition to such plans' benefits for upland farmers.

Author Contributions: Conceptualization, Z.A.G.; methodology, Z.A.G.; data curation, Z.A.G.; formal analysis, Z.A.G.; writing—original draft preparation, Z.A.G.; review and editing, H.R., M.R., C.d.F. and M.A.; supervision, H.R., C.d.F. All authors have read and agreed to the published version of the manuscript.

Funding: This research was funded by Nuffic, Netherlands Initiative for Capacity building in Higher Education of the Netherlands government. It was conducted under the framework of the "Capacity Development of HEIs in Small-Scale Irrigation (and Micro Irrigation)" project (Nuffic/Niche/Eth/197).

Institutional Review Board Statement: Not applicable.

Informed Consent Statement: Not applicable.

Data Availability Statement: Data was obtained from the Ethiopian National Meteorological Agency, Ministry of Water, Irrigation and Energy, the International Soil Reference and Information Center (ISRIC) and are available from the respective organizations.

Acknowledgments: The authors are very grateful to Nuffic, UNESCO-IHE, Delft and Wageningen University, Arba Minch University, Adama Science and Technology University, East Shoa and Arsi Zone offices and Woreda Irrigation Development Authority of the Oromia regional state, the Ethiopian National Meteorology Agency, and the Ministry of Agriculture and Natural Resources for their financial and material support to the study.

Conflicts of Interest: The authors declare no conflict of interest. The funders had no role in the design of the study; in the collection, analyses, or interpretation of data; in the writing of the manuscript, or in the decision to publish the results.

Appendix A

Table A1. Labour input and sediment output of the Arata-Chufa and Ketar irrigation scheme: adopted from Gurmu et al., 2019.

A	B	C	D	E	F	G
Year	Farmers Involved	Working Hours	Working Days	Sediment Removed	Total Input	Out Put
	(Number)	(h/Day)	(Days)	(m^3)	(Days)	(m^3/Day/Farmer)
			Arata-Chufa			
2016	-	-	-	194	-	
2017	260	4.5	6	185	878	0.21
2018	252	4.5	5	163	709	0.23
Average	256	4.5	5.5	181	794	0.22
			Ketar			
2016	-	-	-	2720	-	-
2017	1680	5	3	2690	3150	0.85
2018	1646	5	3	2522	3086	0.81
Average	1663	5	3	264	3118	0.83

Note that 8 h/day of daily working hours is used to estimate labor days and the values from columns A to F are recorded/measured data and columns F and G are calculated values.

Figure A1. Farmers desilting the sediment from the main canal during the annual desilting campaign at the Ketar irrigation scheme.

References

1. Haregeweyn, N.; Balana, B.B.; Melesse, B.; Tsunekawa, A.; Tsubo, M.; Meshesha, D.; Balana, B.B. Reservoir sedimentation and its mitigating strategies: A case study of Angereb reservoir (NW Ethiopia). *J. Soil Sediments* **2012**, *12*, 291–305. [CrossRef]
2. Moridi, A.; Yazdi, J. Sediment flushing of reservoirs under environmental considerations. *Water Resour. Manag.* **2017**, *31*, 1899–1914. [CrossRef]
3. Kondolf, G.M.; Gao, Y.; Annandale, G.W.; Morris, G.L.; Jiang, E.; Zhang, J.; Cao, Y.; Carling, P.; Fu, K.; Guo, Q.; et al. Sustainable sediment management in reservoirs and regulated rivers: Experiences from five continents. *Earth's Future* **2014**, *2*, 256–280. [CrossRef]
4. Aynekulu, E.; Atakliti, S.; Ejersa, A. *Small-Scale Reservoir Sedimentation Rate Analysis for a Reliable Estimation of Irrigation Schemes Economic Lifetime: A Case Study of Adigudom Area Tigray, Northern Ethiopia*; Faculty of Dryland Agriculture and Natural Resources, Mekelle University: Tigray, Ethiopia, 2009.
5. Mekonnen, M.M.; Keesstra, S.D.; Baartman, J.E.M.; Ritsema, C.J.; Melesse, A.M. Evaluating sediment storage dams: Structural off-site sediment trapping measures in Northwest Ethiopia. *Cuad. Investig. Geogr.* **2015**, *41*, 722. [CrossRef]
6. Moges, M.M.; Abay, D.; Engidayehu, H. Investigating reservoir sedimentation and its implications to watershed sediment yield: The case of two small dams in data-scarce upper Blue Nile Basin, Ethiopia. *Lakes Reserv. Res. Manag.* **2018**, *23*, 217–229. [CrossRef]
7. Sumi, T. Reservoir sedimentation management with bypass tunnels in Japan. In Proceedings of the of the Ninth International Symposium on River Sedimentation, Yichang, China, 18–21 October 2004; pp. 1036–1043.
8. Haregeweyn, N.; Poesen, J.; Nyssen, J.; De Wit, J.; Haile, M.; Govers, G.; Deckers, S. Reservoirs in Tigray (Northern Ethiopia): Characteristics and sediment deposition problems. *Land Degrad Dev.* **2006**, *17*, 211–230. [CrossRef]
9. Mekonen, A. Green Accounting Puts Price on Ethiopian Soil Erosion and Deforestation. Available online: http://efdinitiative. org/our-work/policy-interactions/green-accounting-puts-price-ethiopian-soil-erosion-and-deforestation (accessed on 21 February 2021).
10. Young, A. *Land Resources: Now and for the Future*; Cambridge University Press: Cambridge, UK, 1998.
11. Braimoh, A.K.; Vlek, P.L.G. *Land Use and Soil Resources*; Springer: Dordrecht, The Netherlands, 2008. [CrossRef]
12. Amede, T. Technical and institutional attributes constraining the performance of small-scale irrigation in Ethiopia. *Water Resour. Rural Dev.* **2015**, *6*, 78–91. [CrossRef]
13. Awulachew, S.; Ayana, M. Performance of irrigation: An assessment at different scales in Ethiopia. *Exp. Agric.* **2011**, *47*, 57–69. [CrossRef]
14. Gurmu, Z.A.; Ritzema, H.; de Fraiture, C.; Ayana, M. Sedimentation in small-scale irrigation schemes in Ethiopia: It sources and management. *J. Soil Sediments* **2021**. submitted for publication.
15. Abera, A.; Verhoest, N.E.C.; Tilahun, S.A.; Alamirew, T.; Adgo, E.; Moges, M.M.; Nyssen, J. Performance of small-scale irrigation schemes in Lake Tana Basin of Ethiopia: Technical and socio-political attributes. *Phys. Geogr.* **2019**, *40*, 227–251. [CrossRef]
16. Gurmu, Z.A.; Ritzema, H.; de Fraiture, C.; Ayana, M. Stakeholder roles and perspectives on sedimentation management in small-scale irrigation schemes in Ethiopia. *Sustainability* **2019**, *11*, 6121. [CrossRef]
17. Theol, S.; Jagers, B.; Yangkhurung, J.R.; Suryadi, F.X.; de Fraiture, C. Effect of Gate Selection on the Non-Cohesive Sedimentation in Irrigation Schemes. *Water* **2020**, *12*, 2765. [CrossRef]
18. Ganasri, B.P.; Ramesh, H. Assessment of soil erosion by RUSLE model using remote sensing and GIS—A case study of Nethravathi Basin. *Geosci. Front.* **2016**, *7*, 953–961. [CrossRef]
19. Haregeweyn, N.; Tsunekawa, A.; Poesen, J.; Tsubo, M.; Meshesha, D.T.; Fenta, A.A.; Nyssen, J.; Adgo, E. Comprehensive assessment of soil erosion risk for better land use planning in River Basins: Case study of the Upper Blue Nile River. *Sci. Total Environ.* **2017**, *574*, 95–108. [CrossRef]
20. Kumar, T.; Jhariya, D.C.; Pandey, H.K. Comparative study of different models for soil erosion and sediment yield in Pairi watershed, Chhattisgarh, India. *Geocarto Int.* **2019**, 1–22. [CrossRef]
21. Wischmeier, W.H.; Smith, D.D. *Predicting Rainfall Erosion Losses: A Guide to Conservation Planning*; U.S. Department of Agriculture: Bestville, MD, USA, 1978; Volume 537, p. 67.
22. Hurni, H. Erosion-productivity-conservation systems in Ethiopia. In Proceedings of the IV International Conference on Soil Conservation, Maracay, Venezuela, 3–9 November 1985; pp. 654–674.
23. Hurni, H. *Soil Conservation Manual for Ethiopia: A Field Guide for Conservation Implementation*; Ministry of Agriculture: Addis Ababa, Ethiopia, 1985.
24. Gelagay, H.S.; Minale, A.S. Soil loss estimation using GIS and remote sensing techniques: A case of Koga watershed, Northwestern Ethiopia. *Int. Soil Water Conserv.* **2016**, *4*, 126–136. [CrossRef]
25. Williams, J.R. Computer models of watershed hydrology. In *The EPIC Model*; Singh, V.P., Ed.; Water Resources Publications: Highlands Ranch, CO, USA, 1995.
26. Moore, I.D.; Wilson, J.P. Length-slope factors for the revised universal soil loss equation: Simplified method of estimation. *J. Soil Water Conserv.* **1992**, *47*, 423–428.

27. Schmidt, S.; Tresch, S.; Meusburger, K. Modification of the RUSLE slope length and steepness factor (LS-factor) based on rainfall experiments at steep alpine grasslands. *MethodsX* **2019**, *6*, 219–229. [CrossRef]

28. Qin, W.; Guo, Q.; Cao, W.; Yin, Z.; Yan, Q.; Shan, Z.; Zheng, F. A new RUSLE slope length factor and its application to soil erosion assessment in a Loess Plateau watershed. *Soil Till Res.* **2018**, *182*, 10–24. [CrossRef]

29. Hickey, R.; Smith, A.; Jankowski, P. Slope length calculations from a DEM within ARC/INFO grid. *Comput. Environ. Urban. Syst* **1994**, *18*, 365–380. [CrossRef]

30. Wang, M.; Baartman, J.E.M.; Zhang, H.; Yang, Q.; Li, S.; Yang, J.; Cai, C.; Wang, M.; Ritsema, C.J.; Geissen, V. An integrated method for calculating DEM-based RUSLE LS. *Earth Sci. Inform.* **2018**, *11*, 579–590. [CrossRef]

31. Desmet, P.J.J.; Govers, G. A GIS procedure for automatically calculating the USLE LS factor on topographically complex landscape units. *J. Soil Water Conserv.* **1996**, *51*, 427.

32. Almagro, A.; Thomé, T.C.; Colman, C.B.; Pereira, R.B.; Marcato Junior, J.; Rodrigues, D.B.B.; Oliveira, P.T.S. Improving cover and management factor (C-factor) estimation using remote sensing approaches for tropical regions. *Int. Soil Water Conserv. Res.* **2019**, *7*, 325–334. [CrossRef]

33. Panagos, P.; Borrelli, P.; Meusburger, K.; van der Zanden, E.H.; Poesen, J.; Alewell, C. Modelling the effect of support practices (P-factor) on the reduction of soil erosion by water at European scale. *Environ. Sci. Policy* **2015**, *51*, 23–34. [CrossRef]

34. Taye, G.; Vanmaercke, M.; Poesen, J.; Wesemael, B.V.; Tesfaye, S.; Teka, D.; Nyssen, J.; Deckers, J.; Haregeweyn, N. Determining RUSLE P- and C-factors for stone bunds and trenches in rangeland and cropland, North Ethiopia. *Land Degrad Dev.* **2018**, *29*, 812–824. [CrossRef]

35. Nyssen, J.; Clymans, W.; Poesen, J.; Vandecasteele, I.; De Baets, S.; Haregeweyn, N.; Naudts, J.; Hadera, A.; Moeyersons, J.; Haile, M.; et al. How soil conservation affects the catchment sediment budget–A comprehensive study in the north Ethiopian highlands. *Earth Surf. Proc. Landf.* **2009**, *34*, 1216–1233. [CrossRef]

36. Shin, G.J. The Analysis of Soil Erosion Analysis in Watershed Using GIS. Ph.D. Thesis, Gang-won National, Chuncheon, Korea, 1999.

37. Haregeweyn, N.; Poesen, J.; Nyssen, J.; Govers, G.; Verstraeten, G.; de Vente, J.; Deckers, J.; Moeyersons, J.; Haile, M. Sediment yield variability in Northern Ethiopia: A quantitative analysis of its controlling factors. *CATENA* **2008**, *75*, 65–76. [CrossRef]

38. Williams, J.R.; Berndt, H.D. Sediment yield computed with universal equation. *J. Hydrau. Div.* **1972**, *98*, 2087–2098. [CrossRef]

39. Jain, S.K.; Singh, P.; Saraf, A.K.; Seth, S.M. Estimation of sediment yield for a rain, snow and glacier fed river in the Western Himalayan Region. *Water Resour. Manag.* **2003**, *17*, 377–393. [CrossRef]

40. Bhattarai, R.; Dutta, D. Estimation of soil erosion and sediment yield using GIS at catchment scale. *Water Resour. Manag.* **2006**, *21*, 1635–1647. [CrossRef]

41. Onyando, J.O.; Kisoyan, P.; Chemelil, M.C. Estimation of potential soil erosion for river Perkerra catchment in Kenya. *Water Resour. Manag.* **2005**, *19*, 133–143. [CrossRef]

42. Wang, G.; Gertner, G.; Singh, V.; Shinkareva, S.; Parysow, P.; Anderson, A. Spatial and temporal prediction and uncertainty of soil loss using the revised universal soil loss equation: A case study of the rainfall-runoff erosivity R factor. *Ecol. Model.* **2002**, *153*, 143–155. [CrossRef]

43. Biesemans, J.; Van Meirvenne, M.; Gabriels, D. Extending and RUSLE with the Monte Carlo error propagation technique to predict long-term average off-site sediment accumulation. *J. Soil Water Conserv.* **2000**, *55*, 35–42.

44. Falk, M.G.; Denham, R.J.; Mengersen, K.L. Estimating uncertainty in the revised universal soil loss equation via Bayesian melding. *J. Agric. Biol. Environ. Stat.* **2010**, *15*, 20–37. [CrossRef]

45. Herr, A.; Kuhnert, P.M. Assessment of uncertainty in Great Barrier Reef catchment models. *Water Sci. Technol.* **2007**, *56*, 181–188. [CrossRef] [PubMed]

46. Haregeweyn, N.; Tsunekawa, A.; Nyssen, J.; Poesen, J.; Tsubo, M.; Tsegaye Meshesha, D.; Schütt, B.; Adgo, E.; Tegegne, F. Soil erosion and conservation in Ethiopia. *Prog. Phys. Geogr.* **2015**, *39*, 750–774. [CrossRef]

47. Belayneh, M.; Yirgu, T.; Tsegaye, D. Potential soil erosion estimation and area prioritization for better conservation planning in Gumara watershed using RUSLE and GIS techniques'. *Environ. Syst. Res.* **2019**, *8*. [CrossRef]

48. Selassie, Y.G.; Belay, Y. Costs of nutrient losses in priceless soils eroded from the highlands of Northwestern Ethiopia. *J. Agric. Sci.* **2013**, *5*, 1916–9760. [CrossRef]

49. Amsalu, T.; Mengaw, A. GIS based soil soss estimation using RUSLE model: The case of Jabi Tehinan Woreda, ANRS, Ethiopia. *Nat. Resour. J.* **2014**, *5*, 616–626. [CrossRef]

50. Yesuph, A.Y.; Dagnew, A.B. Soil erosion mapping and severity analysis based on RUSLE model and local perception in the Beshillo Catchment of the Blue Nile Basin, Ethiopia. *Environ. Syst. Res.* **2019**, *8*. [CrossRef]

51. Sonneveld, B.G.J.S.; Keyzer, M.A.; Stroosnijder, L. Evaluating quantitative and qualitative models: An application for nationwide water erosion assessment in Ethiopia. *Environ. Modell. Softw.* **2011**, *26*, 1161–1170. [CrossRef]

52. Kebede, W.; Habitamu, T.; Efrem, G.; Fantaw, Y. Soil erosion risk assessment in the Chaleleka wetland watershed, Central Rift Valley of Ethiopia. *Environ. Syst. Res.* **2015**, *4*, 5. [CrossRef]

53. Hui, L.; Xiaoling, C.; Lim, K.J.; Xiaobin, C.; Sagong, M. Assessment of soil erosion and sediment yield in Liao watershed, Jiangxi Province, China, using USLE, GIS, and RS. *J. Earth Sci. China* **2010**, *21*, 941–953. [CrossRef]

54. Mahala, A. Soil erosion estimation using RUSLE and GIS techniques—a study of a plateau fringe region of tropical environment. *Arab. J. Geosci.* **2018**, *11*, 1–18. [CrossRef]
55. FAO. *Ethiopian Highlands Reclamation Study*; Food and Agriculture Organization of the United Nations: Rome, Italy, 1984.
56. Hurni, H. *Ethiopian Highlands Reclamation Study, Soil Formation Rates in Ethiopia*; Food and Agriculture Organization of the United Nations: Addis Ababa, Ethiopia, 1983.
57. Kouli, M.; Soupios, P.; Vallianatos, F. Soil erosion prediction using the revised universal soil loss equation (RUSLE) in a GIS framework, Chania, Northwestern Crete, Greece. *Environ. Geol.* **2009**, *57*, 483–497. [CrossRef]

Article

Can Lumped Characteristics of a Contributing Area Provide Risk Definition of Sediment Flux?

Barbora Jáchymová *, Josef Krása, Tomáš Dostál and Miroslav Bauer

Department of Irrigation, Drainage and Landscape Engineering, Faculty of Civil Engineering, Czech Technical University in Prague, Thakurova 7, 16629 Prague, Czech Republic; josef.krasa@fsv.cvut.cz (J.K.); dostal@fsv.cvut.cz (T.D.); miroslav.bauer@fsv.cvut.cz (M.B.)
* Correspondence: barbora.jachymova@fsv.cvut.cz; Tel.: +42-0-224-354-745

Received: 26 April 2020; Accepted: 20 June 2020; Published: 23 June 2020

Abstract: Accelerated soil erosion by water has many offsite impacts on the municipal infrastructure. This paper discusses how to easily detect potential risk points around municipalities by simple spatial analysis using GIS. In the Czech Republic, the WaTEM/SEDEM model is verified and used in large scale studies to assess sediment transports. Instead of computing actual sediment transports in river systems, WaTEM/SEDEM has been innovatively used in high spatial detail to define indices of sediment flux from small contributing areas. Such an approach has allowed for the modeling of sediment fluxes in contributing areas with above 127,484 risk points, covering the entire Czech Republic territory. Risk points are defined as outlets of contributing areas larger than 1 ha, wherein the surface runoff goes into residential areas or vulnerable bodies of water. Sediment flux indices were calibrated by conducting terrain surveys in 4 large watersheds and splitting the risk points into 5 groups defined by the intensity of sediment transport threat. The best sediment flux index resulted from the correlation between the modeled total sediment input in a 100 m buffer zone of the risk point and the field survey data (R^2 from 0.57 to 0.91 for the calibration watersheds). Correlation analysis and principal component analysis (PCA) of the modeled indices and their relation to 11 lumped characteristics of the contributing areas were computed (average K-factor; average R-factor; average slope; area of arable land; area of forest; area of grassland; total watershed area; average planar curvature; average profile curvature; specific width; stream power index). The comparison showed that for risk definition the most important is a combination of morphometric characteristics (specific width and stream power index), followed by watershed area, proportion of grassland, soil erodibility, and rain erosivity (described by PC2).

Keywords: soil erosion; sediment flux; total soil loss; watershed characteristics; PCA analysis; RUSLE (Revised Universal Soil Loss Equation); WaTEM/SEDEM; Czech Republic; residential areas

1. Introduction

Rainfall-runoff events leading to soil erosion can also cause extensive off-site effects, damage to the urban infrastructure, and can endanger human lives [1,2].

Various models can be used for modeling erosion and sediment transport. In general, these models can be categorized as empirical/statistical, conceptual, and process-based [3]. The models differ in the number of required inputs. Moreover, the quality and the representativeness of the model outputs is very variable. Empirical models based on the universal soil loss equation [1,4–6] are widely used for determining the erosion threat over large areas. In the Czech Republic, the RUSLE-based WaTEM/SEDEM model [7–9] is verified and used in large scale studies [10–13]. This model provides a sufficiently accurate estimate of the erosion intensity and the amount of transported soil material on the basis of a relatively small amount of input data [14].

The spatial resolution and quality of input data for RUSLE-based models in the Czech Republic is rather high, and the method is also used for cross compliance policy application here [15]. Therefore, WaTEM/SEDEM outputs were considered as a relevant basis for definition of sediment flux risk in residential areas for the entire Czech Republic in the framework of research project VG20122015092: "Erosion Runoff—Increased Risk of the Residents and the Water Quality Exposure in the Context of the Expected Climate Change".

Instead of computing actual sediment transports in river systems, WaTEM/SEDEM was innovatively used in high spatial detail, but only to define indices of sediment flux from small contributing areas. Such an approach allowed for the modeling of sediment fluxes in contributing areas with above 127,484 risk points, covering the entire Czech Republic territory (78,866 km^2). Risk points are defined as outlets of contributing areas larger than 1 ha [16], wherein the surface runoff goes into residential areas or vulnerable bodies of water (presented in detail in Section 2.1). Sediment flux indices are calibrated by conducting terrain surveys and splitting the risk points into 5 groups defined by the category of the sediment transport threat (1 to 5). In the following text, the contributing areas of the risk points are called "risk watersheds".

Erosion-related lumped watershed characteristics [17] can be divided into several groups: Morphological, morphometric, land use (presence and state of vegetation), soil quality characteristics, and climatic (precipitation characteristics). The most commonly observed parameter is the slope, which seems to be crucial for the transition from soil cover disturbance to transportation of eroded particles down the slope [18]. The parcel or watershed slope is an important factor for the effectiveness of erosion control measures [19], but this is related to land use [20]. The morphometric parameters, especially the shape of the watershed and the predominant shape of the slopes (convergent/divergent, convex/concave) are important for a description of the rainfall-runoff, erosion, and transport process. The impact of the shape of a watershed, expressed by the specific width (watershed area/watershed length), the planar curvature (describing the convergence/divergence of the slopes), the curvature of the profile (describing the convexity/concavity of the slopes), indices expressing the hydrological behavior, and the erodibility of the watershed and the other morphometric parameters, has been described and assessed in a number of studies [16,21–23]. All watershed characteristics interact, and together they determine the final level of the threat of intensive sediment flux.

Another novelty of this paper lies in correlation analysis and principal component analysis (PCA) of the modeled data and their relation to general watershed (contributing area) characteristics. This way, the sensitivity of model outputs to the general watershed parameters could also be tested. The motivation was the awareness that in many large regions the data of the same spatial resolution and quality (as in the Czech Republic) are not available [24,25]. The research questions are therefore:

- What are typical parameters of a Czech watershed that produces a considerable amount of eroded material and should be modeled in more detail by a process-based model?
- Can single lumped contribution area parameters replace WaTEM/SEDEM modeling if we want to define five classes of the threat of sediment flux (e.g., not having a detailed DEM (Digital elevation model) or spatially detailed land-use maps or soil maps)?
- Can a statistically selected combination of these characteristics provide a better estimate?

If the lumped source area characteristics can define the overall sediment flux risk, the approach can then be used for simplifying the sediment transport assessment methods for regions with a lack of WaTEM/SEDEM input data in a relevant level of detail.

The aim of the study is to use an extensive set of results of the VG20122015092 project to derive a simplified statistical approach. Based on the characteristics of the watershed, which can be easily identified on the basis of open source data, it would then be possible to identify localities where the threat of intensive sediment runoff is high. Measures into the most high-risk areas can be then designed by process-based models.

2. Materials and Methods

2.1. Definition of Source Areas and Risk Points

First, raster-based GIS input data in 10 m spatial resolution were prepared for the entire area of the Czech Republic, consisting of following layers:

- Digital elevation model (DTM) based on 1:10,000 scale vector contours enhanced by a stereophotogrammetrical model in newly developed areas. Model was corrected for artificial sinks in arable areas;
- Land use, defined by the Fundamental Base of Geographic Data of the Czech Republic (http://geoportal.cuzk.cz/(S(aypz0pbaffy4rwohh4fljcu2))/default.aspx?lng=EN&mode=TextMeta&text=dSady_zabaged&side=zabaged&menu=24), and updated by the national register of agricultural areas (Land Parcel Identification System, 1:10,000 scale).

Second, flow accumulation over the entire Czech territory was provided, respecting fragmentation of the DTM by land use (roads, other linear structures, and built-up areas).

Contributing areas larger than 1 ha [26] had defined drainage networks potentially at risk of resulting in concentrated overland flows and sediment transport. By intersecting the drainage network with the boundaries of residential areas, the risk points were defined. Residential areas were identified as "all built-up classes" including gardens up to 50 m from house polygons. Rural gardens (parks) were excluded.

All risk points were considered as potential outlets of sediment flux, so for every point a source area was delineated (called "watershed" in further text). To reduce the number and spatial frequency of the points in the presented results for municipality communities, the risk points closer than 50 m and their watersheds were grouped assuming these outlets are always entering into the same part of any residential area.

The analysis resulted in 127,484 risk points and their watersheds. Further analyses were focused on the definition of the threat (in five classes) to define which points have no risk of sediment flux and which can lead to infrastructure damages. WaTEM/SEDEM was used to define the levels of threat of intensive sediment flux entering residential zones from these watersheds. After extensive terrain surveys and comparison with the model results, the modeled sediment inflow into 100 m buffer zones around residential areas was used as the proper parameter for risk definition.

2.2. Sediment Transport Modeling

For sediment transport modeling and for definitions of indices of erosion threat in risk points, the WaTEM/SEDEM model was used [7–9]. WaTEM/SEDEM is a RUSLE-based model (Equation (1)):

$$A = R \times K \times C \times LS \times P \tag{1}$$

where: A—annual soil erosion rate (Mg/ha·year), R—rainfall erosivity factor (MJ·cm/ha·h·year), K—soil erodibility factor (Mg·h/MJ·cm), LS—topographic factor (-), C—crop management factor (-), and P—erosion control practice factor (-).

Unlike RUSLE, WaTEM/SEDEM calculates the sediment transport capacity based on Equation (2) in each pixel, and then balancing every pixel, it determines the erosion/deposition:

$$TC = K_{TC} \times Ep_{rill} \tag{2}$$

where: TC—transport capacity (Mg/ha·m), K_{TC}—transport capacity coefficient (m), and Ep_{rill}—potential for rill erosion (Mg/ha·year).

A detailed description of the model structure and its parametrization for the Czech Republic is provided by [27].

Distributed R-factor values in 1-km resolution were derived by Hanel [28,29]. Typical C-factor values for land use categories in the Czech Republic are defined by Janeček [30] in accordance with the USDA handbook 537 [4]. The C-factor for arable land was determined as an average value according to the logged crop rotation [31] in each territorial unit (76 districts). A DEM with a spatial resolution of 10 m was used for calculating the LS-factor. K-factor values were determined in accordance with the national methodology [32] based on soil quality maps (BPEJ, 1:5000 scale).

In the Czech Republic and in the 10-m resolution data used in this study, WaTEM/SEDEM was calibrated previously in the Rimov watershed (488 km^2) by [27]. Based on the calibration, the following internal parameters of WaTEM/SEDEM were used in this study: PTEF (arable, forest, grassland = 0, 75, 75); parcel connectivity (arable, others = 40, 75); KTC (arable, others = 35, 55).

Modeling (in tiles) was provided over the entire area of the Czech Republic, considering all surface waters and residential areas as points of delivery (in the terminology of WaTEM/SEDEM, the "river" class of land use). The fully distributed modeling of sediment transports within streams applying river topology maps and reservoirs was out of the scope of the project. Therefore, the model was only used to derive an erosion/deposition map (called "netto erosion") and sediment transport map (called "inflow"). Model output, together with selected model input data, are in Figure 1.

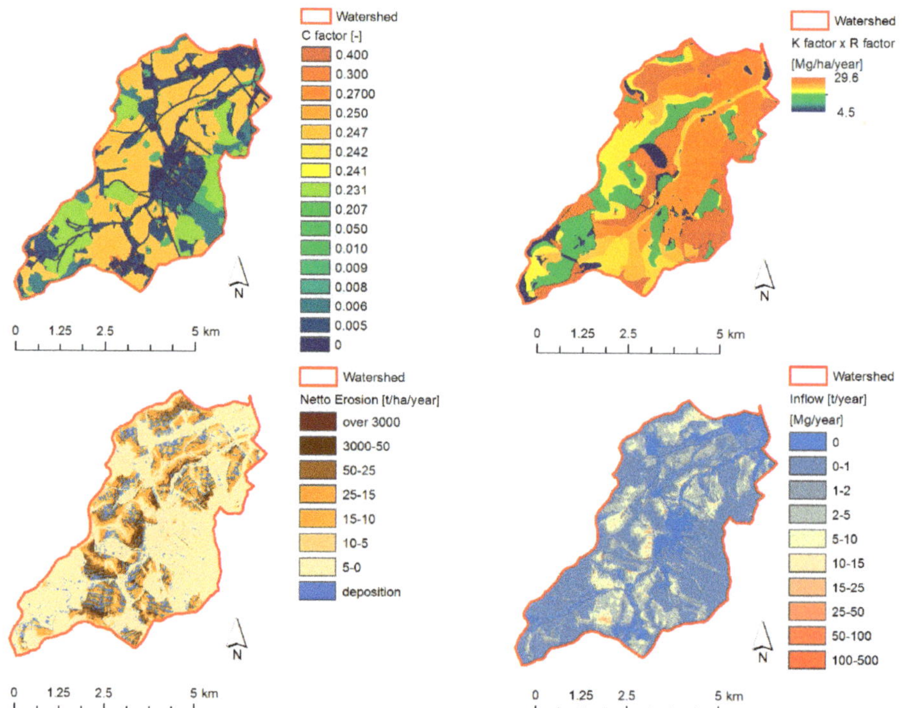

Figure 1. The example of data input (distributed C-factor, K-factor, and R-factor) and WaTEM/SEDEM outputs (netto erosion and inflow).

Raster-based GIS outputs (netto erosion and inflow) were further analyzed by zonal statistics of all 130,000 watersheds to provide risk classification concerning sediment fluxes. Here we should point out that the original calibration could also be used because the actual values of sediment transports in outlet points were not of importance. The only need was to define the high-risk and low-risk classes of the sediment entrance into residential areas, and not to compute the transported sediment volumes.

2.3. Evaluation of the Level of Threat

The level of the threat of sediment transport into residential areas was determined for the risk points. The aim was to classify the risk points into five classes depending on potential sediment fluxes. Since the contributing areas of the risk points were starting only with 1 ha size, for many watersheds we could assume rather high sediment connectivity [33]. Therefore, not only WaTEM/SEDEM sediment delivery to the outlet (inflow) was considered, but also total soil loss and area-specific soil loss in each watershed (example of watershed in Figure 2).

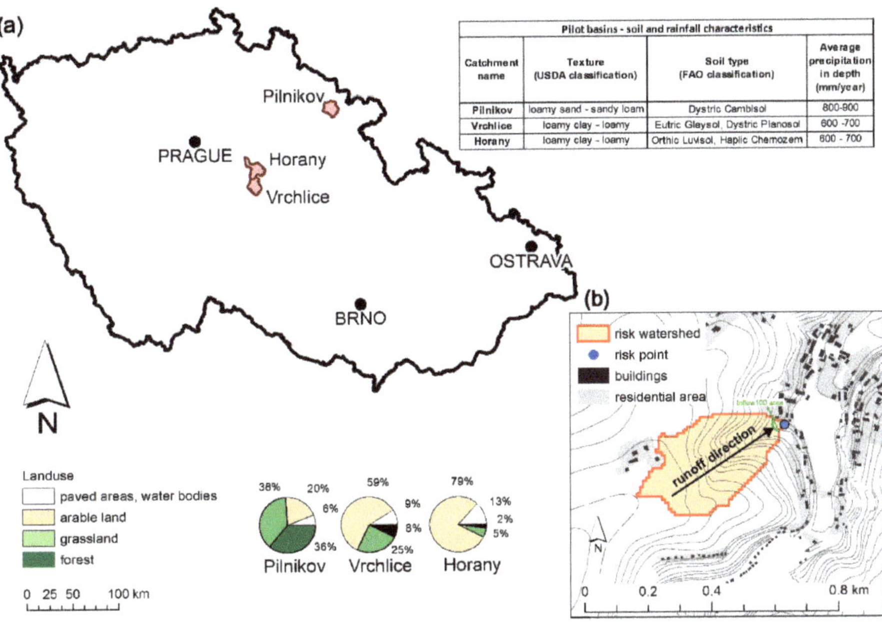

Pilot basins - soil and rainfall characteristics			
Catchment name	Texture (USDA classification)	Soil type (FAO classification)	Average precipitation in depth (mm/year)
Pilnikov	loamy sand - sandy loam	Dystric Cambisol	800-900
Vrchlice	loamy clay - loamy	Eutric Gleysol, Dystric Planosol	600 -700
Horany	loamy clay - loamy	Orthic Luvisol, Haplic Chernozem	600 - 700

Figure 2. Three research catchments selected for the field survey (**a**). Sediment flux from threat watershed flows into residential area through potentially threatened outlet (**b**).

Optimal approach for classifying all risk points would be the terrain survey, but the 79,000 km^2 and 130,000 risk points could not be visited. For that reason, three research catchments (of ca. 100 km^2 each) were selected (Figure 2) to correctly set the five threat categories by terrain survey. The basins represent the most common types of agricultural landscape in the Czech Republic. The Horany Basin represents intensively used lowlands with large parcels, long straight slopes, and intensive crops (corn, sugar beet, and cereals). The Vrchlice Basin represents upland landscapes with morphologically diverse watersheds, steeper slopes, and intensive agriculture, and the Pilnikov Basin represents foothills with steep convergent slopes, and a high proportion of cereals, forage, and grassland. In these basins, the real threat categories (1–5) for the risk points were identified by field surveys. The field survey results were compared with the zonal statistics of the WaTEM/SEDEM outputs for each risk watershed to select a suitable model result for defining the threat categories.

The entire area of the watershed, soil erosion potential and evidence, the runoff trajectory, and the watershed outlets into residential areas were observed. Concurrently, the real sediment transport pathways in pre-selected profiles were surveyed. Information from residents about previous intensive sediment flux was an important aspect of the field survey.

WaTEM/SEDEM modeling provided the output GIS layers for the soil loss, the sediment transport/deposition in each pixel (netto erosion), and the total sediment input in each pixel (inflow).

First, it was necessary to choose a best fitting model output for the correct description of the real threat defined by five classes based on the terrain survey.

The tested model outputs of the model were (Table 1):

- Aspecific (Mg/ha·year)—the specific soil loss in the watershed;
- Atotal (Mg/pixel·year)—the total soil loss in the watershed;
- Inflow100 (Mg/year)—sediment transport to the outlet, the total sediment input in a 100-m buffer zone of the risk point.

The statistical values of the tested model outputs were calculated for threat watersheds in the calibration areas. Then the relationship between the model outputs values and the threat category was evaluated. The Inflow100 was shown to be the most suitable model output for the threat of sediment delivery into the risk point (Table 1).

Table 1. Correlation (correlation coefficient) between tested model outputs value and threat category determined within field survey.

	$A_{specific}$	A_{total}	**Inflow100**
Horany	0.27	0.46	0.57
Pilnikov	0.16	0.63	0.76
Vrchlice	0.34	0.56	0.91
Complete field survey	0.23	0.46	0.70

$A_{specific}$—the specific soil loss in the watershed, A_{total}—the total soil loss in the watershed, Inflow100—sediment transport, the total sediment input in a 100-m buffer zone of the potentially threatened outlet.

In the complete database of threat watersheds for the Czech Republic, the Inflow100 ranges from 0 to 966 Mg/year. The distribution of values shows that the frequency of lower Inflow100 values is higher than the frequency of higher Inflow100 values. The statistical distribution of Inflow100 values in the watershed database was determined in order to set the threshold for the Inflow100 values that define the five threat level categories. Normal distribution was excluded on the basis of the histogram and the Q-Q plot (Figure 3a,b). The statistical distribution of the Inflow100 values corresponds to the log-normal statistical distribution [34] (Figure 3c,d). The expected distribution of the watersheds (in the complete database) in the sediment transport categories indicates that the threat level is not evenly distributed. Watersheds in threat category 4 or 5 appear less frequently than watersheds in category 1 (very low threat level).

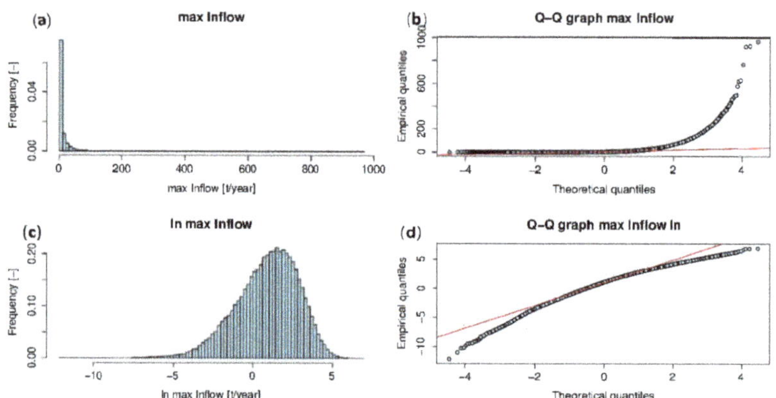

Figure 3. The statistical distribution of Inflow100 values in the watershed threat database represented by histograms and Q-Q plots. It does not corresponds to normal statistical distribution (**a**,**b**). It corresponds to log-normal statistical distribution (**c**,**d**).

2.4. Impact of Watershed Characteristics on the Threat

The following watershed characteristics are assessed for their impacts on the level of threat of sediment flux in comparison with the results of WaTEM/SEDEM modeling.

Soil characteristics (soil texture, soil structure, amount of organic material) are expressed in the K-factor. The precipitation characteristics (average number of intensive rainfall events during the year and their erosivity) are expressed in the R-factor. The average K-factor and R-factor were assessed for each watershed to simulate low-resolution data comparison. The land use was described by the proportion of arable land, forest, and grassland. The morphological characteristics were included in the analysis through the average slope (%) and the watershed area (ha). The analyzed morphometric characteristics were the specific width of the watershed (m), i.e., the ratio between the area of the watershed and the longest runoff line, the curvature of the profile (in the maximum slope direction)—Curveprofile, and the planar curvature (perpendicular to the direction of the maximum slope—Curveplane [35]. The hydrological index stream power index (m rad) (SPI) was considered.

SPI expresses the erosion potential of the surface runoff. It reflects the drainage area and the slope in a specific location in the watershed, on the basis of Equation (3) [36]:

$$SPI = A_s \cdot s \tag{3}$$

where SPI is the local stream power index (m rad), A_s is the local specific drainage area per unit contour length, and s is the local slope (%).

First, the correlation matrix expressing the relationship between the Inflow100 and the analyzed watershed characteristics was set up. Based on our analysis of almost 130,000 potentially threatened points, it can be assumed that there is a higher threat level in watershed with a high proportion of arable land, a steep average slope and a specific width, a large watershed area, and a high value of the SPI coefficient. A multi-variate statistical technique was run to verify this assumption. Within this analysis, we tested the relationships among the watershed characteristics that are important for the Inflow100 (or for the final threat category). Principal component analysis (PCA) is one of the most widely used types of multi-variate data analysis [37]. This method simplifies the complexity in high-dimensional data while retaining trends and patterns. It does this by transforming the data into fewer components, which describe a combination of observed dimensions [38]. In the presented analyses, the PCA method transfers the variables (the threat watershed characteristics) to the principal components. The principal components are a linear combination of the original variables (watershed characteristics). The main aim of this transfer is to reduce the number of variables. R studio software [39] was used for the statistical analyses.

3. Results

The Inflow100 values for the thresholds were set (Table 2) on the basis of the log-normal distribution of the Inflow100 values and required logarithmic representation of the watersheds in the threat categories. The final number of watersheds in the threat categories corresponds to the logarithmic function.

Table 2. Number of threat watersheds in five threat categories.

	Category 1	Category 2	Category 3	Category 4	Category 5	Total
Range (Inflow100 value)	0–2	2–7	7–20	20–55	>55	
Watersheds in category	53,835	32,596	24,389	12,780	3884	127,484

Table 3 shows the average values for the analyzed characteristics in groups of risk watersheds forming the five threat categories.

Table 3. Average values of the analyzed characteristics in risk watersheds representing the five threat categories.

Watershed Characteristics			Category 1	Category 2	Category 3	Category 4	Category 5
Average K-factor	(Mg h/MJ cm)	mean	0.36	0.38	0.39	0.40	0.41
		range	0.60	0.57	0.46	0.53	0.44
		1st and 3rd quartile distance	0.15	0.15	0.14	0.14	0.13
Average R-factor	(MJ cm/ha h year)	mean	63	64	65	67	69
		range	111	111	113	110	112
		1st and 3rd quartile distance	21	19	17	16	15
Average Slope	(%)	mean	9	11	12	13	15
		range	113	83	63	65	72
		1st and 3rd quartile distance	10	11	10	9	9
Area of Arable Land	(%)	mean	35	52	59	62	63
		range	100	100	100	100	100
		1st and 3rd quartile distance	86	96	83	68	61
Area of Forest	(%)	mean	30	28	25	25	27
		range	100	100	100	100	100
		1st and 3rd quartile distance	61	55	43	41	42
Area of Grassland	(%)	mean	35	20	17	14	11
		range	100	100	100	98	98
		1st and 3rd quartile distance	73	33	25	20	15
Total Watershed Area	(ha)	mean	7.12	9.7	14.24	19.77	30.81
		range	1524.96	2238.15	1619.27	1374.07	813.49
		1st and 3rd quartile distance	3.51	6.56	11.50	18.23	30.65
Average Planar Curvature	(-)	mean	0.00	0.00	−0.01	−0.01	−0.02
		range	1.50	1.29	0.96	1.28	0.51
		1st and 3rd quartile distance	0.03	0.03	0.03	0.03	0.03
Average Profile Curvature	(-)	mean	0.00	−0.01	−0.01	−0.01	−0.02
		range	4.21	1.57	1.55	0.89	0.53
		1st and 3rd quartile distance	0.03	0.03	0.03	0.03	0.03
Specific Width	(m)	mean	15	16	16	17	17
		range	419	116	53	88	34
		1st and 3rd quartile distance	5	5	5	5	4
Stream Power Index	(m rad)	mean	1257	2057	2697	3728	5571
		range	488,700	174,800	59,510	105,000	50,820
		1st and 3rd quartile distance	1376.1	2267.6	2785.3	3600	4881

An analysis was made of the simple linear correlation between Inflow100 values and individual analyzed characteristics. The correlation matrix (Table 4) shows a considerable relationship (R > 0.20) only between the Inflow100 and stream power index (SPI). The value of the correlation coefficient between Inflow100 and SPI is 0.30.

Table 3 documents the relationship between the threat category of intensive erosion runoff formation and the average values of the selected characteristics. The SPI coefficient and the proportion of arable land, total area, slope, and specific width increases with higher threat categories. The proportion of grassland decreases and the proportion of forest slightly decreases.

The PCA results for the complete database in Table 5 show the interdependence of the characteristics and the complexity of the relationship between the characteristics and the Inflow100. The individual components explain only a relatively low proportion of the data.

Table 4. Correlation matrix between Inflow100 and studied characteristics.

	Max Inflow100	Average K-Factor	Average R-Factor	Average Slope	Area of Arable Land	Area of Forest	Area of Grassland	Total Watershed Area	Average Curve$_{plane}$	Average Curve$_{profile}$	Specific Width	Stream Power Index
	(Mg/year)	(Mg h/MJ cm)	(MJ cm/ha h year)	(%)	(%)	(%)	(%)	(ha)	(-)	(-)	(m)	(m rad)
Max Inflow100	1.00											
Average K-factor	0.11	1.00										
Average R-factor	0.08	0.03	1.00									
Average Slope	0.13	0.08	0.24	1.00								
Area of Arable Land	0.13	0.02	−0.23	−0.57	1.00							
Area of Forest	−0.02	0.28	0.11	0.57	−0.67	1.00						
Area of Grassland	−0.15	−0.33	0.17	0.09	−0.55	−0.24	1.00					
Total Watershed Area	0.17	0.05	−0.06	−0.03	0.04	0.03	−0.08	1.00				
Average Curve$_{plane}$	−0.10	−0.01	0.07	0.01	−0.03	0.01	0.02	−0.01	1.00			
Average Curve$_{profile}$	−0.06	−0.03	0.02	−0.11	0.02	−0.09	0.06	0.03	0.39	1.00		
Specific Width	0.12	−0.03	−0.08	−0.21	0.11	−0.09	−0.05	0.30	−0.09	0.00	1.00	
Stream Power Index	0.30	0.09	0.11	0.49	−0.29	0.34	−0.02	0.32	−0.08	−0.09	0.38	1.00

Table 5. The variability proportion explained by components (PC1–PC11).

	PC1	PC2	PC3	PC4	PC5	PC6	PC7	PC8	PC9	PC10	PC11
Proportion Explained	0.24	0.16	0.14	0.12	0.08	0.07	0.06	0.05	0.05	0.02	0
Cumulative Proportion	0.24	0.4	0.54	0.66	0.75	0.81	0.87	0.93	0.98	1	1

The correlation coefficients between the studied characteristics and five components are presented in Table 6. The correlation coefficients between the Inflow100 and the components (PC1–PC5) were calculated to identify the importance of the components (and indirectly of the characteristics) in relation to the level of threat (Table 7).

Table 6. Correlation coefficients between the characteristics and components PC1–PC5.

		PC1	PC2	PC3	PC4	PC5
K	(Mg h/MJ cm)	0.17	0.42	−0.53	0.21	0.22
R	(MJ cm/ha h year)	0.36	−0.26	0.06	0.07	0.87
Slope	(%)	0.84	−0.04	−0.11	−0.01	−0.04
Arable land	(%)	−0.8	0.34	−0.16	0.04	0.2
Forest	(%)	0.8	0.19	−0.34	0.12	−0.21
Grassland	(%)	0.2	−0.65	0.58	−0.19	−0.03
Area	(ha)	0.06	0.54	0.44	0.24	−0.02
$Curve_{plane}$	(-)	−0	−0.3	−0.02	0.78	−0.06
$Curve_{profile}$	(-)	−0.1	−0.26	0.13	0.77	−0.08
Spec. width	(m)	−0.1	0.56	0.6	0.07	0.07
SPI	(m rad)	0.6	0.5	0.37	0.06	0.04

Table 7. Correlation coefficients between components (PC1–PC5) and Inflow100.

R between PC and Inflow100	
PC1	0.07
PC2	0.28
PC3	0.05
PC4	0.00
PC5	0.17

PC2 (R = 0.28) and PC5 (R = 0.17) are relatively important. PC2 has positive relationship with the watershed area, the specific width, the SPI, and the K-factor. Conversely, the proportion of grassland has a negative relationship with PC2. PC5 correlates considerably only with the R-factor.

4. Discussion

The accuracy of the modeled Inflow100 value is importantly influenced by the description of watershed connectivity. The index of connectivity based on GIS analysis of landscape was derived by Borselli [33]. Consequently, it was refined by Cavalli [40]. An essential input for determining watershed connectivity is a digital terrain model with high resolution. Therefore, the connectivity based on high-resolution DEM was not evaluated. The connectivity is involved in modeling by respecting parcel boundaries and by setting a sediment transport capacity within WaTEM/SEDEM. Based on our testing [41] and calibrating of the model in numerous previous studies [10–13] we believe in reliable results in defining risk of the sediment fluxes from watersheds of average size of 11.3 ha.

A combination of principal component analysis and correlation analysis between the component values and the Inflow100 shows that the most important watershed characteristics for the threat of sediment flux are morphometric characteristics (the shape of the watershed, expressed by the specific width and SPI), the watershed area, the soil erodibility, and the proportion of grassland. The studies focused on the important factors affecting the value of sediment transport show that the influence of these factors depends on the size of the evaluated watershed. Morphological and morphometric factors are particularly significant for smaller watersheds. The area is a key factor influencing sediment transport in larger watersheds [42]. The presence and state of vegetation cover is also important for runoff generation, erosion intensity, and nutrient transport. [1,43]. The soil quality (organic material content, soil structure and texture) influences infiltration capacity, surface runoff generation, and erosion intensity [44].

Rainfall erosivity also has an important impact on the threat level. According to the results of many studies, rainfall intensity is a key factor that influences not only the total amount of runoff [19] and the erosion event process [18], but also the characteristics of the runoff that is formed and its erosive potential [1]. Rainfall erosivity influences the protective effect of vegetation, and in high erosivity regions the soil conservation techniques have to be adapted [45].

Concerning the land use characteristics, the grassland decreasing accompanied by arable land increasing influences sediment transport. On the other hand, the proportion of a forest is less correlated to the Inflow100 rise. In general, land use has an important influence on the behavior of a watershed in terms of erosion and transport processes [1]. However, land-use characteristics are related to other characteristics (slope length, slope, soil quality, farming methods, etc.) that can have a fundamental effect on runoff behavior [45]. For example, Wu and Wang [20] documented intensive soil erosion on gardens and parcels with shrubs. These situations are consequences of the steepness of the slope on these parcels, or of intensive farming. No direct impact of the average watershed slope on sediment transport was proved by the correlation in our study. A number of studies have demonstrated a direct impact of the parcel slope on erosion intensity [18,46]. In our case, the impact of a slope is related (positively or negatively) to the other characteristics, in the same way as land use is. The multi-variate data analysis presented here shows that the slope has a considerable influence on the erosion threat, particularly in combination with the drainage area. This is expressed by the stream power index (SPI).

5. Conclusions

The presented study deals with the relationship between watershed characteristics and the level of intensive erosion threat in the Czech Republic. Based on our study, we offer the following conclusions relating to the defined scientific questions:

- A typical watershed producing a considerable amount of eroded material is a large convergent area with a steep slope in the lower part and with a low proportion of grassland. The soil erodibility and the frequency of intensive rainfall events are also important factors;
- Morphometric characteristics (the shape of the watershed and the slope in the lower part of the watershed), the area of the watershed, the land use, and soil quality (its susceptibility to erosion) are key factors for the sediment connectivity;
- A simple analysis of a watershed on the basis of widely available data (a digital elevation model, soil characteristics, information about rainfall events in the watershed) can be used for determining the threat level of intensive sediment flux. However, this analysis provides less accurate results than mathematical models provide. The simple analysis presented here is a suitable tool for the initial identification of areas that are susceptible to intensive erosion and transport formation;
- The statistics provided here can form a useful basis for a conceptual model for average conditions in the Czech Republic. However, in different conditions (e.g., parcel sizes, morphology) it would have to be calibrated again.

Author Contributions: Conceptualization, investigation, draft preparation, review, and editing by B.J., J.K., and T.D.; methodology, formal analysis, and visualization by B.J. and M.B. All authors have read and agreed to the published version of the manuscript.

Funding: This research was supported by grant VG20122015092: "Erosion Runoff—Increased Risk of the Residents and the Water Quality Exposure in the Context of the Expected Climate Change" and by grants QK1920224, LTC18030, and H2020 SHui, No. 773903.

Conflicts of Interest: The authors declare no conflict of interest.

References

1. Toy, T.J.; Foster, G.R.; Renard, K.G. *Soil Erosion: Processes, Prediction, Measurement, and Control*; John Wiley and Sons: New York, NY, USA, 2002; ISBN 9780471383697.
2. Borrelli, P.; Van Oost, K.; Meusburger, K.; Alewell, C.; Lugato, E.; Panagos, P. A step towards a holistic assessment of soil degradation in Europe: Coupling on-site erosion with sediment transfer and carbon fluxes. *Environ. Res.* **2018**, *161*, 291–298. [CrossRef] [PubMed]
3. Merritt, W.S.; Letcher, R.A.; Jakeman, A.J. A review of erosion and sediment transport models. *Environ. Model. Softw.* **2003**, *18*, 761–799. [CrossRef]
4. Wischmeier, W.; Smith, D. *Predicting Rainfall Erosion Losses: A Guide to Conservation Planning*; ISBN Agriculture Handbook 537; US Department of Agriculture: Washington, DC, USA, 1978.
5. Renard, K.; Foster, G.; Weesies, G.; McCool, D.; Yoder, D. *Predicting Soil Erosion by Water: A Guide to Conservation Planning with the Revised Universal Soil Loss Equation (RUSLE)*; US Department of Agriculture: Washington, DC, USA, 1997; ISBN 0160489385.
6. Mitasova, H.; Hofierka, J.; Zlocha, M.; Iverson, L.R. Modeling topographic potential for erosion and deposition using GIS. *Int. J. Geogr. Inf. Syst.* **1996**, *10*, 629–641. [CrossRef]
7. Van Oost, K.; Govers, G.; Desmet, P. Evaluating the effects of changes in landscape structure on soil erosion by water and tillage. *Landsc. Ecol.* **2000**, *15*, 577–589. [CrossRef]
8. Van Rompaey, A.; Verstraeten, G.; Van Oost, K.; Govers, G.; Poesen, J. Modelling mean annual sediment yield using a distributed approach. *Earth Surf. Process. Landf.* **2001**, *26*, 1221–1236. [CrossRef]
9. Verstraeten, G.; Van Oost, K.; Van Rompaey, A.; Poesen, J.; Govers, G. Evaluating an integrated approach to catchment management to reduce soil loss and sediment pollution through modelling. *Soil Use Manag.* **2002**, *18*, 386–394. [CrossRef]
10. Van Rompaey, A.; Krasa, J.; Dostal, T.; Govers, G. Modelling sediment supply to rivers and reservoirs in Eastern Europe during and after the collectivisation period. *Hydrobiologia* **2003**, *494*, 169–176. [CrossRef]
11. Krasa, J.; Dostal, T.; Van Rompaey, A.; Vaska, J.; Vrana, K. Reservoirs' siltation measurments and sediment transport assessment in the Czech Republic, the Vrchlice catchment study. *CATENA* **2005**, *64*, 348–362. [CrossRef]
12. Van Rompaey, A.; Krasa, J.; Dostal, T. Modelling the impact of land cover changes in the Czech Republic on sediment delivery. *Land Use Policy* **2007**, *24*, 576–583. [CrossRef]
13. Krása, J.; Dostál, T.; Rosendorf, P.; Borovec, J. Modelling of Sediment and Phosphorus Loads in Reservoirs in the Czech Republic. *Adv. GeoEcol.* **2015**, *44*, 21–34.
14. De Vente, J.; Poesen, J.; Verstraeten, G.; Govers, G.; Vanmaercke, M.; Van Rompaey, A.; Arabkhedri, M.; Boix-Fayos, C. Predicting soil erosion and sediment yield at regional scales: Where do we stand? *Earth-Sci. Rev.* **2013**, *127*, 16–29. [CrossRef]
15. Novotný, I.; Žížala, D.; Kapička, J.; Beitlerová, H.; Mistr, M.; Kristenová, H.; Papaj, V. Adjusting the CPmax factor in the Universal Soil Loss Equation (USLE): Areas in need of soil erosion protection in the Czech Republic. *J. Maps* **2016**, *12*, 58–62. [CrossRef]
16. Chandrashekar, H.; Lokesh, K.V.; Sameena, M.; Roopa, J.; Ranganna, G. GIS –Based Morphometric Analysis of Two Reservoir Catchments of Arkavati River, Ramanagaram District, Karnataka. *Aquat. Procedia* **2015**, *4*, 1345–1353. [CrossRef]
17. Cerdan, O.; Le Bissonnais, Y.; Couturier, A.; Saby, N. Modelling interrill erosion in small cultivated catchments. *Hydrol. Process.* **2002**, *16*, 3215–3226. [CrossRef]

18. Mahmoodabadi, M.; Sajjadi, S.A. Effects of rain intensity, slope gradient and particle size distribution on the relative contributions of splash and wash loads to rain-induced erosion. *Geomorphology* **2016**, *253*, 159–167. [CrossRef]

19. Liu, Q.J.; Shi, Z.H.; Yu, X.X.; Zhang, H.Y. Influence of microtopography, ridge geometry and rainfall intensity on soil erosion induced by contouring failure. *Soil Tillage Res.* **2014**, *136*, 1–8. [CrossRef]

20. Wu, X.; Wang, X. Spatial influence of geographical factors on soil erosion in Fuyang county, China. *Procedia Environ. Sci.* **2011**, *10*, 2128–2133. [CrossRef]

21. Milevski, I. Estimation of Soil Erosion Risk in the Upper Part of Bregalnica Watershed-Republic of Macedonia, Based on Digital Elevation Model and Satellite Imagery. In Proceedings of the 5th International Conference on Geographic Information Systems (ICGIS-2008), Istanbul, Turkey, 2–5 July 2008; pp. 351–358.

22. Conforti, M.; Aucelli, P.P.C.; Robustelli, G.; Scarciglia, F. Geomorphology and GIS analysis for mapping gully erosion susceptibility in the Turbolo stream catchment (Northern Calabria, Italy). *Nat. Hazards* **2011**, *56*, 881–898. [CrossRef]

23. Chaplot, V. Impact of terrain attributes, parent material and soil types on gully erosion. *Geomorphology* **2013**, *186*, 1–11. [CrossRef]

24. European Environment Agency. *Topic Report (ETC LC): CORINE Land Cover—A Key Database for European Integrated Environmental Assessment*; European Environment Agency: Copenhagen, Denmark, 1999.

25. Alatorre, L.C.; Beguería, S.; García-Ruiz, J.M. Regional scale modeling of hillslope sediment delivery: A case study in the Barasona Reservoir watershed (Spain) using WATEM/SEDEM. *J. Hydrol.* **2010**, *391*, 109–123. [CrossRef]

26. Drbal, K.; Štěpánková, P.; Levitus, V.; Říha, J.; Dráb, A.; Satrapa, L.; Horský, M.; Valenta, P.; Valentová, J.; Friedmannová, L. *Methodology for the Creation of Flood Hazard and Flood Risk*; T. G. Masaryk Water Research Institute: Prague, Czech Republic, 2009.

27. Krasa, J.; Dostal, T.; Jachymova, B.; Bauer, M.; Devaty, J. Soil erosion as a source of sediment and phosphorus in rivers and reservoirs—Watershed analyses using WaTEM/SEDEM. *Environ. Res.* **2019**, *171*, 470–483. [CrossRef] [PubMed]

28. Hanel, M.; Máca, P.; Bašta, P.; Vlnas, R.; Pech, P. Rainfall erosivity factor in the Czech Republic and its Uncertainty. *Hydrol. Earth Syst. Sci. Discuss.* **2016**, *20*, 4307–4322. [CrossRef]

29. Krása, J.; Stredova, H.; Dostál, T.; Novotny, I. Rainfall erosivity research on the territory of the Czech Republic. In *Mendel and Bioclimatology*; Mendel University in Brno: Brno, Czech Republic, 2016; pp. 182–196.

30. Janeček, M. *Protection of Agricultural Land from Erosion*; Czech University of Life Science: Prague, Czech Republic, 2012; ISBN 978-80-87415-42-9.

31. Dostál, T.; Krása, J.; Vrána, K. *Methods and Techniques of Prediction of Surface Runoff, Erosion and Transport Processes in Landscape*; CTU in Prague: Prague, Czech Republic, 2006.

32. Vopravil, J.; Janeček, M.; Tippl, M. Revised soil erodibility K-factor for soils in the Czech Republic. *Soil Water Res.* **2007**, *2*, 1–9. [CrossRef]

33. Borselli, L.; Cassi, P.; Torri, D. Prolegomena to sediment and flow connectivity in the landscape: A GIS and field numerical assessment. *CATENA* **2008**, *75*, 268–277. [CrossRef]

34. Becker, R.A.; Chambers, J.M.; Wilks, A.R. *The New S Language*; Wadsworth Brooks/Cole: Pacific Grove, CA, USA, 1988; Volume 1, ISBN 0534091938.

35. Moore, I.D.; Burch, G.J.; Mackenzie, D.H. Topographic Effects on the Distribution of Surface Soil Water and the Location of Ephemeral Gullies. *Trans. ASAE* **1988**, *31*, 1098–1107. [CrossRef]

36. Hengl, T.; Gruber, S.; Shrestha, D.P. *Digital Terrain Analysis in Ilwis: Lecture Notes and User Guide*; International Institute for Geo-Information Science and Earth Observation: Enschede, The Netherlands, 2003.

37. Jambu, M. *Exploratory and Multivariate Data Analysis*; Academia Press: San Diego, CA, USA, 1991.

38. Lever, J.; Krzywinski, M.; Altman, N. Principal component analysis. *Nat. Methods* **2017**, *14*, 641–642. [CrossRef]

39. R Development Core Team. *R: A Language and Environment for Statistical Computing*; R Foundation for Statistical Computing: Vienna, Austria, 2014; ISBN 3900051070.

40. Cavalli, M.; Trevisani, S.; Comiti, F.; Marchi, L. Geomorphometric assessment of spatial sediment connectivity in small Alpine catchments. *Geomorphology* **2013**, *188*, 31–41. [CrossRef]

41. Krasa, J.; Dostal, T.; Vrana, K.; Plocek, J. Predicting spatial patterns of sediment delivery and impacts of land-use scenarios on sediment transport in Czech catchments. *Land Degrad. Dev.* **2009**, *21*, 367–375. [CrossRef]

42. Vanmaercke, M.; Poesen, J.; Verstraeten, G.; de Vente, J.; Ocakoglu, F. Sediment yield in Europe: Spatial patterns and scale dependency. *Geomorphology* **2011**, *130*, 142–161. [CrossRef]
43. Vanmaercke, M.; Poesen, J.; Broeckx, J.; Nyssen, J. Earth-Science Reviews Sediment yield in Africa. *Earth Sci. Rev.* **2014**, *136*, 350–368. [CrossRef]
44. Ferreira, C.S.S.; Ferreira, A.J.D.; Pato, R.L.; Magalhães, M.D.C.; Coelho, C.D.O.; Santos, C. Rainfall-runoff-erosion relationships study for different land uses, in a sub-urban area. *Z. Geomorphol. Suppl.* **2012**, *56*, 5–20. [CrossRef]
45. Maetens, W.; Poesen, J.; Vanmaercke, M. How effective are soil conservation techniques in reducing plot runoff and soil loss in Europe and the Mediterranean? *Earth-Sci. Rev.* **2012**, *115*, 21–36. [CrossRef]
46. Shen, H.; Zheng, F.; Wen, L.; Han, Y.; Hu, W. Impacts of rainfall intensity and slope gradient on rill erosion processes at loessial hillslope. *Soil Tillage Res.* **2016**, *155*, 429–436. [CrossRef]

Article

The Use of Various Rainfall Simulators in the Determination of the Driving Forces of Changes in Sediment Concentration and Clay Enrichment

Judit Alexandra Szabó [1], Csaba Centeri [2], Boglárka Keller [2], István Gábor Hatvani [3], Zoltán Szalai [1,4], Endre Dobos [4] and Gergely Jakab [1,4,5,*]

[1] Geographical Institute, Research Centre for Astronomy and Earth Sciences, 1112 Budapest, Hungary; szabo.judit@csfk.mta.hu (J.A.S.); szalai.zoltan@csfk.mta.hu (Z.S.)

[2] Department of Nature Conservation and Landscape Ecology, Faculty of Agricultural and Environmental Sciences, Szent István University, 2100 Gödöllő, Hungary; Centeri.Csaba@mkk.szie.hu (C.C.); bogi87@gmail.com (B.K.)

[3] Institute for Geological and Geochemical Research, Research Centre for Astronomy and Earth Sciences, 1112 Budapest, Hungary; hatvaniig@gmail.com

[4] Institute of Geography and Geoinformatics, University of Miskolc, 3515 Miskolc, Hungary; ecodobos@uni-miskolc.hu

[5] Department of Environmental and Landscape Geography, ELTE University, 1117 Budapest, Hungary

* Correspondence: jakab.gergely@csfk.mta.hu

Received: 10 September 2020; Accepted: 10 October 2020; Published: 14 October 2020

Abstract: Soil erosion is a complex, destructive process that endangers food security in many parts of the world; thus, its investigation is a key issue. While the measurement of interrill erosion is a necessity, the methods used to carry it out vary greatly, and the comparison of the results is often difficult. The present study aimed to examine the results of two rainfall simulators, testing their sensitivity to different environmental conditions. Plot-scale nozzle type rainfall simulation experiments were conducted on the same regosol under both field and laboratory conditions to compare the dominant driving factors of runoff and soil loss. In the course of the experiments, high-intensity rainfall, various slope gradients, and different soil surface states (moisture content, roughness, and crust state) were chosen as the response parameters, and their driving factors were sought. In terms of the overall erosion process, the runoff, and soil loss properties, we found an agreement between the simulators. However, in the field (a 6 m^2 plot), the sediment concentration was related to the soil conditions and therefore its hydrological properties, whereas in the laboratory (a 0.5 m^2 plot), slope steepness and rainfall intensity were the main driving factors. This, in turn, indicates that the design of a rainfall simulator may affect the results of the research it is intended for, even if the differences occasioned by various designs may be of a low order.

Keywords: comparability; infiltration; rainfall simulation; runoff; soil erosion

---

1. Introduction

Preventing soil erosion is one of the most significant environmental challenges that an increasing global population has to face. Consequently, understanding the infiltration and soil erosion process is a key task, and has generated a wide range of investigations [1].

Concerning interrill erosion, the most commonly regarded drivers are slope gradient, porosity, canopy cover rate (the ratio of the soil surface covered by plants reducing the impact of splash erosion), surface roughness, soil moisture, and organic matter content, including soil texture as representative of surface conditions [2–5]. The other group of driver factors is precipitation-related, including elements such as intensity, uniformity, drop size distribution, and kinetic energy. Taking such driver parameters into account, studies aim to determine response properties, such as the quantity of eroded sediment [6], temporal variation in runoff discharge and sediment concentration [7], or the aggregate size [8] and particle size distribution (PSD) of the eroded sediment [9–11] in the contexts of organic carbon [12] and nutrient loss [13].

A challenge faced by any soil erosion study is that the combined effects of these factors and parameters create a dynamic and complex system of driving factors [14]. Even though the considerable amount of data collected and measured provides a stable basis for the construction of models and the models employed have become more precise, further measurements are still needed to deepen our understanding [15]. As environmental circumstances (soil, canopy cover, tillage operations, rainfall properties, etc.) vary across a wide spectrum, in the interests of comparability, rainfall simulation was introduced. With the application of rainfall simulation, most precipitation-related properties can be standardized, even though the results thus obtained may be affected by the research design selected [16]. In the present study, the term "rainfall simulator" refers to the applied research design including the entire measurement system. Both field and laboratory rainfall simulators are widely used for modeling eroded environments [17–20]. They are useful for understanding the parameters and interactions influencing sediment transport and selective mechanisms under interrill erosion.

Although a large number of rainfall simulation studies have been conducted in the recent past, only a few have compared simulator efficiency [14,21–23], making the comparison of studies conducted using different kinds of simulators difficult. Moreover, the changes in the scale of the research, from point to catchment, typically affect the sediment yield results [1]. In the <m^2 scale, the absence of linear erosion generally mitigates specific soil loss even though the increase of research sites triggers decreasing specific soil loss due to sedimentation within the plot.

The present study, therefore, aimed to examine and compare the measured properties of the same soil under different surface conditions (moisture content, slope gradient, and crust conditions) using both field and laboratory rainfall simulations. Intensive rainfall and location-specific methodology were used at two sites. The main hydrological properties (time to ponding and ponding period, time to runoff, runoff-to-infiltration ratio, and the characteristics of runoff after precipitation), sediment concentration, and clay enrichment were the basis of the comparison. The time to runoff is a key element in the pre-runoff phase, as it decreases with increasing rainfall intensity and, at least generally, decreases with increasing slope gradient [24] and higher initial soil moisture content [7]. Sediment concentration was selected because, according to Iserloh et al. and Chaplot and Bissonnais [22,25], it is able to represent soil loss and susceptibility to erosion well, but is independent of the length of rainfall, and can therefore be used as a comparative value during multi-simulator studies. On the basis of the results of Warrington et al. [9], the preferential loss of clay and its enrichment, compared to the original soil, in runoff was found to be important in erosion control as most of the plant-available nutrients as well as the soil organic matter and thus the buffer capacity and aggregate stability of the soil are related to the clay fraction. Beyond making a simple comparison, the idea was to summarize the results of the single parameters into a complex property (approach), which is relevant to the method applied. The following specific objectives were thus pursued: (i) to determine the similarities and differences between the soil properties measured in the two experiments; (ii) to identify the reasons for any differences; and (iii) beyond the single properties, to attempt to find a complex approach that would make the comparison of the simulators possible.

2. Material and Methods

2.1. Study Area and Soil Properties

The field experiments were conducted near Gerézdpuszta, situated in the Koppány Valley adjacent to the floodplain of the Koppány River (Figure 1). This area is located in the east of the Somogy Region, southwest Hungary.

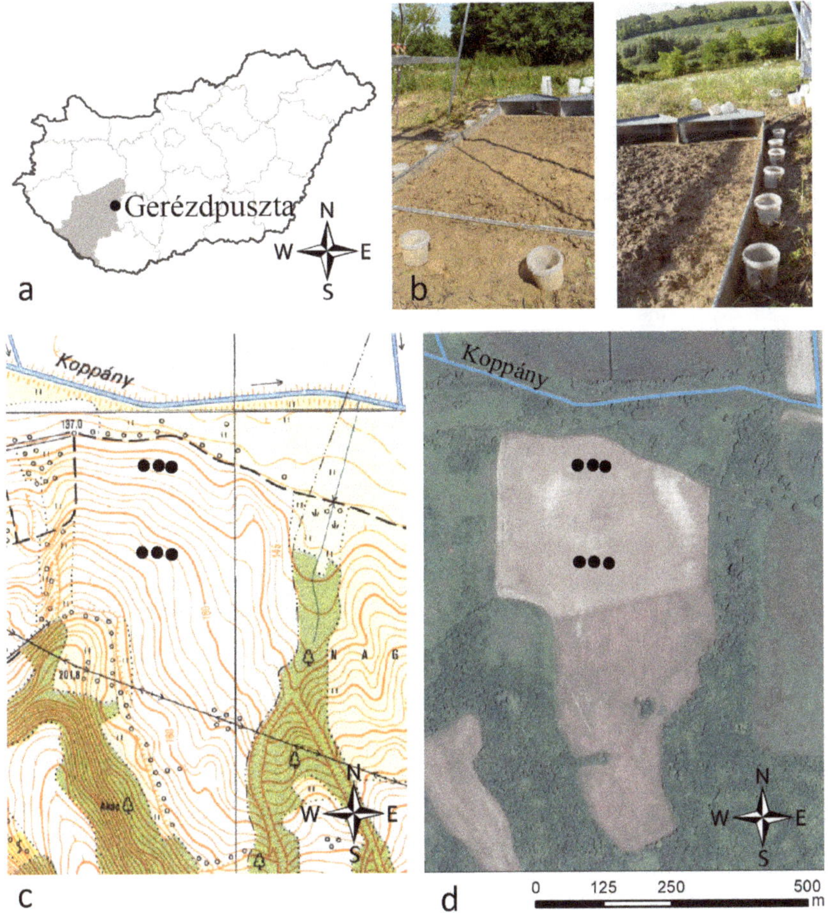

Figure 1. Location of the field experiment and the collection of soil samples for the laboratory simulations in Hungary (Europe) at Gerézdpuszta settlement in Somogy county (area highlighted grey) (**a**). Field rainfall simulation (**b**), the topography (**c**), and the land use (**d**) of the investigated catena. Black dots indicate the locations of the field rainfall simulation plots.

Owing to the loess-like deposits, the area is prone to severe soil erosion; however, a further impact on the soil is the reason why conventional tillage has become less popular in recent decades [26]. The hillslopes that are less eroded due to the lack of tillage are characterized by Cambisols, while mostly Regosols and Colluviums can be found in the cultivated areas [27]. The climate is moderately warm and wet, with an average annual temperature of 10.0–10.2 °C and average annual precipitation of 605–700 mm. Most of the hillsides are characterized by agricultural cultivation, almost half of which is situated on slopes steeper than 12% [28]. The consequent tillage depths have resulted in hard plough-pan with extremely low porosity and infiltration.

The field experiments and the sampling for the laboratory investigations were conducted on recently tilled (seedbed preparation) Regosol, which had a silty clay loam texture, as well as low $CaCO_3$ and organic carbon contents (Table 1).

Table 1. Main properties of the examined topsoil (0–30 cm layer).

	pH_{dw}	pH_{KCl}	$CaCO_3$ (%)	SOC * (%)	Particle Size Distribution (%)		
					Clay ($<8\ \mu$)	Silt (8–$50\ \mu$)	Sand ($>50\ \mu$)
Regosol	8.10	7.40	5.70	1.20	28.93	52.81	18.27

* Soil organic carbon.

Field rainfall simulations were carried out on plots that were 3 m in length and 2 m in width, in triplicate. Altogether, 6 plots were investigated: 3 on the gentler parts of the slope (with an incline of 7–8%), and 3 on steeper parts (17–18%) (see Appendix A). The plots on the same parts of the slope were placed 2–3 m from each other (Figure 1). The soil samples for laboratory simulations (from the topmost 0–25 cm tilled layer) were collected from the site at which the in situ simulations were conducted. In the laboratory, the collected samples were loaded into the flume in the state in which they had arrived. Initially, the flume was covered by woven geotextiles to ensure free drainage through the flume bottom to the collector taps. The soil was compacted solely by the energy of two successive (simulated) rainfall events (30 mm) to achieve the 20 cm soil depth required for the investigations. To exclude the effects of splash erosion, we covered the soil surface by a polyethylene mesh during these compaction pretreatments.

2.2. Rainfall Simulators Used in the Study

Two rainfall simulators were used in the experiment, the Shower Power (SP) and the "ELTE" simulator. SP 02 is a frequently used [29,30], portable field unit that can be employed at the plot-scale with alternating nozzles (Figure 2a,b). The device has a plot of 3×2 m; however, in this instance, the irrigated plot size was 12 m^2 to exclude border effects [30]. The drop forming unit, at a height of 3 m, comprises an alternating axis equipped with two adjacent 80100 Veejet nozzles spaced at 2 m [31]. This distance ensures complete overlap between the two nozzles to gain a uniform spatial intensity and drop size distribution pattern. The intensity can be set to 30–100 mm h^{-1}, depending on the axis alternation frequency. The average kinetic energy of the simulator is 24 J m^{-2} mm^{-1}, ensuring a reasonable drop size distribution and kinetic energy simulation of an intense natural rain [31]. Field measurement results are presented in Table A1. The properties affecting the momentary intensity and drop spectrum during a simulation (pressure regulation, evaporation, wind, pump properties) may well vary in the field, causing anomalies in the preset intensity and kinetic energy [32]. Thus, it was the real, independently measured rainfall intensities that had been collected (an average for each single run) that were used for calculations instead of the preset ones.

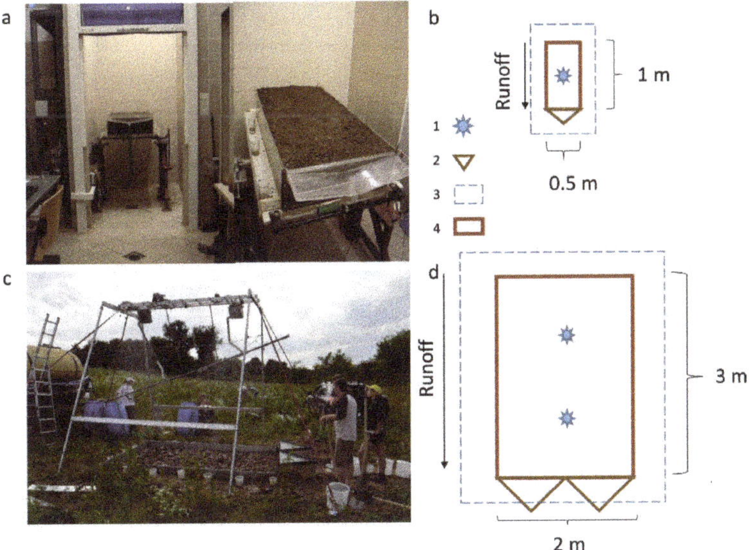

Figure 2. Photographs of the lab (**a**) and field (**c**) simulations and their primary parameters ((**b,d**), respectively) of the simulators used during the study. Marker explanation: (1) nozzle position; (2) triangular collector; (3) irrigated area; (4) plot boundary.

The ELTE simulator [33,34] is a point-scale laboratory device equipped with a fixed 1/2 HH 40 WSQ full-cone nozzle (Figure 2c,d). It was set at a height of 3 m using a constant pressure of 20 kPa, which created a standard intensity of 80 mm h^{-1}, with an average kinetic energy of 17 J m^{-2} mm^{-1}, well representing the effects of an intense natural rainfall [35]. The ELTE simulator has a sample/monolith carrier flume measuring (lwd) 100 × 50 × 20 cm. The steepness of the monolith can be regulated over a range of 0–100%. On the bottom of the flume, under the geotextile cover, a collector system is fitted to drain the leaching water through four taps. In the present investigation, each tap was closed, modelling the quasi impermeable plough-pan layer detected in the field and thus increasing/maximizing the comparability of the research design of the field and laboratory measurements. This research design does not take the effect of splash out (runoff and sediment loss due to raindrop reflex from the surface, which leave the flume across the borders) into account, causing lower runoff and sediment yield values.

2.3. Experimental Design

A total of 18 simulation runs were conducted: 12 in the field and 6 in the laboratory (Table 2). As the soil was the same, the impact of rainfall intensity, slope gradient, and soil surface parameters (roughness, moisture content, and crust state) on soil loss could be examined. The notion of the comparability of the erosion observed in the 2 experiments (comparing an in situ soil to a transported disturbed soil sample) was based on conclusions reached by Thomaz and Pereira [36], who found that the hydro-erosional responses of soils in a conventional tillage system or unstructured soils are similar.

Table 2. Settings of the 18 laboratory and field rainfall simulation experiments. Identification (ID) code fields are as follows: F: field experiment, L: laboratory experiment; G: gentle slope, S: steep slope; P: plain (2% slope). Runs on steep slopes are in italics and runs under seedbed conditions are underlined.

Experiment ID	Location	Rainfall Intensity (mm h^{-1})	Slope Steepness (%)	Amount of Rain (mm)	Rainfall Duration (s)	No. of Sub-Samples
FG1	field plot 1	56	6.7	16.81	1076	10
FG2	field plot 1	84	6.7	10.63	452	10
FG3	field plot 2	70	8	13.53	695	9
FG4	field plot 2	86	8	10.41	435	10
FG5	field plot 3	56	7.7	15.19	961	10
FG6	field plot 3	80	7.7	12.40	555	10
FS1	field plot 4	66	18.3	16.49	889	10
FS2	field plot 4	103	18.3	11.67	406	10
FS3	field plot 5	49	17.2	10.53	760	10
FS4	field plot 5	63	17.2	8.05	459	10
FS5	field plot 6	44	17.6	13.86	1121	10
FS6	field plot 6	76	17.6	11.18	525	10
LG1	laboratory	118	5	93.68	2704	8
LG2	laboratory	141	5	78.65	2008	14
LS1	laboratory	109	12	63.40	2094	16
LS2	laboratory	102	12	23.89	843	16
LP1	laboratory	108	2	35.70	1190	13
LP2	laboratory	134	2	79.80	2144	13

A gentle slope of 5% was set in the laboratory, whereas all values <10% were classified as gentle in the field. The steep slope in the laboratory measurements was 12%, while slopes >10% were considered as steep in the field. Due to the dynamic nature of variable changes in surface roughness and moisture content under the effects of precipitation, we conducted 8 dual experiments, resulting in a total of 16 simulations. The first simulation runs (FG1; FG3; FG5; FS1; FS3; FS5; LG1; LS1) modeled the seedbed (recently tilled, relatively dry) soil with a rough surface, and the second runs (FG2; FG4; FG6; FS2; FS4; FS6; LG2; LS2) were conducted over the sealed and smoothed soil with the evolved crust and higher soil moisture content.

In the remaining two laboratory simulations (LP1; LP2), we simulated extremes of moisture content (modeling "drought" and "inland inundation") on a nearly flat surface (2% slope). A wide moisture content interval was necessary because the soil moisture content is an important factor controlling runoff [37], and its combined effect with the slope gradient could present an increased impact [38].

Four degrees of intensity (55, 70, 80, and 100 mm h^{-1}) were planned and set for the field simulations. To measure the real amount of rainfall reaching, we placed the soil surface collector dishes next to the plot borders. Owing to the changes in evaporation, wind, and fluctuation in water pressure at the nozzles, the intensities as actually measured differed from those that had been set. For calculations, we used the measured amount and intensities. The time was recorded during all simulated rainfalls. All experiments were divided into 4 time periods: (1) time to ponding—in this period the infiltration exceeded rainfall intensity. The length of this period is proportional to the initial infiltration capacity; (2) ponding period (the length of which refers to the water storage capacity of the surface, or, in other words, the surface roughness); (3) runoff; and (4) the post-rain runoff period (runoff after the rain stops)—the length of this period also reflects the volume of water stored on the surface and indicates the degree of spatial connectivity. As the length of the ponding period describes the surface storage capacity, it is therefore an indirect measure of microtopography. During sample collection, all runoff was measured and collected, and runoff volumes were related to time intervals. After precipitation stopped, the runoff was collected into a separate bucket, and its volume was measured. The times at which ponds appeared and the runoff started and ended, as well as the length of runoff time after precipitation ended, were also recorded [29,30].

2.4. Runoff and Soil Loss Characterization

The total amount of soil loss was weighed after drying for 48 h at 60 °C. The sediment concentration (SC) of each sub-sample was calculated as the dry sediment mass per runoff volume (g L^{-1}). The PSD of each sub-sample was determined in triplicate with a laser diffractometer (Horiba LA-950). For the PSD measurements, we selected the Mie theory of light scattering because it is more precise in the determination of clay fractions than Fraunhofer diffraction [39]. The refractive indices (RI) based on the Horiba software calculation were set to 1.63 (real RI) and 0.3 (imaginary RI).

The constant runoff rate was calculated using linear regression on the measured cumulative runoff volumes with time (practically during the second part of period 3) on the basis of the method described by Jakab et al. [30], in which the steepness of the fitted linear function provides the constant runoff rate (Table A2). The runoff coefficient was determined as the percentage of the constant runoff rate related to rainfall intensity. The clay enrichment CE values of each simulation were calculated as the median clay content values of the sub-samples in relation to the median clay content of the in situ soil. The upper limit for the clay fraction was set at 8 μm [40].

2.5. Statistical Analyses

To obtain an overall picture of the dataset, we produced box and whisker plots. Differences between the parameters of the various laboratory and field samples were calculated using independent samples Mann–Whitney U tests [41], while in other cases when more than 2 groups were compared, an independent samples Kruskal–Wallis test [42] was used. It is difficult to determine the combined impact of the most influential independent variables (e.g., SC) in a complex system. However, in many cases, multivariate statistical tools can assist in resolving such problems. For example, variance analysis has been employed in a study on the effect of rain intensity, slope length, and gradient on SC [25]. As in other studies [43–45], principal component analysis (PCA) [46] was used to determine how SC influences response parameters measured in the soil.

In the present study, to explore how soil parameters affect the SC, primary (slope gradient (%), rain intensity (mm h^{-1})) and derived properties (constant runoff/infiltration (%), ponding period length (s), constant runoff rate (mm h^{-1})) of the field and laboratory measurements were selected as input parameters for PCA. Since they are all related to hydrology and runoff formation and, furthermore, independent of soil loss, they are suitable for the measurement and prediction of the likely susceptibility of a given soil to erosion, and thus, also, the sole chosen explanatory parameter, SC. The median diameter of the soil that was lost (μm) was also selected as a response parameter, since it also might reflect SC.

Prior to PCA, we conducted Bartlett's test [47] to investigate the applicability of the input correlation matrices (field and laboratory) to PCA. Both the field ($\chi^2(15) = 56.63, p = 9.47 \times 10^{-7}$) and the laboratory datasets ($\chi^2(15) = 85.87, p = 5.8 \times 10^{-12}$) were found to be statistically suitable for the application of PCA. Thus, the variability of the variables input to PCA was compressed into principal components on the basis of their linear relationships, as presented by Hatvani [48,49]. It should be noted that the observations' principal components (PCs) are referred to as PC scores. The elements of the eigenvectors of the empirical correlation matrix will be referred to as loadings, with these measuring the relationship of the coordinates and the PCs with the Pearson correlation coefficient. Loadings outside the ±0.7 interval are considered as meaningful.

During PCA, we took the PCs into account on the basis of their scree plots [50] and their eigenvalues, which had to be above 1 [51]. Thus, taking into account the previous considerations, the input variables' time series were practically reduced to vectors with uncorrelated coordinates (PC scores) using the first two PCs.

All computations were performed using STATISTICA 10, IBM SPSS 26, R, and visualized in MS Excel 2016.

3. Results

3.1. Variations in Ponding and Runoff Related Properties

Ponding and time to runoff periods did not differ significantly in the lab and the field ($U(N_{Field} = 12$, $N_{Lab} = 6) = 49$, $p = 0.25$) and ($U(N_{Field} = 12$, $N_{Lab} = 6) = 56$, $p = 0.067$) at $\alpha = 0.05$. However, the duration of runoff and post-rain runoff in the field and the lab were significantly different ($U(N_{Field} = 12$, $N_{Lab} = 6) = 69$ and 11, respectively, $p < 0.018$) (Figure 3).

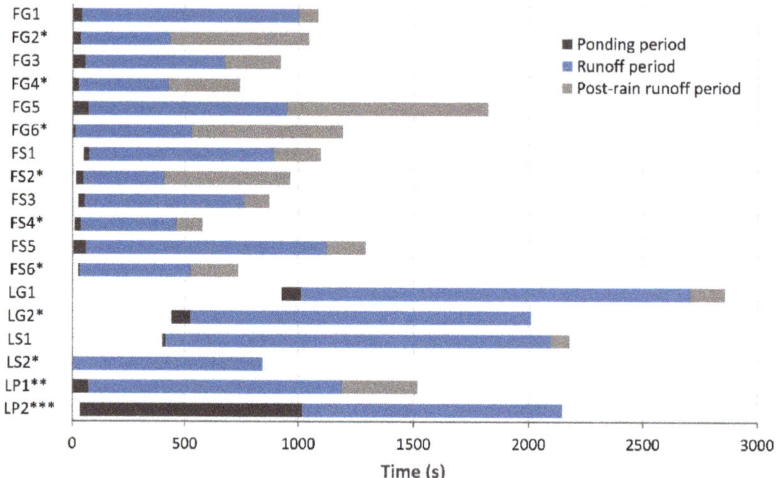

Figure 3. Durations of the ponding, runoff, and post-rain runoff periods at the field and laboratory experiments. Asterisks refer to * crusted, ** wet, and *** dry surface conditions. F—field, L—laboratory; P—plain (2% slope); G—gentle slope; S—steep slope; for details, see Table 2.

In the field, the time to runoff periods were shorter than 2 min. The time to ponding varied from 3 to 77 s, the ponding period varied from 9 to 73 s, and the time to runoff varied from 34 to 119 s. Shorter ponding times corresponded to more intense precipitation (Table A3). The time intervals of the before-runoff periods were shorter, and the post-rain runoff period was lengthened by the inhibited infiltration due to crusting and the improved connectivity on the surface. On the crusted surfaces, the duration of the ponding periods was only half that of the tilled surfaces, except for the following dual experiments: FG1-2, FS1-2, and FS3-4. In these three exceptions, the tilled and crusted surfaces entered the ponding stage for the same duration. The post-rain runoff on crusted surfaces was longer by approximately 1–9 min when compared with that on the tilled surface in the same dual experiment. The effect of slope steepness was only notable during the post-rain runoff period, which was longer on gentle slopes. The median of this closing period was 465 s on gentle slopes and 198 s on steep slopes.

In the laboratory, the runoff in three experiments (LG1-2 and LS1) began after 6–15 min; two of these plots were recently tilled, and one had a crusted surface on a slope of 5%, and which also experienced the highest rainfall intensity of 141 mm h^{-1}. The length of the ponding period decreased with an increasing slope gradient, and the time to runoff decreased with the appearance of the crust. No runoff was detected after rain in experiments where the surface was crusted (LG2, LS2; LF2).

The runoff coefficients did not differ between the slope gradients or locations: $H(18) = 0.711$, $p = 0.701$ (Figure 4). The constant runoff rate was significantly higher in the laboratory ($U(N_{Field} = 12$, $N_{Lab} = 6) = 66$, $p = 0.003$) than in the field. This difference, however, was due to the inclusion of the results of the extreme moisture and dry experiments (Figure 4). It mostly depended on the intensity and moisture content/surface conditions, especially in the field.

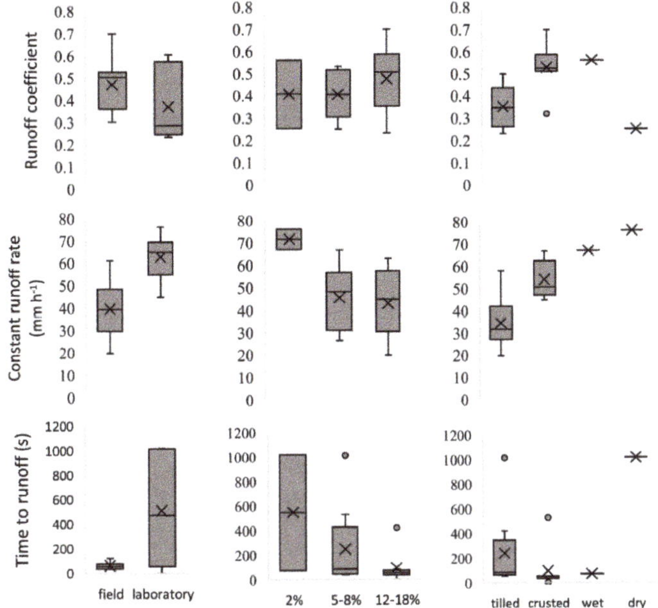

Figure 4. Comparison of the runoff coefficients, constant runoff rates (mm h^{-1}), and time to runoff (s) in accordance with the location, slope gradient, and surface moisture content/roughness. Boxes represent the interquartile interval, the black line in the middle is the median, and the x is the average; the whiskers indicate 1.5 × the interquartile interval. Values outside this are considered outliers.

3.2. Variations in Sediment Concentration and Composition

SC values increased and differed significantly ($H(18) = 11.075$, $p = 0.004$) with the slope gradient, but were unaffected by the location or surface roughness ($p \geq 0.129$; Figure 5). On the gentler (2–8%) slopes, SC values of 9–14 and 8–11 g L^{-1} were recorded, while on the steep (12–18%) slopes, SC values of 10–26 and 15–20 g L^{-1} were observed in the field and laboratory, respectively (Table A1).

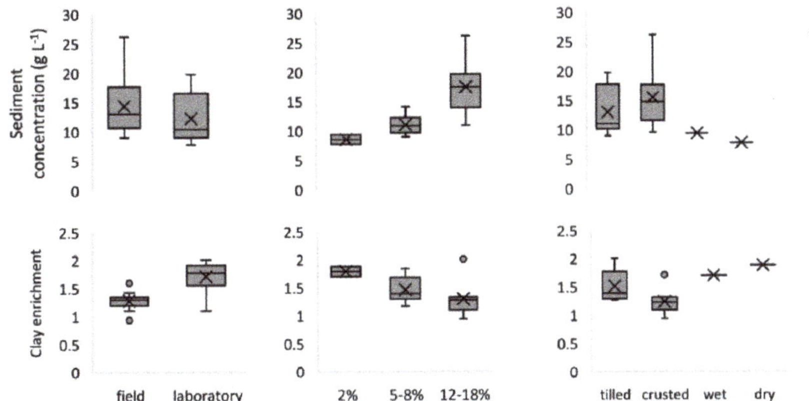

Figure 5. Comparison of the sediment concentration and clay enrichment in relation to the location, slope gradient, and surface properties. The most striking differences were caused by the slope gradient, and the SC increased while the CE decreased with an increasing slope gradient.

The lowest SC values of 5.5 and 9.2 g L^{-1} were measured in the runoff from the experiments with a slope gradient of 2% slope and extremely wet or dry initial moisture contents, respectively. The SC was higher in the field when surfaces were crusted (Table 3).

Table 3. Comparison of sediment concentrations (g L^{-1}) among the different combinations of slope gradients and surface types. LF1 and LF2 were excluded from this comparison as their experimental setups were unique.

Slope	Gentle	Steep
	Sediment concentration (g L^{-1})	
Field tilled	9.93 ± 0.8	14.30 ± 4.2
Field crusted	11.01 ± 1.4	19.69 ± 5.0

The difference in PSD between the in situ soil and soil loss was expressed by the CE ratio. As a general result, all soil loss had a CE > 1, except for the highest SC-related rain with the highest runoff ratio (CE = 0.94). Moreover, CEs were significantly higher in the laboratory ($U(N_{Field} = 12$, $N_{Lab} = 6) = 61$, $p = 0.018$) (Figure 5, Table A1) compared to the field measurements. In contrast, CE did not vary significantly, neither with regards to the slope gradient ($H(18) = 5.088$, $p = 0.079$) nor the surface roughness ($H(18) = 6.868$, $p = 0.076$).

3.3. Regulating Properties of SC for Various Rainfall Simulations

By analyzing the linear relationships of the response variables, we found that the first two PCs (PC1 and PC2) were able to account for approx. 80% of the total original variance in both the laboratory and field experiments together (Table 4a). After correlating the PC scores with the explanatory variable, we found that the main influencing factors of SC differed between the field and laboratory (Table 4).

Table 4. Principal component loadings of the first two PCs and their explanatory power expressed as a percentage in parentheses (**a**). Field and laboratory experiments are both included. Influential loadings (loading outside the +/− 0.70 interval) are in italics, and (**b**) correlation coefficients between the PC scores and the response variable's data series, significant at $\alpha = 0.1$, are marked with an asterisk (*).

	Principal Component/Response Variables	Field		Laboratory	
(a)		1st PC (50.49%)	2nd PC (27.23%)	1st PC (49.55%)	2nd PC (32.55%)
	Slope gradient (%)	0.351	0.694	*0.879*	−0.063
	Constant runoff/infiltration (%)	*0.796*	0.290	−0.111	*0.965*
	Particle size median (μm)	*0.748*	0.567	0.502	*0.746*
	Rain intensity (mm h^{-1})	0.643	*−0.706*	*−0.788*	−0.384
	Ponding period (s)	*−0.760*	0.031	*−0.810*	0.079
	Constant runoff rate (mm h^{-1})	*0.849*	−0.496	*−0.812*	0.555
(b)	Correlation Coefficient/Explanatory Variable				
	Sediment concentration (SC)	0.71 *	0.57 *	0.89 *	−0.27

The most influential parameters in the first PC for the field measurements were constant runoff/infiltration, constant runoff rate, particle size median of soil loss with a positive load, and ponding period with a negative sign indicating a response opposite to the others (Table 4a). Specifically, if the surface was smooth (short ponding period), the constant runoff rate increased, the flow could transport larger particles/aggregates, and the PSD median also increased. The first PC displayed a strong and significant ($p < 0.01$) linear relationship with the explanatory variable SC (Table 4b). The variance compressed into the second PC for the field measurements mostly accounted for the rainfall intensity, which had a smaller linear effect on SC (Table 4b). The loadings of slope gradient did not fall within the chosen ±0.7 interval in neither PC, indicating a secondary effect on SC in the field.

As the slope gradient increased in the laboratory experiments, the rainfall intensity, ponding period, and constant runoff rate decreased according to the first PC (Table 4a); however, the two most determining variables (constant runoff/infiltration and particle size median) in the second PC displayed a parallel change (Table 4a). Significant correlations were observed between the PCs and SC; it was exclusively related to the first PC (Table 4b). Thus, while the actual surface state and development acted together as influential processes on SC in the field, the first PC indicated that, in the laboratory, the constant runoff/infiltration and the particle size median of the soil loss were independent of SC.

4. Discussion

In this study, the soil type, high rainfall intensities, and slope gradients were the matching parameters during the experiments, while initial soil moisture content, surface state (crusting and smoothening), and the device used varied.

No differences were found in time to runoff in either the laboratory or the field. This may suggest that the disturbed soil transported in the laboratory yielded comparable infiltration values with that measured in the field, even though the high variance makes the comparability difficult. In contrast, a larger plot size may trigger poorer surface flow connectivity, resulting in higher ponding capacity in the field, as reported by Langhans et al. [52]. The runoff coefficient values demonstrated the impact of the crust on runoff in both locations, as roughness affected the drainage network [53] and a sealed surface inhibited infiltration [54], although, without the measurement of surface roughness and micro-topography, this is just an indication of its veracity. Initial soil moisture content could also affect the initiation of runoff; it begins earlier on a wet surface, as reported by Jin et al. [7]. Gómez and Nearing [55] also observed delayed time to runoff owing to higher surface roughness, but they did not observe differences in the runoff coefficient. This might have been due to the changes in surface connectivity, which could drastically affect the runoff volume even on rough surfaces [56]. Above and beyond these processes, variations in research design (e.g., differences in kinetic energy, splash loss due to a lack of fences in the laboratory) would surely have contributed to the difference.

As the infiltration–runoff equilibrium of the complex system is affected by several properties, it is impossible to provide a clear explanation of the differences. The point is that the method used will impact the results.

One of the main goals of this study was to test the comparability of field and laboratory SC and CE results. Concerning the constant runoff and constant runoff rate measured in this study, we found that the values from the experiments were similar when conducted on similar slope gradients (Figures 3 and 4). However, the research design (border effect in splash erosion (that is, unmeasured soil loss due to the lack of fences) in the laboratory might decrease SC. At the same time, in the present study, the CE was higher in the laboratory (Figure 5). Zemke [57] also demonstrated the enrichment of fine particles in soil loss on small plots under simulated rainfalls. Presumably, this was due to the smaller plot size, in which deposition of the coarse fraction within a plot is inhibited, as also suggested by Jakab et al. [58], or is simply the result of the higher kinetic energy of the applied rainfall. In contrast, the similar CE from tilled and crusted surfaces of the present study was in contrast to the results of high-intensity small plot experiments on silty loams by Ding and Huang [11].

Concerning the overall erosion process, in the field, both the SC and CE influenced the surface properties, as also reported by Koiter et al. [5]. In the laboratory, the velocity of the moving water and its influencing factors (slope and rainfall intensity) were the main variables driving SC, and the available soil particles and amount of water determined the CE. The above difference, however, is an artefact of the different research designs, including, among others, the plot size, the border effect, and the differences in the drop spectra and kinetic energy of the rain. To achieve more precise and detailed results, additional measurements are needed with a greater degree of standardization in the conduct of the experiments (e.g., using the same nozzles in the field and the laboratory). Moreover, given that the above findings are the results of experiments carried out on a single soil, more investigations on other soil types are required before the results can be generalized.

5. Conclusions

Results gained using the two simulators agree with each other and reflect in similar ways on the soil and surface properties. Therefore, both simulators are suitable for calculations concerning the susceptibility of soils to erosion, and their results are comparable. The slope gradient was found to be an effective regulator only in the laboratory. Rainfall intensity was also more effective in the laboratory than in the field simulations. This, then, suggests that soil-related properties had a prominent role in driving sediment concentration in the field, whereas in the laboratory, slope and rainfall intensity were found to be driving factors independent of soil-regulated sediment concentrations. However, these findings are based on a limited number of simulations and investigate only one type of soil, and some may conclude that different rainfall simulators behave differently, resulting in the overestimation of the role of various sub-processes. Thus, the research design of a rainfall simulator may affect the results even though the differences thus evoked seem to be small in magnitude, probably beneath the sampling error caused by natural spatial heterogeneity of the soil. Ideally, in the future, an optimal research design will be to use paired scenarios for various simulation purposes. To do so, many comparative measurements are needed.

Author Contributions: Conceptualization, J.A.S. and G.J.; methodology, J.A.S., B.K. and G.J.; formal analysis, I.G.H.; investigation, C.C., E.D., and B.K.; writing—original draft preparation, J.A.S.; writing—review and editing, Z.S. and G.J. with contribution from I.G.H.; visualization, J.A.S.; supervision, G.J. All authors have read and agreed to the published version of the manuscript.

Funding: This research was funded by the Development and Innovation Fund of Hungary, grant number NKFIH 123953.

Acknowledgments: The authors are grateful to Paul Thatcher and Lyndre Nel for the English language advice.

Conflicts of Interest: The authors declare no conflict of interest. The funders had no role in the design of the study; in the collection, analyses, or interpretation of data; in the writing of the manuscript; or in the decision to publish the results.

Appendix A

Table A1. Measured drop spectrum of the SP-02 rainfall simulator in the field at a pressure of 0.41 KPa using WATOR (Macherey-Nagel) adsorption papers and digital image analysis.

Drop Diameter (mm)	<0.5	0.5–1.0	1.0–2.0	2.0–3.0	3.0–4.0	4.0–5.0	>5.0	Total
No. of drops	641	504	224	116	50	19	8	1562
Kinetic energy (J m^{-2} mm^{-1})	0.006	0.101	1.396	5.124	6.918	6.328	5.102	24.975

Table A2. Calculation of the final constant runoff (mm h^{-1}) using linear regression on the measured runoff/time dataset. Constant runoff is calculated as the steepness of the function (parameter a in $y = ax - b$). The number of points gives the quantity of measured data used to fit on.

Experiment ID	Fitted Equation	Coefficient of Determination	Number of Points
FG1	$y = 26.448x - 3.7086$	0.998	6
FG2	$y = 52.78x - 2.6902$	0.993	7
FG3	$y = 33.021x - 2.1606$	0.999	7
FG4	$y = 48.244x - 1.4549$	0.999	6
FG5	$y = 30.171x - 2.831$	0.998	6
FG6	$y = 48.066x - 2.4828$	0.999	5
FS1	$y = 33.761x - 3.1841$	0.999	5
FS2	$y = 60.976x - 1.4915$	0.999	7
FS3	$y = 29.21x - 1.2405$	0.999	6
FS4	$y = 46.646x - 0.8294$	0.999	7
FS5	$y = 19.737x - 1.2286$	0.997	7
FS6	$y = 44.645x - 1.2834$	0.999	6
LG1	$y = 58.022x - 20.557$	0.998	30
LG2	$y = 66.651x - 12.452$	0.999	16
LS1	$y = 44.694x - 11.695$	0.994	9
LS2	$y = 62.804x - 0.7752$	0.999	12
LP1	$y = 66.917x - 2.1461$	0.999	9
LP2	$y = 75.973x - 25.359$	0.999	10

Table A3. Main properties and calculated parameters of the simulations.

Experiment ID	Location	Measured Rainfall Intensity (mm h^{-1})	Slope Steepness (%)	Amount of Rain (mm)	Time to Ponding (s)	Ponding Duration (s)	Time to Runoff (s)	Runoff Duration (s)	Runoff after the Rain (s)	Runoff Ratio (%)	Infiltration Ratio (%)	Sediment Concentration (g L^{-1})	Total Soil Loss (g)	Runoff Coefficient	Final Constant Runoff (mm h^{-1})	Clay Enrichment in Soil Loss
FG1	field	56.24	7	16.81	77	42	119	957	85	47	53	10.6	270.6	0.30	26.4	1.60
FG2	field	84.68	7	10.63	18	36	54	398	611	62	38	12.6	306.2	0.53	52.8	1.30
FG3	field	70.19	8	13.53	20	57	77	618	246	47	53	9.0	222.9	0.35	33.0	1.35
FG4	field	86.12	8	10.41	11	27	38	397	319	56	44	11.3	294.6	0.52	48.2	1.17
FG5	field	56.92	8	15.19	16	73	89	872	877	53	47	10.0	318.1	0.45	30.2	1.43
FG6	field	80.44	8	12.40	24	12	36	519	661	60	40	14.1	418.7	0.51	48.1	1.30
FS1	field	66.78	18	16.49	48	28	76	813	208	51	49	10.9	339.4	0.34	33.8	1.28
FS2	field	103.48	18	11.67	16	33	49	357	559	59	41	17.9	582.5	0.53	61.0	1.32
FS3	field	49.89	17	10.53	25	31	56	704	112	59	41	19.2	555.4	0.50	29.2	1.28
FS4	field	63.16	17	8.05	10	26	36	423	120	74	26	26.2	812.1	0.70	46.6	0.94
FS5	field	44.5	17	13.86	3	57	60	1061	170	44	56	13.4	407.1	0.38	19.7	1.33
FS6	field	76.69	17	11.18	25	9	34	491	210	58	42	17.1	537.6	0.51	44.6	1.10
LG1	laboratory	118	5	93.68	925	85	1010	1694	154	49	51	11.2	138.5	0.25	58.0	1.84
LG2	laboratory	141	5	78.65	441	84	525	1483	0	47	53	9.6	111.0	0.32	66.6	1.71
LS1	laboratory	109	12	63.40	399	18	417	1677	83	41	59	19.8	189.7	0.23	44.7	2.01
LS2	laboratory	102	12	23.89	0	1	1	842	0	62	38	15.5	65.9	0.60	62.8	1.10
LF1	laboratory	108	2	35.70	0	72	72	1118	327	62	38	9.4	92.6	0.56	66.9	1.70
LF2	laboratory	134	2	79.80	34	982	1016	1128	0	57	43	7.8	55.4	0.25	76.0	1.88

References

1. García-Ruiz, J.M.; Beguería, S.; Nadal-Romero, E.; González-Hidalgo, J.C.; Lana-Renault, N.; Sanjuán, Y. A meta-analysis of soil erosion rates across the world. *Geomorphology* **2015**, *239*, 160–173. [CrossRef]
2. Balacco, G. The interrill erosion for a sandy loam soil. *Int. J. Sediment Res.* **2013**, *28*, 329–337. [CrossRef]
3. Vahabi, J.; Nikkami, D. Assessing dominant factors affecting soil erosion using a portable rainfall simulator. *Int. J. Sediment Res.* **2008**, *23*, 376–386. [CrossRef]
4. Assouline, S.; Ben-Hur, M. Effects of rainfall intensity and slope gradient on the dynamics of interrill erosion during soil surface sealing. *Catena* **2006**, *66*, 211–220. [CrossRef]
5. Koiter, A.J.; Owens, P.N.; Petticrew, E.L.; Lobb, D.A. The role of soil surface properties on the particle size and carbon selectivity of interrill erosion in agricultural landscapes. *Catena* **2017**, *153*, 194–206. [CrossRef]
6. Bagarello, V.; Di Stefano, C.; Ferro, V.; Kinnell, P.I.A.; Pampalone, V.; Porto, P.; Todisco, F. Predicting soil loss on moderate slopes using an empirical model for sediment concentration. *J. Hydrol.* **2011**, *400*, 267–273. [CrossRef]
7. Jin, K.; Cornelis, W.M.; Gabriels, D.; Schiettecatte, W.; De Neve, S.; Lu, J.; Buysse, T.; Wu, H.; Cai, D.; Jin, J.; et al. Soil management effects on runoff and soil loss from field rainfall simulation. *Catena* **2008**, *75*, 191–199. [CrossRef]
8. Shi, P.; Van Oost, K.; Schulin, R. Dynamics of soil fragment size distribution under successive rainfalls and its implication to size-selective sediment transport and deposition. *Geoderma* **2017**, *308*, 104–111. [CrossRef]
9. Warrington, D.N.; Mamedov, A.I.; Bhardwaj, A.K.; Levy, G.J. Primary particle size distribution of eroded material affected by degree of aggregate slaking and seal development. *Eur. J. Soil Sci.* **2009**, *60*, 84–93. [CrossRef]
10. Asadi, H.; Moussavi, A.; Ghadiri, H.; Rose, C.W. Flow-driven soil erosion processes and the size selectivity of sediment. *J. Hydrol.* **2011**, *406*, 73–81. [CrossRef]
11. Ding, W.; Huang, C. Effects of soil surface roughness on interrill erosion processes and sediment particle size distribution. *Geomorphology* **2017**, *295*, 801–810. [CrossRef]
12. Kuhn, N.J.; Armstrong, E.K.; Ling, A.C.; Connolly, K.L.; Heckrath, G. Interrill erosion of carbon and phosphorus from conventionally and organically farmed Devon silt soils. *Catena* **2012**, *91*, 94–103. [CrossRef]
13. Ouyang, W.; Xu, X.; Hao, Z.; Gao, X. Effects of soil moisture content on upland nitrogen loss. *J. Hydrol.* **2017**, *546*, 71–80. [CrossRef]
14. Stroosnijder, L. Measurement of erosion: Is it possible? *Catena* **2005**, *64*, 162–173. [CrossRef]
15. Poesen, J. Soil erosion hazard and mitigation in the Euro-Mediterranean region: Do we need more research? *Hung. Geogr. Bull.* **2015**, *64*, 293–299. [CrossRef]
16. Iserloh, T.; Ries, J.B.; Arnáez, J.; Boix-Fayos, C.; Butzen, V.; Cerdà, A.; Echeverría, M.T.; Fernández-Gálvez, J.; Fister, W.; Geißler, C.; et al. European small portable rainfall simulators: A comparison of rainfall characteristics. *Catena* **2013**, *110*, 100–112. [CrossRef]
17. Defersha, M.B.; Melesse, A.M. Effect of rainfall intensity, slope and antecedent moisture content on sediment concentration and sediment enrichment ratio. *Catena* **2012**, *90*, 47–52. [CrossRef]
18. Aksoy, H.; Unal, N.E.; Cokgor, S.; Gedikli, A.; Yoon, J.; Koca, K.; Inci, S.B.; Eris, E. A rainfall simulator for laboratory-scale assessment of rainfall-runoff-sediment transport processes over a two-dimensional flume. *Catena* **2012**, *98*, 63–72. [CrossRef]
19. Jakab, G.; Kiss, K.; Szalai, Z.; Zboray, N.; Németh, T.; Madarász, B. Soil Organic Carbon Redistribution by Erosion on Arable Fields. In *Soil Carbon*; Hartemnik, A.E., McSweeney, K., Eds.; Springer International Publishing: Cham, Switzerland, 2014; pp. 289–296. [CrossRef]
20. Rodrigo Comino, J.; Iserloh, T.; Lassu, T.; Cerdà, A.; Keesstra, S.D.; Prosdocimi, M.; Brings, C.; Marzen, M.; Ramos, M.C.; Senciales, J.M.; et al. Quantitative comparison of initial soil erosion processes and runoff generation in Spanish and German vineyards. *Sci. Total Environ.* **2016**, *565*, 1165–1174. [CrossRef]
21. Bryan, R.B.; De-Ploey, J. Comparability of soil erosion measurements with different laboratory rainfall simulators; Comparabilité des mesures d'érosion du sol avec différents simulateurs de pluie de laboratoire. *Catena Suppl.* **1983**, *4*, 33–56.
22. Iserloh, T.; Ries, J.B.; Cerdà, A.; Echeverría, M.T.; Fister, W.; Geißler, C.; Kuhn, N.J.; León, F.J.; Peters, P.; Schindewolf, M.; et al. Comparative measurements with seven rainfall simulators on uniform bare fallow land. *Z. Geomorphol. Suppl. Issues* **2013**, *57*, 11–26. [CrossRef]

23. Mayerhofer, C.; Meißl, G.; Klebinder, K.; Kohl, B.; Markart, G. Comparison of the results of a small-plot and a large-plot rainfall simulator—Effects of land use and land cover on surface runoff in Alpine catchments. *Catena* **2017**, *156*, 184–196. [CrossRef]

24. Sadeghi, S.H.; Harchegani, M.K.; Asadi, H. Variability of particle size distributions of upward/downward splashed materials in different rainfall intensities and Slopes. *Geoderma* **2017**, *290*, 100–106. [CrossRef]

25. Chaplot, V.; Le Bissonnais, Y. Field measurements of interrill erosion under different slopes and plot sizes. *Earth Surf. Process. Landf.* **2000**, *25*, 145–153. [CrossRef]

26. Birkás, M.; Dekemati, I.; Kende, Z.; Pósa, B. Review of soil tillage history and new challenges in Hungary. *Hung. Geogr. Bull.* **2017**, *66*, 55–64. [CrossRef]

27. Szabó, B.; Centeri, C.; Szalai, Z.; Jakab, G.; Szabó, J. Comparison of soil erosion dynamics under extensive and intensive cultivation based on basic soil parameters. *Növénytermelés* **2015**, *64*, 23–26.

28. Dövényi, Z. (Ed.) *Inventory of Micro Regions in Hungary*; MTA Geographical Research Institute: Budapest, Hungary, 2010.

29. Keller, B.; Szabó, J.; Centeri, C.; Jakab, G.; Szalai, Z. Different land-use intensities and their susceptibility to soil erosion. *Agrokémia Talajt. Agrokem* **2019**, *68*, 14–23. [CrossRef]

30. Jakab, G.; Madarász, B.; Szabó, J.A.; Tóth, A.; Zacháry, D.; Szalai, Z.; Kertész, Á.; Dyson, J. Infiltration and soil loss changes during the growing season under ploughing and conservation tillage. *Sustainability* **2017**, *9*, 1726. [CrossRef]

31. Loch, R.J.; Robotham, B.G.; Zeller, L.; Masterman, N.; Orange, D.N.; Bridge, B.J.; Sheridan, G.; Bourke, J.J. A multi-purpose rainfall simulator for field infiltration and erosion studies. *Soil Res.* **2001**, *39*, 599–610. [CrossRef]

32. Zemke, J.J. Set-up and calibration of a portable small scale rainfall simulator for assessing soil erosion processes at interrill scale. *Cuad. Investig. Geográfica* **2017**, *43*, 63. [CrossRef]

33. Szabó, J.; Jakab, G.; Szabó, B. Spatial and temporal heterogeneity of runoff and soil loss dynamics under simulated rainfall. *Hung. Geogr. Bull.* **2015**, *64*. [CrossRef]

34. Szabó, J.A.; Király, C.; Karlik, M.; Tóth, A.; Szalai, Z.; Jakab, G. Rare earth oxide tracking coupled with 3D soil surface modelling: An opportunity to study small-scale soil redistribution. *J. Soils Sediments* **2020**. [CrossRef]

35. Strauss, P.; Cornejo, J.; Strauss, P.; Pitty, J.; Pfeffer, M.; Mentler, A. Rainfall Simulation for Outdoor Experiments. In *Current Research Methods to Assess the Environmental Fate of Pesticides*; Jamet, P., Cornejo, J., Eds.; INRA Editions: Versailles, France, 2000; pp. 329–333.

36. Thomaz, E.L.; Pereira, A.A. Misrepresentation of hydro-erosional processes in rainfall simulations using disturbed soil samples. *Geomorphology* **2017**, *286*, 27–33. [CrossRef]

37. Wang, Y.; You, W.; Fan, J.; Jin, M.; Wei, X.; Wang, Q. Catena Effects of subsequent rainfall events with different intensities on runo ff and erosion in a coarse soil. *Catena* **2018**, *170*, 100–107. [CrossRef]

38. Zhao, X.; Huang, J.; Gao, X.; Wu, P.; Wang, J. Runoff features of pasture and crop slopes at different rainfall intensities, antecedent moisture contents and gradients on the Chinese Loess Plateau: A solution of rainfall simulation experiments. *Catena* **2014**, *119*, 90–96. [CrossRef]

39. Özer, M.; Orhan, M.; Işik, N.S. Effect of particle optical properties on size distribution of soils obtained by laser diffraction. *Environ. Eng. Geosci.* **2010**, *16*, 163–173. [CrossRef]

40. Konert, M.; Vandenberghe, J. Comparison of laser grain size analysis with pipette and sieve analysis: A solution fo rthe underestimation of the clay fraction. *Sedimentology* **1997**, *44*, 523–535. [CrossRef]

41. Mann, H.; Whitney, D. On a Test of Whether one of Two Random Variables is Stochastically Larger than the Other. *Ann. Math. Stat.* **1947**, *18*, 50–60. [CrossRef]

42. Kruskal, W.H.; Wallis, W.A. Use of Ranks in One-Criterion Variance Analysis. *J. Am. Stat. Assoc.* **1952**, *47*, 583–621. [CrossRef]

43. Chaplot, V.; Poesen, J. Sediment, soil organic carbon and runoff delivery at various spatial scales. *Catena* **2012**, *88*, 46–56. [CrossRef]

44. Cerdan, O.; Le Bissonais, Y.; Souchère, V.; Martin, P.; Lecomte, V. Sediment concentration in interill flow: Interactions between soil surface conditions, vegetation and rainfall. *Earth Surf. Process.* **2002**, *205*, 193–205.

45. Maïga-Yaleu, S.; Guiguemde, I.; Yacouba, H.; Karambiri, H.; Ribolzi, O.; Bary, A.; Ouedraogo, R.; Chaplot, V. Soil crusting impact on soil organic carbon losses by water erosion. *Catena* **2013**, *107*, 26–34. [CrossRef]

46. Tabachnick, B.G.; Fidell, L.S. *Using Multivariate Statistics*, 3rd ed.; Harper-Collins: New York, NY, USA, 1996.

47. Bartlett, M.S. The Effect of Standardization on a χ^2 Approximation in Factor Analysis. *Biometrika* **1951**, *38*, 337–344. [CrossRef]

48. Hatvani, I.G.; Kovács, J.; Márkus, L.; Clement, A.; Hoffmann, R.; Korponai, J. Assessing the relationship of background factors governing the water quality of an agricultural watershed with changes in catchment property (W-Hungary). *J. Hydrol.* **2015**, *521*, 460–469. [CrossRef]

49. Hatvani, I.G.; de Barros, V.D.; Tanos, P.; Kovács, J.; Székely Kovács, I.; Clement, A. Spatiotemporal changes and drivers of trophic status over three decades in the largest shallow lake in Central Europe, Lake Balaton. *Ecol. Eng.* **2020**, *151*, 105861. [CrossRef]

50. Cattell, R.B. The Scree Test For The Number Of Factors. *Multivar. Behav. Res.* **1966**, *1*, 245–276. [CrossRef] [PubMed]

51. Kaiser, H.F. The Application of Electronic Computers to Factor Analysis. *Educ. Psychol. Meas.* **1960**, *20*, 141–151. [CrossRef]

52. Langhans, C.; Diels, J.; Clymans, W.; Van den Putte, A.; Govers, G. Scale effects of runoff generation under reduced and conventional tillage. *Catena* **2019**, *176*, 1–13. [CrossRef]

53. Römkens, M.J.M.; Helming, K.; Prasad, S.N. Soil erosion under different rainfall intensities, surface roughness, and soil water regimes. *Catena* **2002**, *46*, 103–123. [CrossRef]

54. Di Prima, S.; Concialdi, P.; Lassabatere, L.; Angulo-Jaramillo, R.; Pirastru, M.; Cerdà, A.; Keesstra, S. Laboratory testing of Beerkan infiltration experiments for assessing the role of soil sealing on water infiltration. *Catena* **2018**, *167*, 373–384. [CrossRef]

55. Gómez, J.A.; Nearing, M.A. Runoff and sediment losses from rough and smooth soil surfaces in a laboratory experiment. *Catena* **2005**, *59*, 253–266. [CrossRef]

56. Antoine, M.; Javaux, M.; Bielders, C. What indicators can capture runoff-relevant connectivity properties of the micro-topography at the plot scale? *Adv. Water Resour.* **2009**, *32*, 1297–1310. [CrossRef]

57. Zemke, J.J. Runoff and soil erosion assessment on forest roads using a small scale rainfall simulator. *Hydrology* **2016**, *3*, 25. [CrossRef]

58. Jakab, G.; Németh, T.; Csepinszky, B.; Madarász, B.; Szalai, Z.; Kertész, A. The influence of short term soil sealing and crusting on hydrology and erosion at balaton uplands, hungary. *Carpathian J. Earth Environ. Sci.* **2013**, *8*, 147–155.

Publisher's Note: MDPI stays neutral with regard to jurisdictional claims in published maps and institutional affiliations.

MDPI
St. Alban-Anlage 66
4052 Basel
Switzerland
Tel. +41 61 683 77 34
Fax +41 61 302 89 18
www.mdpi.com

Water Editorial Office
E-mail: water@mdpi.com
www.mdpi.com/journal/water